NMR in Supramolecular Chemistry

NATO ASI Series

Advanced Science Institute Series

A Series presenting the results of activities sponsored by the NATO Science Committee, which aims at the dissemination of advanced scientific and technological knowledge, with a view to strengthening links between scientific communities.

The Series is published by an international board of publishers in conjunction with the NATO Scientific Affairs Division

A	**Life Sciences**	Plenum Publishing Corporation
B	**Physics**	London and New York
C	**Mathematical and Physical Sciences**	Kluwer Academic Publishers
D	**Behavioural and Social Sciences**	Dordrecht, Boston and London
E	**Applied Sciences**	
F	**Computer and Systems Sciences**	Springer-Verlag
G	**Ecological Sciences**	Berlin, Heidelberg, New York, London,
H	**Cell Biology**	Paris and Tokyo
I	**Global Environment Change**	

PARTNERSHIP SUB-SERIES

1. **Disarmament Technologies**	Kluwer Academic Publishers
2. **Environment**	Springer-Verlag / Kluwer Academic Publishers
3. **High Technology**	Kluwer Academic Publishers
4. **Science and Technology Policy**	Kluwer Academic Publishers
5. **Computer Networking**	Kluwer Academic Publishers

The Partnership Sub-Series incorporates activities undertaken in collaboration with NATO's Cooperation Partners, the countries of the CIS and Central and Eastern Europe, in Priority Areas of concern to those countries.

NATO-PCO-DATA BASE

The electronic index to the NATO ASI Series provides full bibliographical references (with keywords and/or abstracts) to about 50,000 contributions from international scientists published in all sections of the NATO ASI Series. Access to the NATO-PCO-DATA BASE is possible via a CD-ROM "NATO Science and Technology Disk" with user-friendly retrieval software in English, French, and German (©WTV GmbH and DATAWARE Technologies, Inc. 1989). The CD-ROM contains the AGARD Aerospace Database.

The CD-ROM can be ordered through any member of the Board of Publishers or through NATO-PCO, Overijse, Belgium.

Series C: Mathematical and Physical Sciences – Vol. 526

NMR in Supramolecular Chemistry

edited by

M. Pons
Departament de Química Orgànica,
Universitat de Barcelona, Spain

Kluwer Academic Publishers

Dordrecht / Boston / London

Published in cooperation with NATO Scientific Affairs Division

Proceedings of the NATO Advanced Workshop on
Applications of NMR to the Study of Structure and Dynamics of Supramolecular
Complexes
Sitges, Spain
5–9 May 1998

A C.I.P. Catalogue record for this book is available from the Library of Congress.

ISBN 0-7923-5621-7

Published by Kluwer Academic Publishers,
P.O. Box 17, 3300 AA Dordrecht, The Netherlands.

Sold and distributed in North, Central and South America
by Kluwer Academic Publishers,
101 Philip Drive, Norwell, MA 02061, U.S.A.

In all other countries, sold and distributed
by Kluwer Academic Publishers,
P.O. Box 322, 3300 AH Dordrecht, The Netherlands.

Printed on acid-free paper

All Rights Reserved
© 1999 Kluwer Academic Publishers
No part of the material protected by this copyright notice may be reproduced or utilized in any form or by any means, electronic or mechanical, including photocopying, recording or by any information storage and retrieval system, without written permission from the copyright owner.

Printed in the Netherlands

This book contains the proceedings of a NATO Advanced Research Workshop held within the programme of activities of the NATO Special Programme on Supramolecular Chemistry as part of the activities of the NATO Science Committee.

Other books previously published as a result of the activities of the Special Programme are:

WIPFF, G. (Ed.), *Computational Approaches in Supramolecular Chemistry.* (ASIC 426) 1994.
ISBN 0-7923-2767-5

FLEISCHAKER, G.R., COLONNA, S. and LUISI, P.L. (Eds.), *Self-Production of Supramolecular Structures. From Synthetic Structures to Models of Minimal Living Systems.* (ASIC 446) 1994.
ISBN 0-7923-3163-X

FABBRIZZI, L. and POGGI, A. (Eds.), *Transition Metals in Supramolecular Chemistry.* (ASIC 448) 1994.
ISBN 0-7923-3196-6

BECHER, J. and SCHAUMBURG, K. (Eds.), *Molecular Engineering for Advanced Materials.* (ASIC 456) 1995. ISBN 0-7923-3347-0

LA MAR, G.N. (Ed.), *Nuclear Magnetic Resonance of Paramagnetic Macromolecules.* (ASIC 457) 1995.
ISBN 0-7923-3348-9

SIEGEL, JAY S. (Ed.), *Supramolecular Stereochemistry.* (ASIC 473) 1995. ISBN 0-7923-3702-6

WILCOX, C.S. and HAMILTON A.D. (Eds.), *Molecular Design and Bioorganic Catalysis.* (ASIC 478) 1996.
ISBN 0-7923-4024-8

MEUNIER, B. (Ed.), *DNA and RNA Cleavers and Chemotherapy of Cancer and Viral Diseases.* (ASIC 479) 1996. ISBN 0-7923-4025-6

KAHN, O. (Ed.), *Magnetism: A Supramolecular Function.* (ASIC 484) 1996. ISBN 0-7923-4153-8

ECHEGOYEN, L. and KAIFER ANGEL E. (Eds.), *Physical Supramolecular Chemistry.* (ASIC 485) 1996. ISBN 0-7923-4181-3

DESVERGNE J.P. and CZARNIK A.W. (Eds.), *Chemosensors of Ion and Molecule Recognition.* (ASIC 492) 1997. ISBN 0-7923-4555-X

MICHL J. (Ed.), *Modular Chemistry.* (ASIC 499) 1998. ISBN 0-7923-4730-7

ENS W., STANDING K.G. and CHERNUSHEVICH V. (Eds.), *New Methods for the Study of Biomolecular Complexes.* (ASIC 510) 1998. ISBN 0-7923-5003-0

VECIANA J. et al. (Eds.), *Supramolecular Engineering of Synthetic Metallic Materials.* (ASIC 518) 1998. ISBN 0-7923-5311-0

TSOUCARIS G. (Ed.), *Current Challenges on Large Supramolecular Assemblies.* (ASIC 519) 1999 ISBN 0-7923-5314-5

TABLE OF CONTENTS

ORGANIZING COMMITTEE xi

FOREWORD xiii

LIST OF PARTICIPANTS xv

Probing Self-Assembly by NMR 1
    S.J. Cantrill, M.C.T. Fyfe, F.M. Raymo, and J.F. Stoddart

Study of Conformational Behavior of CMP(O)-Cavitands by NMR Spectroscopy
and X-Ray Analysis in Relation to the Extraction Properties 19
    *Harold Boerrigter, Willem Verboom, Ron Hulst, Gerrit J. van Hummel,
    Sybolt Harkema, and David N. Reinhoudt*

NMR Studies of Molecular Recognition by Metalloporphyrins 37
    *Nick Bampos, Zoë Clyde-Watson, Joanne C. Hawley, Chi Ching Mak,
    Anton Vidal-Ferran, Simon J. Webb, and Jeremy K.M. Sanders*

Reversible Dimerization of Tetraureas Derived from Calix[4]arenes 45
    V. Böhmer, O. Mogck, M. Pons, and E.F. Paulus

Self-Assembling Peptide Nanotubes 61
    Juan R. Granja and M. Reza Ghadiri

Symmetry: Friend or Foe? 67
    *M. Pons, M.A. Molins, O. Millet, V. Böhmer, P. Prados, J. de Mendoza,
    J. Veciana, and J. Sedó*

Quenching Spin Diffusion in Transferred Overhauser Studies and Detection of
Octupolar Order in Gaseous Xenon-131 83
    Michaël Deschamps and Geoffrey Bodenhausen

Applications of NMR Spectroscopy to the Study of the Bound Conformation of
O- and C-Glycosides to Lectins and Enzymes 99
    *J. Jiménez-Barbero, J. Cañada, J.L. Asensio, J.F. Espinosa,
    M. Martín-Pastor, E. Montero, and A. Poveda*

Application of NMR to Conformational Studies of an HIV Peptide Bound
to a Neutralizing Antibody 117
    Anat Zvi and Jacob Anglister

Paramagnetic NMR Effects of Lanthanide Ions as Structural Reporters of
Supramolecular Complexes 133
    Carlos F.G.C. Geraldes

Characterization of Reaction Complex Structures of ATP-Utilizing Enzymes
by High Resolution NMR 155
    B.D. Nageswara Rao

Spin Relaxation Methods for Characterizing Picosecond-Nanosecond and
Microsecond-Millisecond Motions in Proteins 171
    Arthur G. Palmer III and Clay Bracken

Structural Diversity of the Osmoregulated Periplasmic Glucans of Gram-
Negative Bacteria by a Combined Genetics and Nuclear Magnetic Resonance
Approach 191
    Guy Lippens and Jean-Pierre Bohin

NMR Studies of the 269 Residue Serine Protease PB92 from *Bacillus
Alcalophilus* 227
    Yasmin Karimi-Nejad, Frans A.A. Mulder, John R. Martin,
    Axel T. Brünger, Dick Schipper, and Rolf Boelens

Intermolecular Interactions of Proteins Involved in the Control of Gene
Expression 247
    G. Wagner, H. Matsuo, H. Li, C.M. Fletcher, A.M. McGuire,
    A.-C. Gingras, and N. Sonenberg

NMR-Based Modeling of Protein-Protein Complexes and Interaction of
Peptides and Proteins with Anisotropic Solvents 255
    H. Kessler and G. Gemmecker

Protein Surface Recognition 267
    X. Salvatella, T. Haack, M. Gairí, J. de Mendoza, M.W. Peczuh,
    A.D. Hamilton, and E. Giralt

NMR Studies of Protein-Ligand and Protein-Protein Interactions Involving
Proteins of Therapeutic Interest 281
    J. Feeney

NMR Diffusion Measurements in Chemical and Biological Supramolecular
Systems 301
  Yoram Cohen, Orna Mayzel, Ayelet Gafni, Moshe Greenwald,
  Dana Wessely, Limor Frish, and Yaniv

Determination of Calixarene Conformations by Means of NMR Techniques 307
  A. Casnati, J. de Mendoza, D.N. Reinhoudt, and R. Ungaro

Structure, Molecular Dynamics, and Host-Guest Interactions of a Water-Soluble
Calix[4]arene 311
  Jürgen H. Antony and Andreas Dölle

Cation Complexation and Stereochemistry of Macrocycles Studied by ^{13}C NMR
Relaxation-Time Measurements 315
  Çakil Erk

A Comparative Study on Liquid State CSA Methods 319
  Gy. Batta, K.E. Kövér, and J. Kowalewski

Hydrogen Bonding and Cooperativity Effects on the Assembly of
Carbohydrates 325
  M. López de la Paz, J. Jiménex-Barbero, and C. Vicent

Influence of Substituents on the Edge-to-Face Aromatic Interaction: Halogens ... 331
  G. Chessari, C.A. Hunter, J.L. Jiménez-Blanco, C.R. Low, and
  J.G. Vinter

SUBJECT INDEX **335**

Miquel Pons	Universitat de Barcelona, Spain
Geoffrey Bodenhausen	Ecole Normale Supérieure, France
James Feeney	National Institute for Medical Research, UK
Ernest Giralt	Universitat de Barcelona, Spain
Horst Kessler	Technische Universität München, Germany
David Reinhoudt	University of Twente, The Netherlands

FOREWORD

The challenge of developing large but well characterized discrete systems for technological or other applications is being met by exploiting non-covalent interactions between smaller subunits to form unique supramolecular structures. The weakness of the forces involved in the stabilization of the desired structures make them inherently dynamic and the desired uniqueness has to be built in during the design of the subunits. Careful experimental characterization of the final equilibrium structure is mandatory.

The power of NMR as a tool for the structural determination of complex systems has been proven in the field of biomolecules. In addition, NMR has a unique capability for the study of dynamics in different time scales that is being extensively exploited at present in the field of molecular biology. The strong interaction between NMR specialists and structural and molecular biologists has been crucial for the success of NMR in this field : it has guided the development of new NMR techniques needed for specific problems but has also led to the development of labeling techniques that have expanded the range of feasible NMR experiments.

Supramolecular systems have obvious connections with biomolecules. Most processes inside a living organisms are based on non-covalent interactions between different macromolecules and many designed supramolecular systems have been inspired by natural models. On the other hand, the range of possible supramolecular systems extends well beyond the biological models. The Advanced Research Workshop "Applications of NMR to the study of structure and dynamics of supramolecular systems" was conceived as a way of stimulating the interaction between Supramolecular Chemists and NMR specialists, including biomolecular NMR experts.

This idea was encouraged by the steering committee of the NATO Supramolecular Chemistry Program and benefitted from the advice and suggestions of the colleagues, and friends, that accepted invitations to sit on the international organizing committee. Additional financial support was provided by the Ministerio de Educación y Cultura and the Generalitat de Catalunya. The meeting was held in Sitges, near Barcelona (Spain) from the 4th to the 9th of May, 1998.

The presentations could be broadly classified under three general categories: the first class illustrated the state of the art in the design of supramolecular systems and included examples of different classes of supramolecular complexes: catenanes, rotaxanes, hydrogen-bonded rosettes, tubes, capsules, dendrimers, and metal-containing hosts. A second class comprised contributions to NMR methods that can be applied to address the main structural problems that arise in supramolecular chemistry. The third class included biological supramolecular systems studied by state of the art NMR techniques.

Specific structural problems in supramolecular chemistry were identified. These include determining the conformation of a bound molecule in equilibrium with a free state; avoiding the degeneracy problems that complicate the study of symmetrical molecules; avoiding the spin-diffusion problems arising from chemical exchange; coping with ensembles of different conformations in fast exchange; characterizing the dynamics of the systems at different time scales or determining the size or aggregation state of the system.

In addition to the traditional NMR techniques already used extensively, the meeting highlighted some special NMR methods that are not exploited very often in "ordinary" chemistry but that may find increased applications in supramolecular systems: the use of paramagnetic ions, the creation of coherence due to the special properties of quadrupolar nuclei in anisotropic environments, the exploitation of the interference between different relaxation mechanisms, the use of optical pumping to increase NMR sensitivity or the combination of multiple-quantum filtration with diffusion measurements.

The true interdisciplinary atmosphere that developed during the Sitges meeting is difficult to reflect in the pages of a book but browsing through the titles of the contributions may provide a hint. If some of the readers of this volume are encouraged to walk out of their labs and talk to the colleagues in the Department next door, the main purpose of this book of Proceedings and of the Advanced Research Workshop itself will have been achieved.

Miquel Pons

LIST OF PARTICIPANTS

KEY SPEAKERS

Feng Ni
Biotechnology Research Institute
6100 Royalmount Ave
Montreal Quebec H4P 2R2 Canada

Geoffrey Bodenhausen
Ecole Normale Supérieure
24 rue Lhormond
75231 Paris cedex 05, France

Guy Lippens
Institute Pasteur Lille
1, Rue Calmette
59019-Lille CEDEX, France

Volker Böhmer
Institut für Organische Chemie, Johannes
Guttenberg Universität Mainz
J.J. Becher-Weg 18-20
D-55099 Mainz, Germany

Horst Kessler
Institute für Organische Chemie und
Biochemie II
T. U. München
Lichtenbergstrasse, 4
85747 Garching, Germany

Fritz Vögtle
Institute für Organische Chemie und
Biochemie
Rheinischen Friedrich-Wilhelms Universität
Bonn
Gerhard-Domagk Str. 1
53121 Bonn, Germany

Rolf Boelens
Department of Chemistry
University of Utrecht
Padualaan 8,
3584 CH Utrecht, The Netherlands

David N. Reinhoudt
Faculty of Chemical Technology
University of Twente
P.O Box 217
7500 AE Enschede, The Netherlands

Carlos Geraldes
Departamento de Bioquimica
Universidade de Coimbra
Apartado 3126,
3000-Coimbra, Portugal

Ernest. Giralt
Departament de Química Orgànica
Universitat de Barcelona
Marti i Franques, 1
08028-Barcelona, Spain

Jesus Jimenez-Barbero
Instituto de Química Orgánica, CSIC
Juan de la Cierva, 3
28006 Madrid, Spain

Javier de Mendoza
Departamento de Quimica Organica
Universidad Autonoma de Madrid
Cantoblanco
28049-Madrid, Spain

Miquel Pons
Departament de Química Orgànica
Universitat de Barcelona
Marti i Franques, 1
08028-Barcelona, Spain

James Feeney
Division of Molecular Structure
National Institute for Medical Research
The Ridgeway, Mill-Hill
London NW7 1AA, United Kingdom

Jeremy Sanders
Department of Chemistry
University of Cambridge
Lensfield Road,
Cambridge CB2 1EW, United Kingdom

Art Palmer
Department of Biochemistry and Molecular Biophysics
Columbia University
630 W. 168th St.
New York, NY 10032, USA

Alex Pines
Department of Chemistry
University of California at Berkeley
CA 94720, USA

B.D. Nageswara Rao
Department of Physics, IUPUI
402 N. Blackford Street
Indianapolis, IN 46202-3273, USA

J Fraser Stoddart
Department of Chemistry and Biochemistry
University of California, Los Angeles
405 Hilgard Avenue,
Los Angeles, CA 90095-1569, USA

Gerhard Wagner
Department of Biological Chemistry and Molecular Pharmacology, Harvard Medical School
240 Longwood Ave,
Boston, MA 02115, USA

Jacob Anglister
Department of Structural Biology
Weizmann Institute of Science
76100 Rehovot, Israel

Gil Navon
School of Chemistry, Tel Aviv University
Tel Aviv 69978, Israel

Anil Kumar
Department of Physics and Sophisticated Instruments Facility
Indian Institute of Science
Bangalore-560012
India

OTHER PARTICIPANTS

Pascale Delangle
CEA-Grenoble
/DRFMC/SCIB/Reconnaissance Ionique
17 avenue des martyrs
38054 Grenoble cedex 9, France

Andreas Doelle
Institute of Physical Chemistry, RWTH
Aachen
52056 Aachen, Germany

Margit Gruner
Institute of Organic Chemistry of the TU
Dresden
Mommsenstrasse 13
01062 Dresden, Germany

Tanja Kortemme
European Molecular Biology Laboratory
Meyerhofstr. 1
69117 Heidelberg, Germany

Alessandro Casnati
Dept. of Organic and Industrial Chemistry
Viale delle Scienze 78
43100 Parma, Italy

Enrico Redenti
Chiesi Farmaceutici
Via S. Leonardo 96
43100 Parma, Italy

Gert Jan Boender
University of Nijmegen
Toernooiveld 1
6525 ED, Nijmegen, The Netherlands

Willem Verboom
Laboratory of Suparmolecular Chemistry
and Technology
University of Twente
P.O.Box 217
7500 AE Enschede, The Netherlands

Juan Espinosa
Instituto de Química Orgánica
Juan de la Cierva 3
28006 Madrid, Spain

Juan Granja
Departamento de Química
Universidad de Santiago
Santiago de Compostela, Spain

Mª Antonia Molins
University of Barcelona
Martí i Franqués 1
08028 Barcelona, Spain

Pedro Nieto
Instituto de Investigaciones Químicas
c/ Américo Vespucio s/n
41092 Isla de la Cartuja
Sevilla, Spain

Eduardo Rubio
Universidad de Oviedo
Dept. Química Orgànica e Inorgànica
Facultad de Química
33071 Oviedo, Spain

Francesc Sánchez-Ferrando
Universitat de Barcelona
Departament de Química
08193 Bellaterra, Spain

Cristina Vicent
Instituto de Química Orgánica, CSIC
c/ Juan de la Cierva 3
28006 Madrid, Spain

Anton Vidal
Dept. Química Orgánica
Universitat de Barcelona
Martí i Franquès 1
08028 Barcelona, Spain

Erk Çakil
Istanbul Technical University
Maslak
80626 Istanbul, Turkey

Nick Bampos
Department of Chemistry
University of Cambridge
Lensfield Road
CB2 1EW Cambridge, United Kingdom

Eric Brouwer
Department of Chemistry
University of Durham
Science Laboratories, South Road
DH1 3LE Durham, United Kingdom

Gianni Chessari
University of Sheffield
Chemistry Department, Brook Hill
S3 7HF Sheffield, United Kingdom

Stefan Freund
Centre for Protein Engeneering
Chemical Laboratories, Lensfield Road
CB4 3PP Cambridge, United Kingdom

Phil Douglas
School of Chemistry
University of Birmingham
Edgbaston
Birmingham B15 2TT, United Kingdom

Nuria Assa-Munt
Structural Biology Program
NMR Facility Faculty Supervisor
The Burnham Institute
10901 North Torrey Pines Road
La Jolla, CA 92037, U.S.A.

Brian J. Stockman
Structural, Analytical and Medicinal
Chemistry
Pharmacioa & Upjohn
301 Henrietta St.
Kalamazoo, MI 49001, U.S.A.

Gyula Batta
L. Kossuth University
Egyetem ter 1
4010 Debrecen, Hungary

Yoram Cohen
Tel Aviv University
Ramat-Aviv
69978 Tel-Aviv, Israel

Konstantin Pervouchine
Institut für Molekularbiologie und
Biophysik
Eidgenossische Technische Hochshule
Hoggerberg
8093 Zurich, Switzerland

PROBING SELF-ASSEMBLY BY NMR

S.J. CANTRILL, M.C.T. FYFE, F.M. RAYMO, J.F. STODDART
Department of Chemistry and Biochemistry
University of California, Los Angeles
405 Hilgard Avenue, Los Angeles, CA 90095-1569, USA

1. NMR Spectroscopy in Supramolecular Chemistry

One of the major challenges in supramolecular chemistry [1] is the characterization of the (super)structural properties of the (supra)molecular species resulting from self-assembly [2,3] processes. Although the stereoelectronic properties of the supramolecular synthons are responsible for determining the geometries of the final self-assembled (super)strucures, in most instances the experimental conditions (*e.g.*, solvent, temperature, pressure, and stoichiometry) play a dominant role. As a result, the prediction of the geometries of supramolecular entities is difficult and, more often than not, impossible even by means of computational methods [4]. Experimental techniques that are able to establish the (super)structural properties of (supra)molecular species unequivocally are much in demand. X-Ray crystallography has played a major role in determining the superstructures of supermolecules and supramolecular arrays, as well as the structures of large and complex molecular assemblies, in the solid state [5]. The characterization of self-assembled monolayers at the air/water interface — and on solid supports — has been achieved employing sophisticated microscopic techniques {*e.g.*, Brewster angle microscopy [6], scanning and transmission electron microscopies (SEM and TEM) [7], atomic force microscopy (AFM) [8] and scanning tunneling microscopy (STM) [9]}. The combination of absorption and emission spectroscopies [10], along with electrochemical techniques [11], has permitted the characterization of self-assembled products in solution. Chromatographic methods {*e.g.*, high performance liquid chromatography (HPLC) [12] and gel permeation chromatography (GPC) [13]}, often coupled with mass spectrometric techniques {*e.g.*, fast atom bombardment and liquid secondary ion mass spectrometries (FABMS and LSIMS) [14], electrospray mass spectrometry (ESMS) [15], and matrix-assisted laser desorption ionization time-of-flight mass spectometry (MALDI–TOF–MS) [16]}, are becoming more and more powerful analytical methods for the characterization of (supra)molecular species. However, the most informative and widely used analytical method employed for the study of such species in solution is nuclear magnetic resonance (NMR) spectroscopy [17]. By relying on the evaluation of simple parameters (*e.g.*, chemical shifts, relative signal intensities, and spin–spin couplings) inter-component interactions, thermodynamic (*e.g.*, association constants) and kinetic (*e.g.*, rate constants) data, as well as (co)conformational changes, can be elucidated easily. However, in many cases, advanced NMR spectroscopic techniques [*e.g.*, nuclear Overhauser effects (nOe), variable temperature nuclear magnetic resonance (VT–NMR), and line-shape analysis] are required to supplement the conventional ^{1}H- and ^{13}C-NMR spectroscopies.

We have employed a combination of ^{1}H-NMR spectroscopic techniques to probe thermodynamically- and kinetically-controlled self-assembly processes [18] that lead to pseudorotaxanes, catenanes, and rotaxanes [19]. The association constants (K_a) of numerous pseudorotaxane complexes have been measured [20] by single-point determinations or by dilution/titration experiments when the complexation/ de-complexation processes are slow or fast, respectively, on the ^{1}H-NMR timescale. Furthermore, the kinetic and thermodynamic parameters, associated with dynamic processes involving the relative movements of the mechanically-interlocked components of catenanes and rotaxanes, have been estimated [21] by variable temperature ^{1}H-NMR spectroscopy. These results have paved the way for the design and realization of molecular and supramolecular switches [22], in the shape of catenanes, rotaxanes, and pseudorotaxanes. In this Chapter, some representative examples of pseudorotaxanes, rotaxanes, and 'daisy chain' supermolecules, whose structural and functional properties have been studied by ^{1}H-NMR spectroscopy, are discussed.

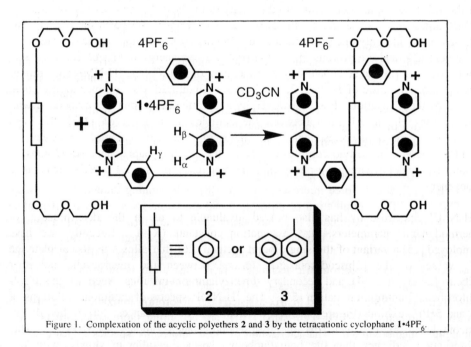

Figure 1. Complexation of the acyclic polyethers **2** and **3** by the tetracationic cyclophane **1•4PF$_6$**.

2. Pseudorotaxanes

The π-electron deficient bipyridinium-based tetracationic cyclophane 1•4PF$_6$ binds [23] (Figure 1) the π-electron rich 1,4-dioxybenzene- and 1,5-dioxynaphthalene-based polyethers 2 and 3, respectively. The resulting pseudorotaxanes are held together by *(i)* [C–H···O] hydrogen bonds (between the α-bipyridinium hydrogen atoms and the polyether oxygen atoms), *(ii)* [π···π] stacking (between the complementary π-electron deficient and π-electron rich aromatic units), and *(iii)* [C–H···π] interactions (between the dioxyarene hydrogen atoms and the *p*-phenylene spacers of 1•4PF$_6$). The ^{1}H-NMR spectrum (CD$_3$CN, 233 K) of 1•4PF$_6$ shows (Figure 2a) sharp and well-resolved signals for H$_\alpha$, H$_\beta$, and H$_\gamma$. On mixing equimolar amounts of 1•4PF$_6$ and 2 in CD$_3$CN, these resonances shift (Figure 2b) as a result of complex formation. When 1•4PF$_6$ is combined with equimolar amounts of 3 in CD$_3$CN, all the signals for constitutionally identical protons in the tetracationic host separate (Figure 2c) into two sets of resonances on complexation, as a consequence of the local C_{2h} symmetry imposed by the 1,5-dioxynaphthalene unit of the guest. Upon warming, the rate of the dynamic process, involving the loss of the local C_{2h} symmetry — *i.e.*, *(i)* extrusion of the 1,5-dioxynaphthalene unit from the cavity of the host, followed by *(ii)* either a 180° rotation of the bipyridinium rings, with respect to their [N···N] axis, or a 180° rotation of the 1,5-dioxynaphthalene ring system, with respect to its [O···O] axis, and *(iii)* reinsertion of the 1,5-dioxynaphthalene unit into the cavity of the host — becomes fast on the ^{1}H-NMR timescale. Consequently, the two sets of resonances, observed for H$_\alpha$, H$_\beta$, and H$_\gamma$, coalesce into only one set in each case. By employing the approximate coalescence treatment [24] and the protons H$_\alpha$ as the probe, a free energy barrier of *ca.* 14.1 kcal mol^{-1} was determined at a coalescence temperature of 304 K. The ^{1}H-NMR spectrum of an equimolar mixture of 1•4PF$_6$, 2, and 3 shows (Figure 2d) the complexation of both guests by the tetracationic host. However, by integrating the resonances associated with the protons H$_\beta$, a 5:1 ratio between the complexes, in favor of [1:3]•PF$_6$, was estimated. This ratio corresponds to a difference of *ca.* 1.6 kcal mol^{-1} between the binding energies of the two complexes. Thus, the 1,5-dioxynaphthalene-based guest is bound by cyclobis(paraquat-*p*-phenylene) in preference to its 1,4-dioxybenzene-based congener.

^{1}H-NMR Spectroscopy has also proved invaluable to us in the determination of thermodynamic parameters, *viz.*, association constants (K_as). Recently, we have employed [25] a variant of the single-point method [26] that allows us to calculate the K_a values of the [2]pseudorotaxanes formed between the macrocyclic polyether dibenzo[24]crown-8 (4) and secondary dibenzylammonium ions, such as the bis(4-chlorobenzyl)ammonium cation (5$^+$). The ^{1}H-NMR spectra of equimolar solutions of 4 and 5•PF$_6$ exhibit (Figure 3) three different sets of resonances: for *(i)* free 4, *(ii)* uncomplexed 5•PF$_6$, and *(iii)* the [2]pseudorotaxane complex [4:5]•PF$_6$. This observation indicates that the bound/unbound species equilibrate slowly with one another on the ^{1}H-NMR timescale by virtue of the large steric barrier experienced when the cation's relatively bulky benzenoid rings attempt to pass through/out of **4**'s cavity. This phenomenon permits the calculation of the absolute concentrations of *(i)* uncomplexed **4/5**•PF$_6$ and *(ii)* pseudorotaxane from the known initial **4/5**•PF$_6$ concentrations and the ^{1}H-NMR signals' relative intensities. Ultimately, the K_a value is calculated using the relationship K_a = [[**4**:**5**]•PF$_6$] / [**4**][**5**•PF$_6$].

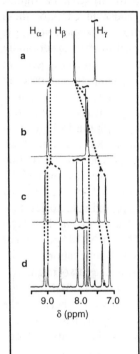

Figure 2. Partial ^{1}H-NMR spectra [CD$_3$CN, 233 K] of (a) **1**•4PF$_6$, and equimolar solutions of (b) **1**•4PF$_6$ and **2**, (c) **1**•4PF$_6$ and **3**, and (d) **1**•4PF$_6$, **2**, and **3**.

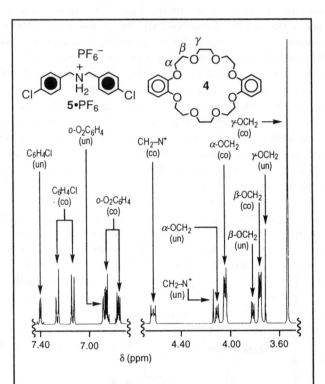

Figure 3. Partial ^{1}H-NMR spectrum [CDCl$_3$ / CD$_3$CN (1:1), 304 K] of an equimolar **4**–**5**•PF$_6$ solution (each 1.0 × 10^{-2} M), demonstrating that uncomplexed (un) and complexed (co) species interconvert slowly with each other on the ^{1}H-NMR timescale.

3. The Slippage Approach to Rotaxanes

If good size complementarity between the cavity of a macrocycle and the stoppers of a dumbbell-shaped compound is achieved, the slipping-on of the macrocycle over one of the two stoppers of the dumbbell occurs (Figure 4) when a solution of the two components is heated at an appropriate temperature. If macrocyclic and dumbbell-shaped components incorporate complementary recognition sites, the resulting [2]rotaxane is stabilized, relative to its free components, by noncovalent bonding interactions. Upon cooling the solution down to an appropriate temperature, both slipping-on and slipping-off processes are 'frozen' and the kinetically-stable [2]rotaxane can be isolated from the free macrocyclic and dumbbell-shaped components. When an MeCN solution of the macrocyclic polyether **6** and the dumbbell-shaped compound **7**•4PF$_6$ (Figures 5 and 6) is heated at 323 K, both the [2]rotaxane **8**•4PF$_6$ and the [3]rotaxane **9**•4PF$_6$ self-assemble [27] after the slipping-on of **6** over the stoppers of **7**•4PF$_6$. When the solution is cooled down to ambient temperature after 10 d, the rotaxanes become kinetically-stable and can be isolated by column chromatography. Their yields are related to the starting ratios of the macrocyclic and dumbbell-shaped compounds. When 4.0 molar equivalents of **6**, with respect to **7**•4PF$_6$, are employed, the yields of **8**•4PF$_6$ and **9**•4PF$_6$ are 31 and 8%, respectively. When 10.0 molar equivalents of **6** are employed, under otherwise identical conditions, the yields of **8**•4PF$_6$ and **9**•4PF$_6$ change to 20 and 55%, respectively.

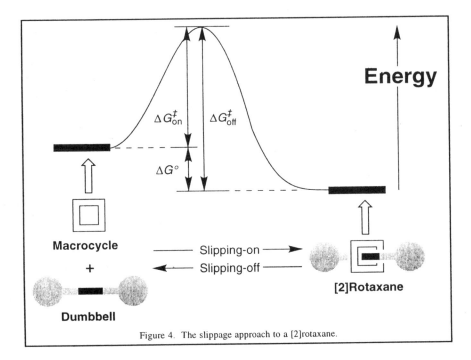

Figure 4. The slippage approach to a [2]rotaxane.

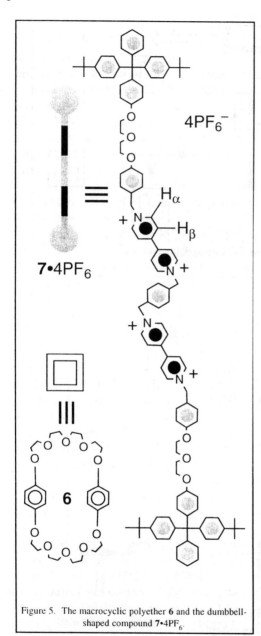

Figure 5. The macrocyclic polyether **6** and the dumbbell-shaped compound **7**•4PF$_6$.

When a (CD$_3$)$_2$SO solution of the pure [3]rotaxane **9**•4PF$_6$ is heated at 100°C, the macrocyclic components slip-off (Figure 6) over the stoppers of the dumbbell-shaped component to afford the [2]rotaxane **8**•4PF$_6$, and ultimately, the free dumbbell-shaped compound **7**•4PF$_6$. This process can be followed (Figure 7) by ^{1}H-NMR spectroscopy. The partial ^{1}H-NMR spectrum, illustrated in Figure 7a, shows the resonances associated with the pure [3]rotaxane **9**•4PF$_6$ in (CD$_3$)$_2$SO at 298 K. When this solution is heated up to 373 K, some of the [3]rotaxane **9**•4PF$_6$ is converted into the [2]rotaxane **8**•4PF$_6$ — the ^{1}H-NMR spectrum recorded at this temperature reveals (Figure 7b) signals attributable to both rotaxanes. After 600 s, the resonances of the free dumbbell-shaped compound **7**•4PF$_6$ are also evident (Figure 7c) in the ^{1}H-NMR spectrum. After 7200 s, the only species present in solution are (Figure 7e) the free dumbbell-shaped compound **7**•4PF$_6$ and the free macrocyclic polyether **6**. By measuring the relative intensities of the resonances associated with **7**•4PF$_6$–**9**•4PF$_6$ in the ^{1}H-NMR spectra illustrated in Figures 7b–7e, the rate constants k_A and k_B (Figure 6) were determined. When the protons H$_\alpha$ were employed as the probe, the values of k_A and k_B were 18.9×10^{-4} and 4.91×10^{-4} s^{-1}, respectively. Similarly, when the protons H$_\beta$ were employed as the probe, k_A and k_B were 20.3×10^{-4} and 4.92×10^{-4} s^{-1}, respectively.

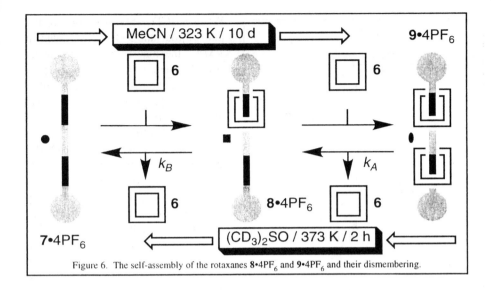

Figure 6. The self-assembly of the rotaxanes 8•4PF$_6$ and 9•4PF$_6$ and their dismembering.

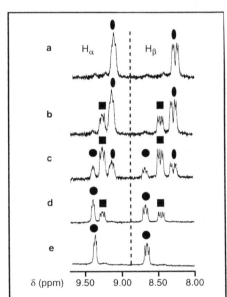

Figure 7. Partial ^{1}H-NMR spectra [(CD$_3$)$_2$SO] of the [3]rotaxane 9•4PF$_6$ (a) at 298 K and (b)–(e) at 373 K: (b) at initial time, and after (c) 600 s, (d) 2400 s, and (e) 7200 s.

The bipyridinium-based compound 10•6PF$_6$ incorporates tetra-arylmethane-based stoppers of appropriate sizes for the slipping-on of the macrocyclic polyether 6. When a solution of 10•6PF$_6$ and 6 is heated (Figure 8) at 323 K for 10 d, the [2]rotaxane 11•6PF$_6$, the [3]rotaxane 12•6PF$_6$, and the [4]rotaxane 13•6PF$_6$ self-assemble [28]. When the solution is cooled down to ambient temperature, these rotaxanes become kinetically-stable and can be isolated by column chromatography. When only 2.0 molar equivalents of 6, with respect to 10•6PF$_6$, are employed, the yields of the rotaxanes are 46, 26, and 6% for 11•6PF$_6$, 12•6PF$_6$, and 13•6PF$_6$, respectively. By employing a large amount of 6 (15.0 molar equivalents), the yields of 11•6PF$_6$, 12•6PF$_6$, and 13•6PF$_6$ change to 22, 41, and 19 %, respectively. In 10•6PF$_6$, the three 'free' bipyridinium units are equivalent and only two sets of signals are observed (Figure 9a) for the protons H'$_\alpha$ and H"$_\alpha$ in the ^{1}H-NMR

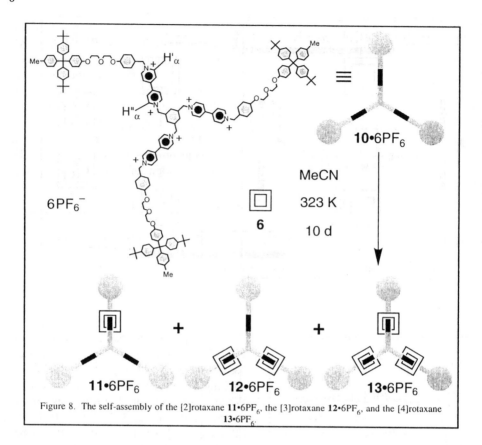

Figure 8. The self-assembly of the [2]rotaxane **11·6PF$_6$**, the [3]rotaxane **12·6PF$_6$**, and the [4]rotaxane **13·6PF$_6$**.

spectrum [(CD$_3$)$_2$CO, 298 K]. The ^{1}H-NMR spectrum of the [2]rotaxane **11·6PF$_6$** shows (Figure 9b) two sets of signals for the protons H'$_\alpha$ and H"$_\alpha$ of the two equivalent 'free' bipyridinium units. In addition, another two sets of signals, associated with the protons H'$_\alpha$ and H"$_\alpha$ of the 'occupied' bipyridinium unit, appear at 'higher fields'. The chemical shift change observed for these protons is a result of the shielding effects exerted by the sandwiching 1,4-dioxybenzene rings. Similarly, four sets of signals are observed for the protons H'$_\alpha$ and H"$_\alpha$ of the three bipyridinium units of the [3]rotaxane **12·6PF$_6$**. However, in this instance, the two equivalent bipyridinium units are 'occupied' and the signals appearing at 'higher fields' are more intense than those appearing at 'lower fields'. In the [4]rotaxane **13·6PF$_6$**, the three 'occupied' bipyridinium units are equivalent and only two sets of signals are observed.

The amphiphilic compound **10·6PF$_6$** possesses a hexacationic core surrounded by hydrophobic tetra-arylmethane-based stoppers and undergoes aggregation in nonpolar solvents. Indeed, while the ^{1}H-NMR spectra of CD$_3$CN and (CD$_3$)$_2$CO solutions of **10·6PF$_6$** show (Figure 10a and 10b, respectively) sharp and well-resolved signals, those

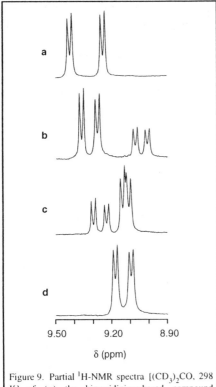

Figure 9. Partial ^{1}H-NMR spectra [(CD$_3$)$_2$CO, 298 K] of (a) the bipy-ridinium-based compound **10·6PF$_6$**, (b) the [2]rotaxane **11·6PF$_6$**, (c) the [3]rotaxane **12·6PF$_6$**, and (d) the [4]rotaxane **13·6PF$_6$**.

recorded in CDCl$_3$, under otherwise identical conditions (1.0 mM, 298 K), reveal (Figure 10c) broad resonances. However, sharpening of the signals is observed (Figures 10d and 10e) upon the grad-ual addition of (CD$_3$)$_2$CO. The plot, shown in Figure 11, illustra-tes the chemical shift change obser-ved for the protons H$_\alpha$ upon varying the concentration of a CDCl$_3$/(CD$_3$)$_2$CO

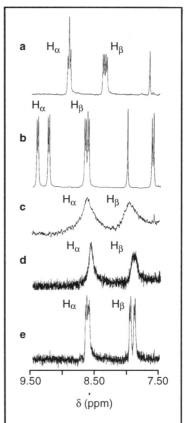

Figure 10. Partial ^{1}H-NMR spectra [298 K, concentration = 1.0 mM] of the amphiphilic compound **10·6PF$_6$** in (a) CD$_3$CN, (b) (CD$_3$)$_2$CO, (c) CDCl$_3$, (d) CDCl$_3$/(CD$_3$)$_2$CO (85:15), and (e) CDCl$_3$/(CD$_3$)$_2$CO (70:30).

(85:15) solution of **10·6PF$_6$** at 298 K. Significant changes in the chemical shifts are observed at low concentra-tions (<10^{-3} M), as a result of the equilibration of **10·6PF$_6$** in its 'free' and aggregate forms. On the contrary, no appreciable changes in the chemical shifts values are observed at higher concentrations (>10^{-3} M).

When a CH$_2$Cl$_2$ solution of the macrocyclic polyether **4** and the secondary dialkylammonium salt **14·PF$_6$** is heated at 313 K for 36 d, the complex **[4:14]·PF$_6$** is obtained [29] (Figure 12) in a yield of 90 %, after the slipping-on of **4** over

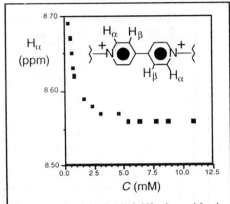

Figure 11. Plot of the chemical shifts observed for the protons H_α in the ^{1}H-NMR spectra [CDCl$_3$/(CD$_3$)$_2$CO (85:15), 298 K] of **10·6PF$_6$** against the concentration.

the cyclohexyl groups of **14·PF$_6$**. The process can be followed (Figure 13) by ^{1}H-NMR spectroscopy. The partial ^{1}H-NMR spectrum [recorded in CDCl$_3$/CD$_3$CN (3:1) at 293 K] of an equimolar mixture of **4** and **14·PF$_6$** shows (Figure 13a) the signals of the uncomplexed species (un). The gradual appearance of signals associated with the complexed species (co) is observed (Figure 13b and 13c) when the same solution is heated at 313 K. By measuring the relative intensities of the complexed and uncomplexed species, the rate constants for the slipping-on (k_{on}) and slipping-off (k_{off}) processes were determined employing the [CH$_2$–N$^+$] protons as the probe. At 313 K in CDCl$_3$/CD$_3$CN (3:1), k_{on} and k_{off} are 2.9 × 10^{-5} M^{-1} s^{-1} and 2.6 × 10^{-7} s^{-1}, respectively. On dissolving the pure complex [**4:14**]·PF$_6$ in (CD$_3$)$_2$SO, [**4:14**]·PF$_6$ dismembers (Figure 14) into its constituent components gradually. The ^{1}H-NMR spectrum of a (CD$_3$)$_2$SO solution of [**4:14**]·PF$_6$, recorded after 3 min, shows (Figure 14a), for the most part, only the resonances of the complex (co). The gradual disappearance of the resonances of the complex is accompanied (Figures 14b and 14c) by the appearance of the signals for the free species. After 18 h, the resonances of free **4** and **14·PF$_6$** are observed (Figure 14d) in the ^{1}H-NMR spectrum solely.

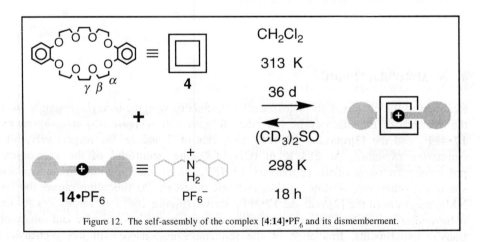

Figure 12. The self-assembly of the complex [**4:14**]·PF$_6$ and its dismemberment.

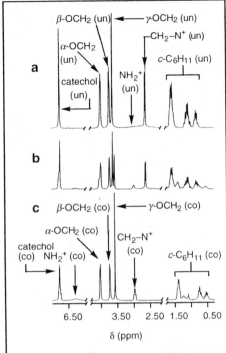

Figure 13. Partial ¹H-NMR spectra [CDCl₃/CD₃CN (3:1)] of an equimolar mixture of **4** and **14**•PF₆ (a) at 293 K and (b) at 313 K after 112 d, as well as (c) of the pure complex [**4**:**14**]•PF₆ at 293 K [(un) and (co) stand for uncomplexed and complexed, respectively].

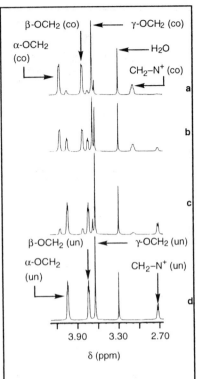

Figure 14. Partial ¹H-NMR spectra [(CD₃)₂SO (3:1), 298 K] of [**4**:**14**]•PF₆ after (a) 3 min, (b) 48 min, (c) 4 h, and (d) 18 h.

4. A Molecular Shuttle

Reaction of the bis(hexafluorophosphate) salt **15**•2PF₆ with the benzylic bromide **16** in the presence of the macrocyclic polyether **6** gave [30] (Figure 15) the [2]rotaxane **17**•4PF₆ and the [3]rotaxane **18**•4PF₆ in yields of 5 and 23 %, respectively, after counterion exchange. At 213 K in (CD₃)₂CO, the 'shuttling' of the macrocyclic polyether component of the [2]rotaxane **13**•4PF₆, from one bipyridinium recognition site to the other one, is slow on the ¹H-NMR timescale. At this temperature, the ¹H-NMR spectrum of the [2]rotaxane **17**•4PF₆ shows (Figure 16d) four resonances for the α-bipyridinium protons. These signals were assigned on the basis of nOe and saturation transfer experiments. Irradiation of the resonances associated with the *p*-phenylene protons H–5 (Figures 17 and 18) at 213 K results in an nOe experienced by the signals

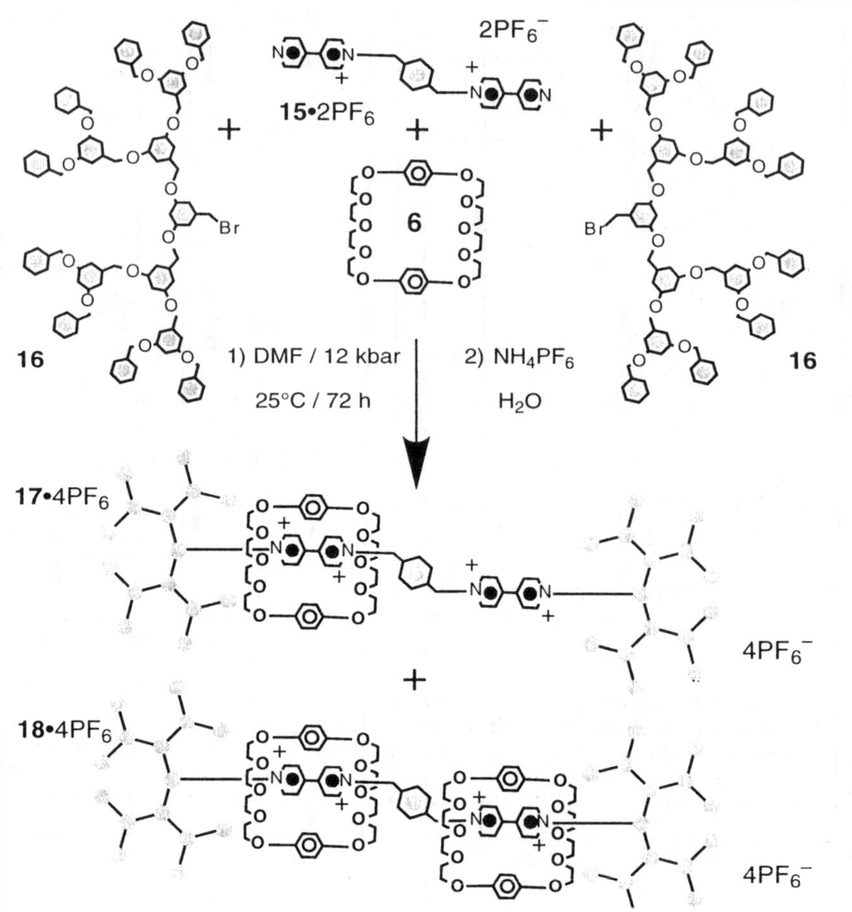

Figure 15. The self-assembly of the [2]rotaxane **17·4PF$_6$** and of the [3]rotaxane **18·4PF$_6$**.

centered on δ 9.42 and 9.23 ppm, which correspond to the adjacent α-protons. However, the protons associated with an 'occupied' bipyridinium unit resonate at 'higher fields' than those of a 'free' one. Thus, the resonances, centered on δ 9.42 and 9.23 ppm, must correspond to H–3 and H–2, respectively, while the remaining two resonances, centered on δ 9.35 and 9.04 ppm, must correspond to H–4 and H–1, respectively. As a result of the 'shuttling' process, H–1 and H–4 are in slow exchange on the ^{1}H-NMR timescale at 213 K. Consistently, irradiation of the signal at δ 9.04 ppm at this temperature results (Figure 19) in saturation transfer to the resonance centered on δ 9.35 ppm. On warming a (CD$_3$)$_2$CO solution of **17·4PF$_6$** up, the resonances associated with H–1 and H–4, as well as those associated with H–2 and H–3, coalesce (Figure 16b and 16c) and, at 273 K, only two partially overlapping sets of

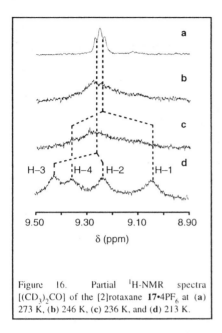

Figure 16. Partial ^{1}H-NMR spectra [$(CD_3)_2CO$] of the [2]rotaxane **17•4PF$_6$** at (**a**) 273 K, (**b**) 246 K, (**c**) 236 K, and (**d**) 213 K.

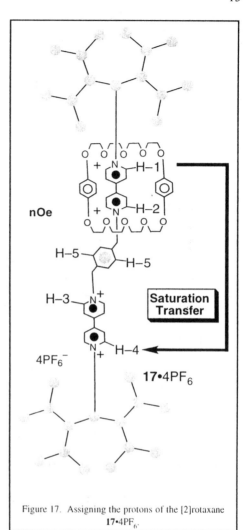

Figure 17. Assigning the protons of the [2]rotaxane **17•4PF$_6$**.

signals are observed (Figure 16a) for the four sets of α-protons. By employing the approximate coalescence treatment, the rate constants and derived free energy barriers for the 'shuttling' process were determined in a range of solvents of decreasing polarity [$(CD_3)_2CO$, THF-d$_8$, CD_2Cl_2, and $CDCl_3$]. At 298 K, the rate of 'shuttling' decreases from *ca.* 33000 to 200 s^{-1} on going from $(CD_3)_2CO$ to $CDCl_3$ and the free energy barrier increases by *ca.* 3 kcal mol^{-1}.

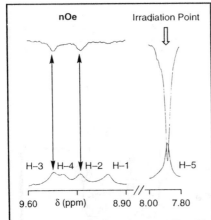

Figure 18. Partial ^{1}H-NMR spectra [$(CD_3)_2CO$, 213 K] of the [2]rotaxane **17**•4PF$_6$ illustrating the nOe occurring between the protons attached to the *p*-xylene spacer H–5 and the bipyridinium α-protons H–2 and H–3.

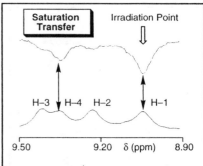

Figure 19. Partial ^{1}H-NMR spectra [$(CD_3)_2CO$, 213 K] of the [2]rotaxane **17**•4PF$_6$ illustrating the saturation transfer from H–1 to H–4.

5. Supramolecular Daisy Chains

The quest for ever more elaborate mechanically-interlocked (super)structures led us to synthesize (Figure 20) the self-complementary monomer **24**•HPF$_6$. It was hoped that this self-complementary molecule would self-assemble into linear and/or cyclic 'daisy chain' arrays [31] in appropriate non-competitive solvents [20a]. Indeed, three stereochemically distinct head-to-tail dimeric supermolecules (the enantiomers **A/C** and the *meso* form **B**, see Figure 21) can be formed by **24**•HPF$_6$. In these dimeric complexes, the NH$_2^+$-containing sidearms of each component are threaded simultaneously through the complementary macrocyclic recognition sites of their counterparts. In the solid state, a racemic mixture of the two C_2 symmetric supermolecules **A** and **C** is observed (Figure 21): no *meso* form **B** is present in the crystal. The ^{1}H-NMR spectrum of a $(CD_3)_2SO$ solution of **24**•HPF$_6$ reveals (Figure 22a) signals that can be assigned to the uncomplexed monomer **24**•HPF$_6$. By contrast, an extremely complicated ^{1}H-NMR spectrum was obtained (Figure 22b) in CD$_3$CN. Since it is known [20a] that the complexation of secondary dialkylammonium salts by macrocyclic polyethers is highly disfavored in $(CD_3)_2SO$, the simplicity of the spectrum shown in Figure 22a is hardly surprising. Instead, in CD$_3$CN, relatively strong hydrogen bonds favor [20a] the insertion of the NH$_2^+$-containing sidearms through the cavity of the macrocyclic polyether. Thus, the complexity of the ^{1}H-NMR spectrum, illustrated in Figure 22b, is presumably a result of the equilibration of several supermolecules and/or supramolecular arrays. Interestingly, however, the ^{1}H-NMR spectrum of a CD$_3$CN solution of **24**•HPF$_6$ recorded at 353 K, is remarkably similar to that recorded in $(CD_3)_2SO$ at 298 K, indicating that, at high temperatures, the uncomplexed monomer **24**•HPF$_6$ is the predominant species.

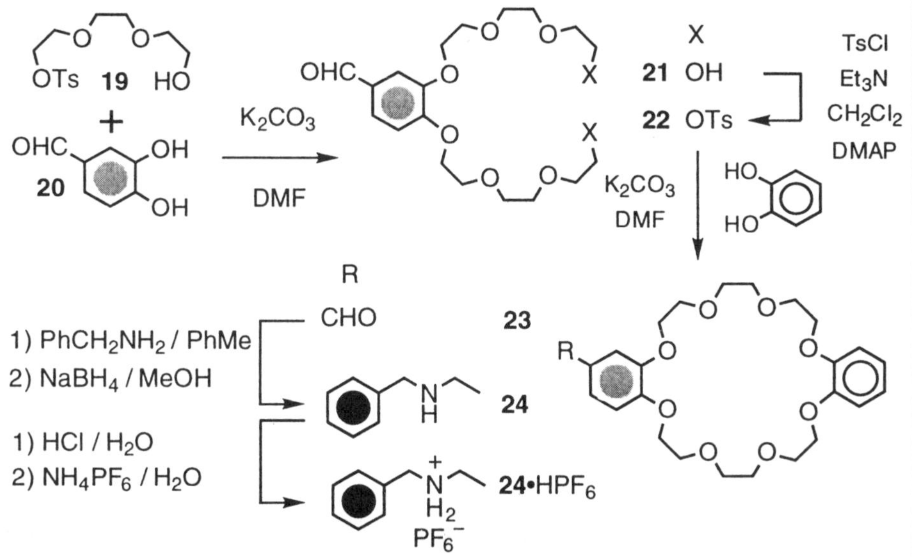

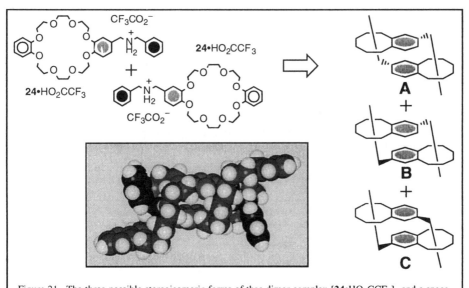

Figure 20. The synthesis of the self-complementary monomer 24•HPF$_6$.

Figure 21. The three possible stereoisomeric forms of thee dimer complex [24•HO$_2$CCF$_3$]$_2$ and a space-filling representation of its solid state geometry.

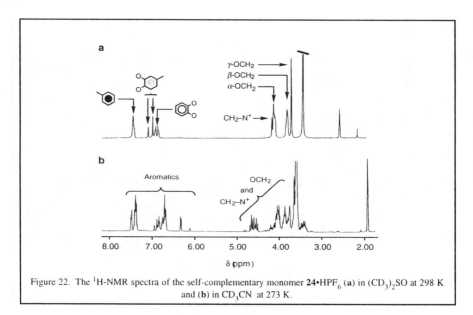

Figure 22. The ^{1}H-NMR spectra of the self-complementary monomer **24**•HPF$_6$ (a) in (CD$_3$)$_2$SO at 298 K and (b) in CD$_3$CN at 273 K.

6. Conclusions

The elucidation of the (super)structural properties of large and complex (supra)molecular species has been achieved using NMR spectroscopy by evaluating simple parameters such as chemical shifts, relative signal intensities, and spin–spin couplings, as well as by nOe experiments. The association constants (K_a) and the corresponding free energies of complexation ($\Delta G°$) of numerous pseudorotaxane complexes have been determined. An evaluation of the relative intensities of appropriate signals is sufficient to be able to derive the K_a and $\Delta G°$ values when the complex and its separate host/guest components are in slow exchange on the NMR timescale. On the other hand, when the complex and its separate host/guest components are in fast exchange on the ^{1}H-NMR timescale, K_a and $\Delta G°$ values can be derived by monitoring the chemical shift changes of probe resonances on varying the absolute/relative concentrations of the host and/or guest. In addition to their stabilities, the rates of association and disassociation of pseudorotaxane complexes have also been determined, employing NMR spectroscopy, by monitoring the changes in the relative intensities of appropriate probe resonances with time. Similarly, the rate constants and free energy barriers associated with the relative movements of the mechanically-interlocked components of a number of rotaxanes and catenanes have been determined by variable-temperature NMR spectroscopy employing the approximate coalescence treatment. In conclusion, structural and superstructural information, as well as thermodynamic and kinetic data, that would otherwise be unattainable, can be derived by NMR spectroscopy, rendering this analytical method an indispensable tool in contemporary chemistry — be it molecular or supramolecular in nature.

7. References

1. a) Vögtle, F. (1991) *Supramolecular Chemistry*, Wiley, New York. b) Lehn, J.-M. (1995) *Supramolecular Chemistry*, VCH, Weinheim. c) Eds. Atwood, J.L., Davies, J.E.D., MacNicol, D.D., and Vögtle, F. (1996) *Comprehensive Supramolecular Chemistry*, Pergamon, Oxford.
2. a) Lehn, J.-M. (1990) *Angew. Chem., Int. Ed. Engl.*, **29**, 1304–1319. b) Lindsey, J.S. (1991) *New J. Chem.*, **15**, 153–180. c) Lawrence, D.S., Jiang, T., and Levett, M. (1995) *Chem. Rev.*, **95**, 2229–2260. d) Conn, M.M. and Rebek, J., Jr. (1997) *Chem. Rev.*, **97**, 1647–1668. e) Linton, B. and Hamilton, A.D. (1997) *Chem. Rev.*, **97**, 1669–1680. f) Stupp, S.I., LeBonheur, V., Walker, K., Li, L.S., Huggins, K.E., Keser, M., and Amstutz, A. (1997) *Science*, **276**, 384–389.
3. a) Philp, D. and Stoddart, J.F. (1991) *Synlett*, 445–458. b) Philp, D. and Stoddart, J.F. (1996) *Angew. Chem., Int. Ed. Engl.*, **35**, 1154–1196. c) Raymo, F.M. and Stoddart, J.F. (1996) *Curr. Op. Coll. Interf. Sci.*, **1**, 116–126. d) Fyfe, M.C.T. and Stoddart, J.F. (1997) *Acc. Chem. Res.*, **30**, 393–401. e) Raymo, F.M. and Stoddart, J.F. (1998) *Curr. Op. Coll. Interf. Sci.*, **3**, 150–159.
4. a) Roterman, I., Rybarska, J., Konieczny, L., Skowronek, M., Stopa, B., Piekarska, B., and Bakalarski, G. (1998) *Computers Chem.*, **22**, 61–70. b) Cervello, E., and Jaime, C. (1998) *J. Mol. Struct. (Theochem)*, **428**, 195–201. c) Goncalves, H., Robinet, G., Barthelat, M., and Lattes, A. (1998) *J. Phys. Chem. A*, **102**, 1279–1287. d) Kaminski, G.A. and Jorgensen, W.L. (1998) *J. Phys. Chem. B*, **102**, 1787–1796.
5. a) Power, K.N., Hennigar, T.L., and Zaworotko, M.J. (1998) *Chem. Commun.*, 595–596. b) Suarez, M., Branda, N., Lehn, J.-M., Decian, A., and Fischer, J. (1998) *Helv. Chim. Acta*, **81**, 1–13. c) Thalladi, V.R., Brasselet, S., Weiss, H.C., Blaser, D., Katz, A.K., Carrell, H.L., Boese, R., Zyss, J., Nangia, A., and Desiraju, G.R. (1998) *J. Am. Chem. Soc.*, **120**, 2563–2577. d) MacGillivray, L.R., Groeneman, R.H., and Atwood, J.L. (1998) *J. Am. Chem. Soc.*, **120**, 2676–2677.
6. a) Werkman, P.J., Schouten, A.J., Noordegraaf, M.A., Kimkes, P., and Sudholter, E.J.R. (1998) *Langmuir*, **14**, 157–164. b) Matsuzawa, Y., Seki, T., and Ichimura, K. (1998) *Langmuir*, **14**, 683–689. c) Maruyama, T., Lauger, J., Fuller, G.G., Frank, C.W., and Robertson, C.R. (1998) *Langmuir*, **14**, 1836–1845. d) Liley, M., Gourdon, D., Stamou, D., Meseth, U., Fischer, T.M., Lautz, C., Stahlberg, H., Vogel, H., Burnham, N.A., and Duschl, C. (1998) *Science*, **280**, 273–275.
7. a) Li, S., Clarke, C.J., Lennox, R.B., and Eisenberg, A. (1998) *Coll. Surf. A, Physicochem. Eng. Asp.*, **133**, 191–203; b) Ichinose, I., Tagawa, H., Mizuki, S., Lvov, Y., and Kunitake, T. (1998) *Langmuir*, **14**, 187–192. c) Samori, P., Francke, V., Mangel, T., Mullen, K., and Rabe, J.P. (1998) *Optic. Mater.*, **9**, 390–393. d) Pedersen, E.V., Shiryaev, S.Y., Jensen, F., Hansen, J.L., and Petersen, J.W. (1998) *Surf. Sci.*, **399**, L351–L356.
8. a) Toncheva, V., Wolfert, M.A., Dash, P.R., Oupicky, D., Ulbrich, K., Seymour, L.W., and Schacht, E.H. (1998) *Biochim. Biophys. Acta., General Subjects*, **1380**, 354–368. b) Xu, S., Cruchon Dupeyrat, S.J.N., Garno, J.C., Liu, G.Y., Jennings, G.K., Yong, T.H., and Laibinis, P.E. (1998) *J. Chem. Phys.* **108**, 5002–5012. c) Loiacono, M.J., Granstrom, E.L., and Frisbie, C.D. (1998) *J. Phys. Chem. B*, **102**, 1679–1688. d) Meine, K., Vollhardt, D., and Weidemann, G. (1998) *Langmuir*, **14**, 1815–1821.
9. a) Dominguez, O., Echegoyen, L., Cunha, F., and Tao, N.J. (1998) *Langmuir*, **14**, 821–824. b) Yamada, R., and Uosaki, K. (1998) *Langmuir*, **14**, 855–861. c) Charra, F., and Cousty, J. (1998) *Phys. Rev. Lett*, **80**, 1682–1685. d) Blumentritt, S., Burghard, M., Roth, S., and Nejo, H. (1998) *Surf. Sci.*, **397**, L280–L284.
10. a) Balzani, V., Campagna, S., Denti, G., Juris, A., Serroni, S., and Venturi, M. (1998) *Acc. Chem. Res.*, **31**, 26–34. b) Fabbrizzi, L., Licchelli, M., Pallavicini, P., and Parodi, L. (1998) *Angew. Chem. Int. Ed.*, **37**, 800–802. c) Faidherbe, P., Wieser, C., Matt, D., Harriman, A., DeCian, A., and Fischer, J. (1998) *Eur. J. Inorg. Chem.*, 451–457. d) Connolly, S., Fullam, S., Korgel, B., and Fitzmaurice, D. (1998) *J. Am. Chem. Soc.*, **120**, 2969–2970.
11. a) Beer, P.D. (1998) *Acc. Chem. Res.*, **31**, 71–80. b) Boulas, P.L., Gómez-Kaifer, M., and Echegoyen, L. (1998) *Angew. Chem. Int. Ed.*, **37**, 216–247. c) Wang, Y., Mendoza, S., and Kaifer, A.E. (1998) *Inorg. Chem.*, **37**, 317–320. d) Fall, M., Aaron, J.J., Sakmeche, N., Dieng, M.M., Jouini, M., Aeiyach, S., Lacroix, J.C., and Lacaze, P.C. (1998) *Synthetic Metals*, **93**, 175–179.
12. a) Kaida, Y., Okamoto, Y., Chambron, J.-C., Mitchell, D. K., Sauvage, J.-P. (1993) *Tetrahedron Lett*, **34**, 1019–1022. b) Asakawa, M., Pasini, D., Raymo, F.M., and Stoddart, J.F. (1996) *Anal. Chem.*, **68**, 3879–3881. c) Yamamoto, C., Okamoto, Y., Schmidt, T., Jäger, R., Vögtle, F. (1997) *J. Am. Chem. Soc.*, **119**, 10547–10548. d) Ashton, P.R., Matthews, O.A., Menzer, S., Raymo, F.M., Spencer, N., Stoddart, J.F., and Williams, D.J. (1997) *Liebigs Ann./Recueil*, 2485–2494.
13. a) Zimmerman, S.C., Zeng, F.W., Reichert, D.E.C., and Kolotuchin, S.V. (1996) *Science*, **271**, 1095–1098. b) Mattei, S., Wallimann, P., Kenda, B., Amrein, W., and Diederich, F. (1997) *Helv. Chim. Acta*, **80**, 2391–2417. c) Simanek, E.E., Isaacs, L., Li, X.H., Wang, C.C.C., and Whitesides, G.M. (1997) *J. Org. Chem.*, **62**, 8994–9000. d) Balogh, L., Samuelson, R., Alva, K.S., and Blumstein, A. (1998) *J. Polym. Sci. A, Polym. Chem.*, **36**, 703–712.
14. a) Armaroli, N., Diederich, F., Dietrich-Buchecker, C.O., Flamigni, L., Marconi, G., Nierengarten, J.-F., and Sauvage, J.-P. (1998) *Chem. Eur. J.*, **4**, 406–416. b) Ashton, P.R., Balzani, V., Credi, A., Kocian,

O., Pasini, D., Prodi, L., Spencer, N., Stoddart, J.F., Tolley, M.S., Venturi, M., White, A.J.P., and Williams, D.J. (1998) *Chem. Eur. J.,* **4**, 590–607. c) Hamilton, D.G., Davies, J.E., Prodi, L., and Sanders, J.K.M. (1998) *Chem. Eur. J.,* **4**, 608–620. d) Amabilino, D.B., Ashton, P.R., Boyd, S.E., Lee, J.Y., Menzer, S., Stoddart, J.F., and Williams, D.J. (1998) *J. Am. Chem. Soc.*, **120**, 4295–4307.
15. a) Ashton, P.R., Brown, C.L., Chapman, J.R., Gallagher, R.T., and Stoddart, J.F. (1992) *Tetrahedron Lett.*, **33**, 7771–7774. b) Marquis-Rigault, A., Dupont-Gervais, A., Van Dorsselaer, A., and Lehn, J.-M. (1996) *Chem. Eur. J.*, **2**, 1395–1398. c) Marquis-Rigault, A., Dupont-Gervais, A., Baxter, P.N.W., Van Dorsselaer, A., and Lehn, J.-M. (1996) *Inorg. Chem.*, **35**, 2307–2310. d) Scherer, M., Sessler, J.L., Moini, M., Gebauer, A., and Lynch, V. (1998) *Chem. Eur. J.*, **4**, 152–158.
16. a) Fischer, C., Nieger, M., Mogck, O., Bohmer, V., Ungaro, R., and Vögtle, F. (1998) *Eur. J. Org. Chem.*, 155–161. b) Dopke, N.C., Treichel, P.M., and Vestling, M.M. (1998) *Inorg. Chem.*, **37**, 1272–1277. c) Hempenius, M.A., Langeveld Voss, B.M.W., van Haare, J.A.E.H., Janssen, R.A.J., Sheiko, S.S., Spatz, J.P., Moller, M., and Meijer, E.W. (1998) *J. Am. Chem. Soc.*, **120**, 2798–2804. d) Cabezon, B., Irurzun, M., Torres, T., and Vázquez, P. (1998) *Tetrahedron Lett.*, **39**, 1067–1070.
17. a) Kraft, A., and Osterod, F. (1998) *J. Chem. Soc., Perkin Trans. 1*, 1019–1025. b) Martin, N., Bunzli, J.C.G., McKee, V., Piguet, C., and Hopfgartner, G. (1998) *Inorg. Chem.*, **37**, 577–589. c) Murner, H., von Zelewsky, A., and Hopfgartner, G. (1998) *Inorg. Chim. Acta*, **271**, 36–39. d) Khoury, R.G., Jaquinod, L., and Smith, K.M. (1998) *Tetrahedron*, **54**, 2339–2346.
18. Gillard, R.E., Raymo, F.M., and Stoddart, J.F. (1997) *Chem. Eur. J.*, **3**, 1933–1947. d) Raymo, F.M. and Stoddart, J.F. (1998) *Chemtracts*, **11**, In press.
19. a) Chambron, J.C., Dietrich-Buchecker, C.O., and Sauvage, J.-P. (1993) *Top. Curr. Chem.*, **165**, 131–162. b) Gibson, H.W., Bheda, M.C., and Engen, P.T. (1994) *Prog. Polym. Sci.*, **19**, 843–945. c) Amabilino, D.B. and Stoddart, J.F. (1995) *Chem. Rev.*, **95**, 2725–2828. d) Jäger, R. and Vögtle, F. (1997) *Angew. Chem., Int. Ed. Engl.*, **36**, 930–944.
20. a) Ashton, P.R., Chrystal, E.J.T., Glink, P.T., Menzer, S., Schiavo, C., Spencer, N., Stoddart, J.F., Tasker, P.A., White, A.J.P., Williams, D.J. (1996) *Chem. Eur. J.*, **2**, 709–728. b) Asakawa, M., Ashton, P.R., Menzer, S., Raymo, F.M., Stoddart, J.F., White, A.J.P., and Williams, D.J. (1996) *Chem. Eur. J.*, **2**, 877–893. c) Asakawa, M., Brown, C.L., Menzer, S., Raymo, F.M., Stoddart, J.F., and Williams, D.J. (1997) *J. Am. Chem. Soc.*, **119**, 2614–2627.
21. a) Ashton, P.R., Boyd, S.E., Claessens, C.G., Gillard, R.E., Menzer, S., Stoddart, J.F., Tolley, M.S., White, A.J.P., and Williams, D.J. (1997) *Chem. Eur. J.*, **3**, 788–798. b) Ballardini, R., Balzani, V., Brown, C.L., Credi, A., Gillard, R.E., Montalti, M., Philp, D., Stoddart, J.F., Venturi, M., White, A.J.P., Williams, B.J., and Williams, D.J. (1997) *J. Am. Chem. Soc.*, **119**, 12503–12513. c) Asakawa, M., Ashton, P.R., Boyd, S.E., Brown, C.L., Gillard, R.E., Kocian, O., Raymo, F.M., Stoddart, J.F., Tolley, M.S., White, A.J.P., and Williams, D.J. (1997) *J. Org. Chem.*, **62**, 26–37. d) Ballardini, R., Balzani, V., Gandolfi, M.T., Gillard, R.E., Stoddart, J.F., and Tabellini, E. (1998) *Chem. Eur. J.*, **4**, 449–459.
22. a) Bissell, R.A., Córdova, E., Kaifer, A.E., and Stoddart, J.F. (1994) *Nature*, **369**, 133–137. b) Martínez-Díaz, M.-V., Spencer, N., and Stoddart, J.F. (1997) *Angew. Chem., Int. Ed. Engl.*, **36**, 1904–1907. c) Asakawa, M., Ashton, P.R., Balzani, V., Credi, A., Hamers, C., Mattersteig, G., Montalti, M., Shipway, A.N., Spencer, N., Stoddart, J.F., Tolley, M.S., Venturi, M., White, A.J.P., and Williams, D.J. (1998) *Angew. Chem. Int. Ed.*, **37**, 333–337. d) Balzani, V., Gómez-López, M., and Stoddart, J.F. (1998) *Acc. Chem. Res.*, **31**, In press.
23. Asakawa, M., Dehaen, W., L'abbé, G., Menzer, S., Nouwen, J., Raymo, F.M., Stoddart, J.F., and Williams, D.J. (1996) *J. Org. Chem.*, **61**, 9591–9595.
24. Sandström, J. (1982) *Dynamic NMR Spectroscopy*, Academic Press, London.
25. Ashton, P.R., Fyfe, M.C.T., Hickingbottom, S.K., Stoddart, J.F., White, A.J.P., and Williams, D.J. (1998) *J. Chem. Soc., Perkin Trans. 2*, In press.
26. Adrian, J.C., Jr., and Wilcox, C.S. (1991) *J. Am. Chem. Soc.* **113**, 678–680.
27. Ashton, P.R., Ballardini, R., Balzani, V., Belohradsky, M., Gandolfi, M.T., Philp, D., Prodi, L., Raymo, F.M., Reddington, M.V., Spencer, N., Stoddart, J.F., Venturi, M., and Williams, D.J. (1996) *J. Am. Chem. Soc.*, **118**, 4931–4951.
28. Amabilino, D.B., Asakawa, M., Ashton, P.R., Ballardini, R., Balzani, V., Belohradsky, M., Credi, A., Higuchi, M., Raymo, F.M., Shimizu, T., Stoddart, J.F., Venturi, M., and Yase, K. (1998) *New J. Chem.*, In press.
29. Ashton, P.R., Baxter, I., Fyfe, M.C.T., Raymo, F.M., Spencer, N., Stoddart, J.F., White, A.J.P., and Williams, D.J. (1998) *J. Am. Chem. Soc.*, **120**, 2297–2307.
30. Amabilino, D.B., Ashton, P.R., Balzani, V., Brown, C.L., Credi, A., Fréchet, J.M.J., Leon, J.W., Raymo, F.M., Spencer, N., Stoddart, J.F., and Venturi, M. (1996) *J. Am. Chem. Soc.*, **118**, 12012–12020.
31. Ashton, P.R., Baxter, I., Cantrill, S.J., Fyfe, M.C.T., Glink, P.T., Stoddart, J.F., White, A.J.P., and Williams, D.J. (1998) *Angew. Chem., Int. Ed.*, **37**, 1294–1297.

STUDY OF CONFORMATIONAL BEHAVIOR OF CMP(O)-CAVITANDS BY NMR SPECTROSCOPY AND X-RAY ANALYSIS IN RELATION TO THE EXTRACTION PROPERTIES

HAROLD BOERRIGTER,[a] WILLEM VERBOOM,[b] RON HULST,[c] GERRIT J. VAN HUMMEL,[d] SYBOLT HARKEMA,[d] AND DAVID N. REINHOUDT[b,*]

Netherlands Energy Research Foundation (ECN),[a] P.O. Box 1, 1755 ZG Petten, The Netherlands and Laboratories of Supramolecular Chemistry and Technology (SMCT),[b] Chemical Analysis,[c] and Chemical Physics,[d] University of Twente, P.O. Box 217, 7500 AE Enschede, The Netherlands; Phone +31 53 4892980, Fax +31 53 4894645, E-mail smct@ct.utwente.nl

Abstract

Tetra-CMP(O)-functionalized cavitands are very effective europium(III) extractants. Cavitand **1b** is the strongest ligand and has the highest extraction constant for 1:1 complexation with Eu(picrate)$_3$ (*viz.* $K_{ex}^1 = 2.7 \times 10^{12}$ M^{-4}). The conformational behavior of the CMP(O) cavitands **1a,b** and **2a,b** was studied by means of NMR spectroscopy. The introduction of an *N*-propyl substituent imposes large conformational differences due to inhibition of the free rotation around the amide *N*-C(O) bond. The Gibbs free energy of activation is about 75 kJ/mol for a rotation around the amide moiety (*inward-outward* motion). The *N*-propyl-substituted cavitands form clusters; this aggregation originates from the association of propyl and pentyl chains of the cavitand molecules. The ability of bromomethylcavitand **3** to complex solvent molecules (*e.g.* CH$_2$Cl$_2$ or toluene) in the cavity is illustrated by a single crystal X-ray structure (with CH$_2$Cl$_2$).

1. Introduction

At present time, nearly all commercial nuclear fission plants are using uranium as their basic fuel. After burn-up the spent fuel, which contains uranium plus the side products of the fission processes, has to be treated to allow handling and storage in (geologic) depositories. In the reprocessing step in the nuclear fuel cycle, uranium and plutonium can be quantitatively recovered in the PUREX (*Plutonium URanium EX*traction) process [1]. After the extraction of uranium and plutonium the high-level liquid waste (HLLW) stream of the reprocessing process still contains higher actinides (mainly americium and curium). To date, a major point of interest is the efficient removal of trivalent actinides from the HLLW as the long term radiotoxicity (and heat production) of the waste is mainly due to these elements.

The most commonly used process for the recovery of the trivalent trans-plutonium actinides, the TRUEX (*TRansUranium EX*traction) process [2], utilizes carbamoylmethyl-phosphoryl [CMP(O)] derivatives as organic extractants. The trivalent actinides are generally extracted as complexes with three CMP(O) molecules. Therefore, preorganization of the complexing CMP(O) groups will favor the complexation as the coordinating moieties are already grouped and positioned [3].

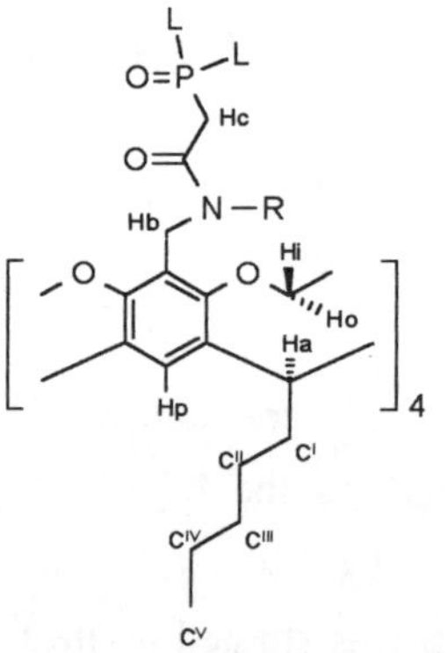

Chart 1. Structure of CMP(O)-Cavitands.

Recently [4], we demonstrated that ligands based on functionalized resorcinarene cavitands [5] allow a high degree of preorganization of the complexing moieties due to the rigidity of the cavitand frame (see Chart 1). Attachment of CMP(O) sites to the cavitand frame resulted in a new type of ligands which very efficiently extract europium (selected as representative for the trivalent actinides americium and curium) [6]. However, the conformational behavior of the derivatives with the *N*-propyl-substituted amide moieties (cavitands **1b** and **2b**) and the cavitands **1a** and **2a** with unsubstituted amides is very different. The ^{1}H NMR spectra of **1b** and **2b** are very complicated and show multiple signals for nearly all hydrogens, while cavitands **1a** and **2a** yield sharp and well resolved NMR spectra. In this paper we present a detailed study to the conformational behavior of the CMP(O)-cavitands by NMR and X-Ray spectroscopy. The relation with the extraction properties and the conformational behavior of the structural related precursors of **1** and **2** are also discussed.

2. Synthesis

The CMP(O)-based cavitands have been prepared starting from tetrakis(bromomethyl)-cavitand **3** via the (alkyl)aminomethylcavitands **4a** and **4b**. Aminomethylcavitand **4a** was obtained after substitution of the bromo atoms of **3** with phthalimido functionalities and, subsequent deprotection with hydrazine hydrate to yield the free amine. Propyl-aminomethylcavitand **4b** was synthesized directly from **3**, by reaction of **3** with *n*-propylamine. Acylation of the amines **4a** and **4b** with chloroacetyl chloride and Et$_3$N as a base in CH$_2$Cl$_2$ afforded chloroacetamidomethylcavitand **5a** and *N*-propylchloroacetamido-methylcavitand **5b** (Scheme 1) [4a].

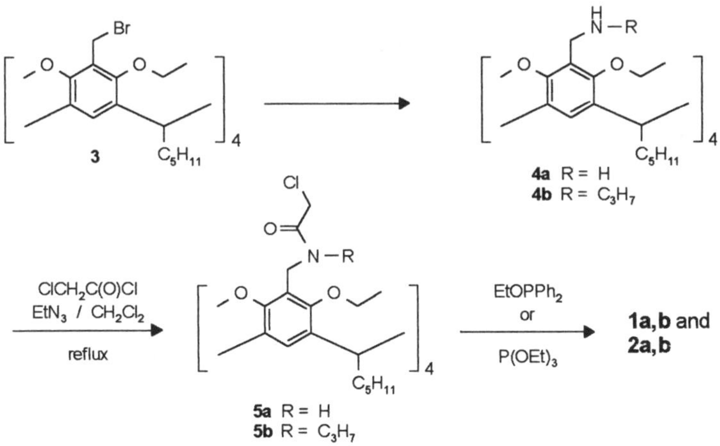

Scheme 1. Synthesis of the CMP(O)-Cavitands.

The CMP(O) ligating sites are introduced in Arbusov reactions on **5a** and **5b** with ethyl diphenylphosphinite to give (diphenyl-*N*-methylcarbamoylmethylphosphine oxide)cavitand **1a** and (diphenyl-*N*-methyl-*N*-propylcarbamoylmethylphosphine oxide)cavitand **1b**, respectively. Similar reactions of **5a** and **5b** with triethyl phosphite afforded (diethyl-*N*-methylcarbamoylmethylphosphinate)cavitand **2a** and (diethyl-*N*-methyl-*N*-propylcarbamoylmethylphosphinate)cavitand **2b**, respectively [4a].

A related cavitand with a 'simple' acetamide function was prepared for reason of comparison. Reaction of bromomethylcavitand **3** with *N*-methylacetamide in the presence of NaOH, K_2CO_3, and tetrabutylammonium hydrogensulfate in toluene afforded tetrakis(*N*-methylacetamidomethyl)cavitand **6** in 82% yield (Scheme 2).

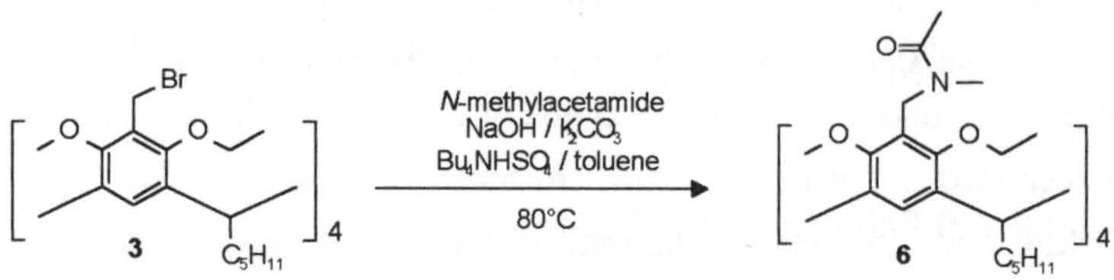

Scheme 2. Synthesis of Acetamidocavitand **6**.

3. Extraction properties

Initially, europium(III) picrate (2,4,6-trinitrophenolate) extractions were carried out with the new ligands **1** and **2**, in order to study the influence of different phosphorus and amide substituents [4a]. In a follow-up study the extraction properties were studied from acidic nitrate solutions which are more similar to the HLLW conditions [7]. Under all conditions ligand **1b** is the most efficient and selective extractant for Eu(III). The CMP(O)-resorcinarene ligands, and also the related calix[4]arene-based CMPO-ligands, are much more efficient than the simple CMP(O) derivatives [3].

Table 1. Extraction Constants of 1:1-Complexes.[a]

	1a	1b	2a	2b
$E^b \times 10^2$	60	94	4.9	84
K_{ex} [M^{-4}]	3.7×10^{10}	2.7×10^{12}	5.4×10^8	3.4×10^{11}

[a] $[L]_{o,i} = 10^{-4}$ M; $[Eu^{3+}]_{w,i} = 10^{-4}$ M; $[LiPic]_w = 10^{-2}$ M; $[HNO_3]_w = 10^{-3}$ M; pH = 3.0.
[b] E = percentage Eu(III) extracted.

Under the assumptions that in the case of $[L]_{o,i} = [Eu^{3+}]_{w,i} = 10^{-4}$ M, the cation is extracted in a 1:1 stoichiometry, and that the variation in the aqueous picrate concentration, $[Pic^-]_w$, is negligible as $[Pic^-]_w \approx [Pic^-]_{w,i} = 10^{-2}$ M. The extraction coefficients for the 1:1 complexes (K_{ex}^1) have been calculated and are summarized in Table 1.

Infrared data confirmed that in the 1:1 complex both the phosphoryl oxygen and amide carbonyl interact with the cation [8], and the presence of a substituent on the amide nitrogen largely influences the value of the extraction constants. Ligand **1b** is a ~75 times stronger extractant than its unsubstituted analogue **1a**, while the K_{ex} value of **2b** is a factor of ~650 higher than that of **2a**. Besides the increasing effect on the basicity of the amide due to the introduction of a propyl group in **1b** and **2b**, also intramolecular 'self-complexation' within the CMP(O) moieties of **1a** and **2a** might be an important effect. The phosphine oxide and phosphonate phosphorus oxygens of **1a** and **2a** can coordinate with the amide hydrogen, forming a six-membered ring, which has to adjust the geometry prior to complexation.

4. Conformational Behavior of CMP(O) Cavitands

4.1. GENERAL NMR SPECTROSCOPIC OBSERVATIONS

Cavitands **1a**, **2a**, and **5a** with unsubstituted amide moieties yield sharp and well resolved NMR spectra over a large temperature interval (243-318 K) when recorded in CDCl$_3$ (at 5 mM concentration). The ^{1}H NMR spectrum of cavitand **1a** exhibits characteristic signals for the aromatic cavitand hydrogen (at 7.06 ppm), the doublets for the outer and inner bridge hydrogens (at 5.55 and 4.43 ppm) and the doublet for the CMPO methylene hydrogens (H$_c$) at 3.24 ppm with a phosphorus coupling constant ($^2J_{PH}$) of 13.7 Hz. The ^{1}H NMR signals for the corresponding cavitand **2a** were found at 7.14 (H$_p$), 5.95 and 4.18 (H$_o$ and H$_i$) and at 2.90 ppm with $^2J_{PH}$ = 21.1 Hz for H$_c$.

The ^{1}H NMR spectra of the derivatives with the *N*-propyl-substituted amide moieties (cavitands **1b**, **2b**, and **5b**) at ambient temperature are very complicated and show multiple signals for nearly all hydrogens due to the internal hindered rotation around the amide *N-C(O)* bonds. In particular this behavior is very pronounced for the absorption of the aromatic (H$_p$) and the methyleneoxy bridge hydrogens (H$_o$ and H$_i$), since these hydrogens are highly sensitive to (small) conformational changes. The multiple signals of **1b**, **2b**, and **5b** (partly) coalesced to a single, though broad, signal when the spectra were recorded at elevated temperatures (393 K in C$_2$D$_2$Cl$_4$ or DMSO-d_6). These effects are due to several rotational isomers or conformers resulting from hindered rotation around the amide *N-C(O)* bond. In Figure 1 the ^{1}H NMR spectra of CMPO-cavitands **1a** and **1b** are shown.

From these experiments it became clear that the conformations observed are largely influenced by changes in temperature and/or solvents used. It has to be noted that also a very strong concentration dependency was observed leading to a sudden increase of the line broadening upon increase of the concentration (> 5 mM). These effects largely take place in CDCl$_3$ but are markedly less effective in toluene-d_8 or DMSO-d_6.

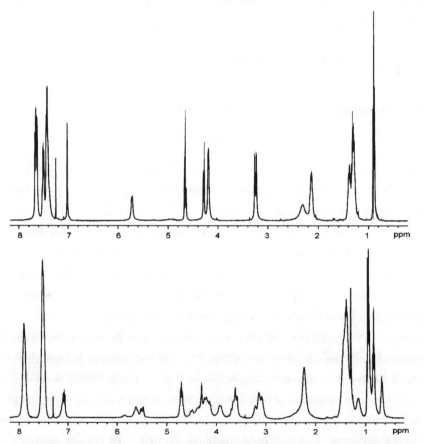

Figure 1. ^{1}H NMR Spectra in CDCl$_3$ at Ambient Temperature of *NH*-free Cavitand **1a** (top) and *N*-propyl-substituted Cavitand **1b** (bottom).

4.2. BROMOMETHYLCAVITAND 3

4.2.1. Conformational Analysis

Surprisingly, also starting compound bromomethylcavitand 3, without a(n) (substituted) amide moiety, shows broadened signals when recorded in CDCl$_3$ (303 K) at 400 MHz (with 250 MHz NMR this was not observed). The outer hydrogen (H$_o$) shows three large and one smaller scalar coupling with the signals belonging to the inner hydrogen (H$_i$), indicating the existence of at least three and possibly four different conformers and/or complexed forms (*vide infra*) at ambient temperatures. This behavior is probably due to the hindered rotation around the methylene bridge due to the large bromo atom (Ar-*CH$_2$*Br). The ^{1}H NMR spectra recorded in toluene-d_8 and DMSO-d_6 show no line broadening, suggesting the presence of a single conformer.

4.2.2. Solvent Inclusion

NMR Spectroscopy. The introduction of one equivalent of CH$_2$Cl$_2$ to a solution of bromomethylcavitand 3 does not give rise to significant induced shifts, although in toluene-d_8 the absorption of CH$_2$Cl$_2$ is shifted upfield to 4.43 ppm in the ^{1}H NMR spectrum compared with the 'normal' position at 5.33 ppm in toluene-d_8. ROESY experiments revealed several interesting contacts between the CH$_2$Cl$_2$ molecule and the cavitand molecule. The most important is a contact with the benzyl group (H$_b$) at 4.39 ppm, the outer methylene hydrogen (H$_o$) at 5.92 ppm, and a *very weak* contact with the bridge CH (H$_a$) and CH$_2$-group (C^{I}H$_2$) of the pentyl moiety. These data point to complexation of CH$_2$Cl$_2$ in the cavity of cavitand 3 in solution.

Surprisingly, in toluene-d_8 a very large contact of CH$_2$Cl$_2$ is found with a signal arising at 2.84 ppm (this signal is only clearly observed in the ROESY spectra). Also a large contact is visible between this signal and the methyl moiety of the solvent toluene (2.09 ppm). This indicates the (possible) complexation of toluene in cavitand 3 in competition with the complexation of CH$_2$Cl$_2$. The inclusion of toluene in the cavity has been observed in several X-ray structures of cavitands [9,10] (with the Ar*CH$_3$* pointing inwards, similar to other included -*CH$_3$* containing solvent molecules [11]).

X-Ray Analysis. The stoichiometric complexation of CH$_2$Cl$_2$ in the cavity of bromomethylcavitand 3 is also observed in the single crystal X-ray structure of 3 (Figure 2). The crystal structure was determined by X-ray diffraction (crystallographic data are collected in Table 3; see Experimental Section). The unit cell contains two independent molecules **3** and a CH_2Cl_2 molecule is included in the aromatic cavities of both molecules.

In the crystal packing (not shown) the molecules are approximately arranged in layers with the polar groups (Br, Cl, O atoms) of all molecules in one plane and the hydrocarbon chains in another plane. This arrangement, with little interaction between the hydrocarbon chains, may explain the disorder found in some of the hydrocarbon chains.

In the solid state one $ClCH_2$- of the CH_2Cl_2 molecule is positioned 'perpendicular into the cavity' (perpendicular to the plane through the benzylic groups). The second chloro atom is pointing to one benzylic methylene (Ar-CH_2Br) group of which the bromo atom is rotated away from the cavity, to accommodate the Cl atom of the guest. The other three bromo atoms are positioned on top of the cavity and directed to the methylene of the CH_2Cl_2.

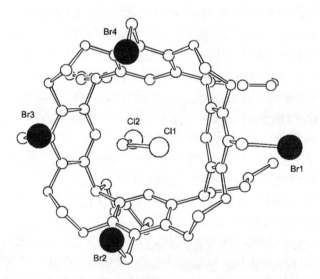

Figure 2. Single Crystal X-Ray Structure of Bromomethylcavitand **3**.CH_2Cl_2 (Top View).

Also by other single crystal X-ray structures the influence is illustrated of (large) (halo) substituents on the cavitand aromatic units on the inclusion of CH_2Cl_2 [10]. In an X-ray structure of a bromocavitand (however, with ethylene- instead of methyleneoxy bridges) the included CH_2Cl_2 pointed perpendicularly into the cavity with the $ClCH_2$-.

In X-ray structures of unsubstituted cavitands with CH_2Cl_2 also positioning with the -CH_2- pointing inwards the cavity was found, besides perpendicular inclusion [10,12].

4.3. (CHLORO)ACETAMIDOCAVITANDS 5a AND 5b

The chloroacetamidocavitands **5a** and **5b** are the precursors in the synthesis of the CMP(O) cavitands **1** and **2** (*vide supra*) [4a]. Chloroacetamidocavitand **5a**, which contains no propyl substituent attached to the amide nitrogen, yields sharp ^{1}H NMR spectra even at lowered temperatures (243 K). The 2D NOESY and ROESY spectra revealed relatively limited information about the conformation. Contacts between the aromatic (H_p) and methylene hydrogen H_a and parts of the pentyl moiety (*viz.* C^I, C^{III} and C^{IV}) are clearly visible. The contacts between the aromatic hydrogen and the pentyl groups up to carbon C^{IV} suggest *inter*molecular contacts, however, they may alternatively arise from *intra*molecular contacts due to backfolding.

The ^{1}H NMR spectrum of *N*-propyl-substituted cavitand **5b** recorded at low temperature (243 K, in CDCl$_3$) showed multiple broadened signals, which tend to coalesce at elevated temperature. At temperatures typically around 393 K (in DMSO-d_6) well resolved spectra were obtained except for the resonances arising from the benzylic hydrogens (H_b), which remain relatively broad due to the hindered rotation around the amide *N-C(O)* bond. From the signal group at about 5.80 ppm (H_o) it becomes evident that at least three different conformers are present at ambient temperature. The exchange connectivities in the NOESY/ROESY experiments suggest a dynamic equilibrium between several other conformers. Several of the signals are in the slow exchange region with other spin systems, allowing the determination of the Gibbs free energy of activation.

Based on the exchange signals of the *N*-propyl moiety the Gibbs free energy is estimated to be 74 kJ/mol (± 4 kJ/mol) for the *inward-outward* rotation around the amide *N-C(O)* [13]. Unfortunately, due to limited resolution and intensity of the signal groups of interest, the Gibbs free energy of rotation of the other possible rotamers (double, triple, and four times rotational products) could not be established. The 2D NOESY and ROESY spectra revealed very limited additional information. The contacts between H_p and the pentyl moiety, however, indicate close spatial orientation, either due to back-folding of the pentyl group or, alternatively, due to contacts with other cavitand molecules (*vide infra*).

To illustrate the conformational behavior of the *N*-substituted cavitands more clearly also the cavitand (*N*-methylacetamidocavitand **6**; Scheme 2) with a 'simple' acetamide function was included. In the COSY spectrum of cavitand **6** in DMSO-d_6 (Figure 3) the signals for the different conformers are nicely visible (*e.g.* H_o hydrogens at 5.7 ppm).

4.4. CMP(O)-CAVITANDS 1 AND 2

4.4.1. *Conformational Analysis*

The ^{1}H decoupled ^{31}P NMR spectrum of cavitand **1b** recorded at ambient temperature indicated the presence of one major and two minor conformers. Due to limited resolution, no quantitative analysis could be obtained from the ^{31}P NMR spectra. The presence of several conformers was also concluded from the ^{1}H NMR spectra recorded at ambient temperature [14]. Similar to the *N*-propyl-substituted chloroacetamido-cavitand **5b**, limited rotation around the *N-C(O)* bond was observed. This is most elegantly illustrated by the appearance of an *AB* spin system for the methylene group H_c besides the clearly visible $^2J_{PH}$ coupling. Also methylene group H_b gives rise to an *AB* spin system, although severe exchange line broadening and overlap with other signal groups did not allow proper quantitative analysis.

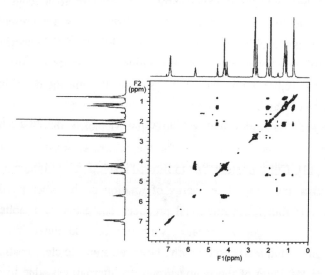

Figure 3. COSY Spectrum of Cavitand **6** in CDCl$_3$ at Ambient Temperature.

From the *N*-propyl exchange resonances the Gibbs free energies of activation were estimated to be 70 (± 3.8) and 78 (± 4.3) kJ/mol, respectively, for a single and a double rotation around the amide moiety (*inward-outward* motion) [13]. Although proof was generated that also triple and four times rotated conformers must exist, overlap of signals prevented obtaining quantitative information about the dynamic behavior.

Comparison with the free-NH equivalent cavitand **1a** again indicates the large influence of the propyl chain at the conformational behavior and hence the spectral appearance. For cavitand **1a** a clear-cut ^{1}H NMR spectrum could be obtained in CDCl$_3$ throughout the entire temperature range (243-318 K). In the 2D NOESY and ROESY spectra of cavitand **1a** contacts were observed for the aromatic hydrogens of the P-phenyl moiety with the inner bridge hydrogens (H$_i$), the methylene group (H$_c$), and, surprisingly, with the methyl moieties (at 0.90-1.00 ppm). This indicated that close contacts are viable between these pending groups and the more or less rigid structure of the cavitand units. All other expected contacts were also clearly observed from these spectra.

4.4.2. *Aggregation*

Since the 2D spectra allowed the visualization of only part of the exchange dynamics in detail, the relaxation behavior of the materials was also investigated. Due to limitations in solubility, only ^{1}H and ^{31}P T$_1$ relaxation measurements were conducted (in Table 2 selected T$_1$ relaxation times are shown). The lower T$_1$ times indicate that the *N*-propyl-substituted cavitands **1b** and **5b** behave as somewhat larger clusters compared to the free-NH cavitands **1a**, **2a**, and **5a**, respectively, which could be attributed to a tendency to associate (*viz.* dimerize or oligomerize) to a certain extent.

Table 2. Longitudinal Relaxation Times T$_1$ of Selected Signals.

Compd	Longitudinal Relaxation Times T$_1$ [s] Nuclei			
	H$_a$	H$_b$	H$_c$	^{31}P
1a	0.98	0.35	0.46	1.29
1b	0.27	0.05	0.13	0.06
2a	0.99	0.31	0.29	1.09
2b	0.86	0.31	0.38	1.84
5a	0.82	0.43	0.30	--
5b	0.51	0.23	0.12	--

This conclusion is also strengthened by a comparison of the T_1 times from similar molecules going from the free to the propyl-substituted compounds, *viz.* **1a** *vs.* **1b** and **5a** *vs.* **5b**. The largest differences were observed between cavitands **1a** and **1b**, of which **1b** also shows the most considerable exchange in the 2D *EXSY* experiments. Cavitand **2b** shows T_1 relaxation times that are not easily interpreted and do not fit nicely into the proposed explanation, probably due to the balance of the T_1 relaxation behavior between the conformational flexibility and the tendency to form larger clusters [15].

For all CMP(O) ligands **1** and **2** contacts between the aromatic hydrogen (H_p) and the pentyl moieties were observed, indicating back-folding of the pentyl group or interactions with other cavitand molecules. However, from the T_1 relaxation times it was concluded that only *N*-propyl substituted cavitands form (larger) clusters in solution. The large extent of aggregation, therefore, most likely originates from the presence of the substituent on the amide and *not solely* from the mutual association of pentyl side chains [16]. By aggregation of cavitands with the mutual pentyl groups (only) a dimer or layers can be formed. On the other hand, clustering by interactions of propyl and pentyl groups of cavitands allows the formation of larger aggregates. The *N*-propyl CMP(O) cavitands can be visualized as a three-dimensional micel with a polar 'box' formed by the eight P=O and C=O groups with on the apolar 'outside' four propyl and four pentyl side chains (and the eight phosphorus substituents, *e.g.* phenyl or ethoxy) [17].

4.4.3. *Relation with Extraction Properties*

Besides the difference in dynamic behavior, the ligands with *N*-propyl-substituted amide functions (**1b** and **2b**) are significantly more effective extractants for europium(III) than their unsubstituted counterparts (**1a** and **2a**), cavitand **1b** being the most effective one. The stronger extraction of **1b** compared to its unsubstituted counterpart **1a** (~75 times) was attributed to the influence of the increased basicity of the amide carbonyl due to the propyl substituent. In addition to the decreasing effect due to the *intra*molecular auto-association of the CMP(O) groups in **1a** (and **2a**), the phosphine oxide groups of **1a** (and **2a**) might coordinate with the amide hydrogen forming a six-membered ring, which has to adjust its geometry prior to complexation (*vide supra*) [4a].

Based on the current NMR study no direct information was obtained to confirm this behavior. However, contacts in the 2D NOESY and ROESY spectra of **1a** for the aromatic hydrogens of the P-phenyl moiety with the inner bridge hydrogens (H_i) indicate close geometric orientation. These contacts of the phosphorus substituents are expected when the P=O folds back to associate with the amide. The extraction is not affected by the (*inter*molecular) aggregation behavior, as the complexing P=O and C=O sites are far away from the side chains.

5. Conclusions

Bromomethylcavitand **3** shows broadened signals in $CDCl_3$ due to hindered rotation around the methylene bridge (*Ar-CH$_2$Br*) induced by complexation of **3** with solvent molecules (*e.g.* CH_2Cl_2 or/and toluene). The ability of cavitand **3** to complex CH_2Cl_2 in the cavity is illustrated by a single crystal X-ray structure.

The introduction of an *N*-propyl substituent imposes large dynamic changes in the behavior of the CMP(O)- and chloroacetamidocavitands **1**, **2**, and **5** due to the hindered rotation around the amide *N-C(O)* bond. The Gibbs free energy of activation is around 75 kJ/mol for a rotation around the amide moiety (*inward-outward* motion).

Based upon the T_1 relaxation time measurements the tendency of the *N*-propyl-substituted cavitands to form larger clusters is suspected. This aggregation originates in the association of propyl and the pentyl chains of the cavitand molecules. The extraction properties of **1b** for the association with europium(III) are not affected by this aggregation behavior. The contacts in the 2D NOESY and ROESY spectra of **1a** support the proposed behavior of intramolecular auto-association of the CMPO groups, resulting in a decrease of the extraction efficiency.

6. Experimental Section

Apparatus. All spectra were recorded at ambient temperature (293 K) unless stated otherwise. The ^1H (400 MHz) and ^{13}C NMR (100 MHz) spectra were recorded using a Unity 400 WB instrument. Unless otherwise indicated the chemical shifts are expressed relative to the residual solvent signals of $CDCl_3$ for ^1H NMR (at 7.26 ppm) and ^{13}C NMR (at 76.91 ppm). *NOESY* [18], *ROESY* [19], *TOCSY* (MLEV17) [20], and *COSY* [21] spectra were performed using standard Varian pulse programs. The *TOCSY* (MLEV17) experiments were performed with mixing times of 30 ms. The mixing times of the *NOESY* experiments ranged fom 10 to 80 ms. The mixing time of the *ROESY* experiments consisted of a spin lock pulse of 2 KHz field strength with a duration of 30 ms, typically. All 2D experiments were collected using 2D hypercomplex data [22] and Fourier transformed in the phase-sensitive mode after weighting with shifted square sine-bells or shifted Gaussian functions. NMR data were processed by the standard VnmrS software packages on the Unity 400 WB host computers (SUN IPX and Sparc stations). The T_1 measurements were performed using standard Varian Pulse programmes and the calculations were carried out using the standard Varian Microprogrammes.

Materials. CDCl$_3$ was dried over an Al$_2$O$_3$ (activity I) column and DMSO-d_6 was distilled from CaH$_2$ under an argon atmosphere just prior to use. The samples used for T_1 measurements were carefully deoxygenated by three free pump thaw cycles using Schlenck line techniques. Concentrations of the samples used were typically 5 mM. Only the chemical shifts of the major conformations and some selected signals for the minor conformers are given. The preparation and characterization (melting points, Mass-FAB spectra, and elemental analyses) of cavitands **1-5** have been described previously [4a].

Tetrakis(diphenyl-N-methylcarbamoylmethylphosphine oxide)cavitand (1a); ^{1}H NMR (one conformer, recorded at 243 K) δ 7.65-7.7 (m, 16 H, P-phenyl), 7.35-7.50 (m, 24 H, P-phenyl), 7.22 (br s, 4 H, H$_p$), 7.06 (br s, 4 H, NH), 5.72 (d, 4 H, J = 5.2 Hz, H$_o$), 4.66 (t, 4 H, J = 7.9 Hz, H$_a$), 4.28 (d, 4 H, J = 7.6 Hz, H$_i$), 4.20 (d, 8 H, J = 4.1 Hz, H$_b$), 3.24 (d, 8 H, $^2J_{PH}$ = 13.8 Hz, H$_c$), 2.2-2.05 (m, 8 H, C^{I}H$_2$), 1.2-1.4 (m, 24 H, C^{II}H$_2$-C^{IV}H$_2$), 0.91 (t, 12 H, J = 6.8 Hz, C^{V}H$_3$); ^{13}C NMR δ 164.2, 164.1 (C=O), 153.9 (ArCOCH$_2$O), 138.0 (ArCCH), 132.5-128.7 (P-phenyl), 123.7 (ArCCH$_2$), 119.9 (ArCH), 100.1 (OCH$_2$O), 39.3 [CH$_2$P(O)], 36.8 (ArCHAr), 34.0 (ArCH$_2$NH); ^{31}P NMR δ 29.66.

Tetrakis(diphenyl-N-methyl-N-propylcarbamoylmethylphosphine oxide)cavitand (1b); ^{1}H NMR (major conformer) δ 7.9-7.8 (m, 16 H, P-phenyl), 7.45-7.3 (m, 20 H, P-phenyl), 7.1-6.9 (dd, 4 H, J = 6.8 Hz, J = 2.3 Hz, H$_p$), 5.55 (d, 4 H, J = 7.6 Hz, H$_o$), 4.65-4.55 (m, 4 H, H$_a$), 4.4-4.3 (m, 4 H, H$_b$), 4.16 (d, 4 H, J = 7.6 Hz, H$_i$), 4.1-4.0 (m, 4 H, H$_b$), 3.75-3.6 (m, 8 H, H$_c$), 3.1-3.0 (m, 8 H, NCH$_2$CH$_2$), 2.2-2.1 (m, 8 H, C^{I}H$_2$), 1.4-1.3 (m, 8 H, NCH$_2$CH_2), 1.4-1.3 (m, 24 H, C^{II}H$_2$-C^{IV}H$_2$), 1.0-0.8 (m, 12 H, C^{V}H$_3$), 0.8-0.7 [m, 12 H, N(CH$_2$)$_2$CH$_3$]; ^{1}H NMR (minor conformer I, recorded at 243 K, selected signals) δ 5.65-5.58 (br d, 4 H, H$_o$), 4.20-4.15 (m, 4 H, H$_i$), 3.98-3.90 (br dd, 4 H, H$_b$); ^{1}H NMR (minor conformer II, recorded at 243 K, selected signals) δ 5.40 (d, 4 H, J = 6.8 Hz, H$_o$), 4.05 (d, 4 H, J = 6.8 Hz, H$_i$), 3.95-3.88 (br dd, 4 H, H$_b$); ^{13}C NMR δ 165.0 (C=O), 154.1 (ArCOCH$_2$O), 137.8 (ArCCH), 132.0-128.3 (P-phenyl), 122.6 (ArCCH$_2$), 119.9 (ArCH), 99.0 (OCH$_2$O), 46.5 (NCH$_2$CH$_2$), 38.7, 38.1 [CH$_2$P(O)], 36.7 (ArCHAr), 31.9, 30.0 (ArCH$_2$NH); ^{31}P NMR δ 28.67.

Tetrakis(diethyl-N-methylcarbamoylmethylphosphonate)cavitand (2a); ^{1}H NMR (one conformer, recorded at 243 K) δ 7.92 (br s, 4 H, NH), 7.42 (br s, 4 H, H$_p$), 5.95 (d, 4 H, J = 7.9 Hz, H$_o$), 4.58 (t, 4 H, J = 8.0 Hz, H$_a$), 4.24 (d, 4 H, J = 7.9 Hz, H$_i$), 4.08 (d, 8 H, J = 3.6 Hz, H$_b$), 4.0-3.9 (m, 16 H, OCH$_2$CH$_3$), 2.90 (d, 8 H, $^2J_{PH}$ = 21.1 Hz, H$_c$), 2.24 (q, 8 H, J = 6.2 Hz, C^{I}H$_2$), 1.45-1.2 (m, 8 H, C^{II}H$_2$-C^{IV}H$_2$), 1.25-0.9 (m, 24 H, OCH$_2$CH$_3$), 0.89 (t, 12 H, J = 6.8 Hz, C^{V}H$_3$); ^{13}C NMR δ 163.4 (C=O), 154.2 (ArCOCH$_2$O), 138.1 (ArCCH), 124.8 (ArCCH$_2$), 119.8 (ArCH), 101.6 (OCH$_2$O), 62.6 (OCH$_2$CH$_3$), 36.8 (ArCHAr), 35.8, 33.6 [CH$_2$P(O)], 31.9 (ArCH$_2$NH), 15.6 (OCH$_2$CH$_3$); ^{31}P NMR δ 23.09.

Tetrakis(diethyl-N-methyl-N-propylcarbamoylmethylphosphonate)cavitand (2b); ^{1}H NMR (major conformer) δ 7.10 (d, 4 H, J = 2.1 Hz, H$_p$), 5.91 (d, 4 H, J = 7.2 Hz, H$_o$), 4.78 (t, 4 H, J = 6.0 Hz, H$_a$), 4.5-4.2 (m, 8 H, H$_b$), 4.4-4.3 (m, 8 H, H$_i$), 4.3-4.1 (m, 16 H, OCH$_2$CH$_3$), 3.2-3.15 (m, 8 H, NCH$_2$CH$_2$), 3.06 (d, 8 H, $^2J_{PH}$ = 22.0 Hz, H$_c$), 2.2-2.15 (m, 8 H, C^{I}H$_2$), 1.6-1.4 (m, 8 H, NCH$_2$CH_2, 24 H, OCH$_2$CH_3), 1.5-1.4 (m, 24 H, C^{II}H$_3$-C^{IV}H$_2$), 0.95-0.75 [m, 12 H, N(CH$_2$)$_2$CH$_3$, 12 H, C^{V}H$_3$]; ^{1}H NMR (minor conformer I, recorded at 243 K, selected signals) δ 5.95-5.89 (br d, 4 H, H$_o$), 4.76-4.74 (m, 4 H, H$_a$), 4.41-4.35 (br d, 4 H, H$_i$); ^{1}H NMR (minor conformer II, recorded at 243 K, selected signals) δ 5.98-5.93 (br d, 4 H, H$_o$), 4.81-4.78 (m, 4 H, H$_a$), 4.40-4.38 (br d, 4 H, H$_i$); ^{13}C NMR δ 164.2 (C=O), 154.0 (ArCOCH$_2$O), 137.7

(ArCCH), 122.6 (ArCCH$_2$), 119.6 (ArCH), 99.2 (OCH$_2$O), 62.3 (OCH$_2$CH$_3$), 49.0 (NCH$_2$CH$_2$), 38.9 (NCH$_2$CH$_2$), 36.5 (ArCHAr), 33.6 [CH$_2$P(O)], 31.7 (ArCH$_2$NH), 16.0 (OCH$_2$CH$_3$); ^{31}P NMR δ 21.81.

Tetrakis(bromomethyl)cavitand (3); ^{1}H NMR (major conformer) δ 7.13 (br s, 4 H, H$_p$), 6.02 (d, 4 H, J = 6.7 Hz, H$_o$), 4.77 (t, 4 H, J = 8.0 Hz, H$_a$), 4.55 (d, 4 H, J = 6.8 Hz, H$_i$), 4.41 (s, 8 H, H$_b$), 2.19 (q, 8 H, J = 8.0 Hz, J = 6.1 Hz, C^{I}H$_2$), 1.50-1.25 (m, 24 H, C^{II}H$_2$-C^{IV}H$_2$), 0.90 (t, 12 H, J = 6.8 Hz, C^{V}H$_3$); ^{1}H NMR (major conformer, toluene-d_8) δ 7.20 (br s, 4 H, H$_p$), 5.96 (br d, 4 H, H$_o$), 5.09 (t, 4 H, J = 7.2 Hz, H$_a$), 4.82 (br d, 4 H, H$_i$), 4.41 (s, 8 H, H$_b$), 2.3-2.25 (m, 8 H, C^{I}H$_2$), 1.4-1.25 (m, 24 H, C^{II}H$_2$-C^{IV}H$_2$), 0.98 (t, 12 H, J = 6.6 Hz, C^{V}H$_3$); ^{13}C NMR δ 153.6 (ArCOCH$_2$O), 138.1 (ArCCH), 124.5 (ArCCH$_2$Br), 121.0 (ArH), 99.1 (OCH$_2$O), 36.9 (ArCHAr), 26.9, 23.0 (ArCH$_2$Br).

Tetrakis(chloroacetamido)cavitand (5a); ^{1}H NMR (one conformer, recorded at 243 K) δ 7.15 (s, 4 H, H$_p$), 7.01 (t, 4 H, J = 5.2 Hz, NH), 6.03 (d, 4 H, J = 7.3 Hz, H$_o$), 4.81 (t, 4 H, J = 7.9 Hz, H$_a$), 4.41 (d, 4 H, J = 7.3 Hz, H$_i$), 4.37 (d, 8 H, J = 4.5 Hz, H$_b$), 4.02 (s, 8 H, H$_c$), 2.23 (q, 8 H, J = 5.9 Hz, C^{I}H$_2$), 1.4-1.35 (m, 24 H, C^{II}H$_2$-C^{IV}H$_2$), 0.98 (t, 12 H, J = 6.8 Hz, C^{V}H$_3$); ^{13}C NMR δ 165.6 (C=O), 153.6 (ArCOCH$_2$O), 138.2 (ArCCH), 123.0 (ArCCH$_2$), 120.1 (ArH), 99.9 (OCH$_2$O), 42.6 (CH$_2$Cl), 36.9 (ArCHAr).

Tetrakis(*N*-propylchloroacetamidomethyl)cavitand (5b); ^{1}H NMR (major conformer) δ 7.13 (d, 4 H, J = 2.0 Hz, H$_p$), 5.80 (d, 4 H, J = 6.4 Hz, H$_o$), 4.72 (t, 4 H, J = 7.9 Hz, H$_a$), 4.48 and 4.40 (2d, 4 H, J = 6.5 Hz, H$_i$), 4.42 (dd, 8 H, 2 × J = 4.3 Hz, H$_b$), 4.05 (s, 8 H, H$_c$), 3.2-3.1 (m, 8 H, NCH$_2$CH$_2$), 2.25-2.2 (m, 8 H, C^{I}H$_2$), 1.5-1.3 (m, 24 H, C^{II}H$_3$-C^{IV}H$_2$), 1.45-1.4 (m, 8 H, NCH$_2$CH$_2$), 0.95-0.8 [m, 24 H, N(CH$_2$)$_2$CH$_3$, C^{V}H$_3$]; ^{13}C NMR δ 166.1 (C=O), 154.2 (ArCOCH$_2$O), 137.9 (ArCCH), 122.5 (ArCCH$_2$), 119.8 (ArCH), 99.5 (OCH$_2$O), 48.9 (NCH$_2$CH$_2$), 41.6 (CH$_2$Cl), 36.8 (ArCHAr).

Synthesis. The reaction was carried out under an argon atmosphere and flash column chromatography was performed with Merck silica gel 60 (0.040-0.063 mm, 230-400 mesh). Toluene was distilled from sodium and kept on molecular sieves (3/4 Å), and other chemicals were of reagent grade and used without further purification. The melting point was determined with a Reichert melting point apparatus and is uncorrected, the Mass (FAB) spectrum was recorded with a Finnigan MAT 90 spectrometer using *m*-nitrobenzyl alcohol as a matrix, and the elemental analysis was performed with a Model 1106 Carlo Erba Strumentazione Elemental Analyzer. The presence of solvent molecules in the analytical sample was confirmed by ^{1}H NMR spectroscopy.

Tetrakis(*N*-methylacetamidomethyl)cavitand (6); A solution of tetrakis(bromomethyl)cavitand 3 (100 mg, 84.1 μmol), *N*-methylacetamide (34 mg, 0.47 mmol), pulvered NaOH (50 mg, 2.0 mmol), K$_2$CO$_3$ (86 mg, 0.62 mmol), and a catalytic amount of tetrabutylammonium hydrogensulfate in toluene (5 mL) was stirred at 80 °C for 4.5 h. After removal of the solvent the residue was dissolved in CH$_2$Cl$_2$ (10 mL), washed with water (10 mL), 1 M NaOH (10 mL), water (10 mL), 1 M HCl (10 mL), water (2 × 10 mL), brine (10 mL), and dried over Na$_2$SO$_4$. The residue was purified by flash column chromatography (SiO$_2$, CH$_2$Cl$_2$/MeOH/Et$_3$N, gradient 95/5/0 to 90/9.5/0.5) to afford **6** as a white powder. Yield 80 mg (82%); mp 136-138 °C; MS *m/z* 1179.5 (100%, [M+Na]$^+$, calcd 1179.6); ^{1}H NMR for 2D COSY spectrum see Figure 3; ^{13}C NMR δ 170.5 (s, C=O), 154.3 (s, ArCOCH$_2$O), 137.8 (s, ArCCH), 122.6 (ArCCH$_2$), 119.6 (d, ArCH), 99.7 (t, OCH$_2$O), 45.8 (t, ArCH$_2$N), 36.9 (d, ArCHAr). Anal. Calcd for C$_{68}$H$_{92}$N$_4$O$_{12}$·0.5H$_2$O: C, 70.02; N, 4.80; H, 8.04. Found: C, 69.96; N, 4.54; H, 8.03.

X-Ray Spectroscopy. Intensities were measured, using graphite monochromated MoK$_\alpha$ radiation (λ = 0.7107 Å). The crystal structure was solved using direct methods and heavy atom techniques.

Table 3. Crystallographic Data of the Single Crystal X-Ray Structure of $3 \cdot CH_2Cl_2$.

Crystal data		Data Collection	
formula	$C_{56}H_{68}Br_4O_8 \cdot CH_2Cl_2$	radiation	MoK_α ($\lambda = 0.7107$ Å)
F_w	1273.7	scan type	$\omega/2\theta$
crystal system	triclinic	θ range	3.0 - 22.5°
space group	P-1	refl. measured	15452
a	11.901(3) Å	refl. obs. I > 3.0σ(I)	5831
b	21.262(5) Å	h range	-12 - 13
c	24.261(6) Å	k range	-22 - 23
α	72.75(2)°	l range	0 - 26
β	79.42(3)°		
γ	82.05(4)°	**Refinement**	
V	5740(3) Å^3	Final R	0.083
Z	4	wR	0.100
D	1.47 Mg.m^{-3}	S	2.67
μ	2.92 mm^{-1}	reflections	5831
T	293(1) K	parameters	661

7. References and Notes

1. (a) McKay, H. A. C., Miles, J. H., and Swanson, J. L. (1984) in Schulz, W. W. and Navratil, J. D. (eds.) *Science and Technology of Tributyl Phosphate*, CRC Press, Boca Raton, pp. 1-10. (b) Miles, J. H. (1984) in Schulz, W. W. and Navratil, J. D. (eds.) *Science and Technology of Tributyl Phosphate*, CRC Press, Boca Raton, p. 11.

2. For example see: (a) Horwitz, E. P., Kalina, D. G., Diamond, H., and Vandegrift, G. F. (1985) *Solv. Extr. Ion Exch.* **3**, 75. (b) Horwitz, E. P. and Schulz, W. W. (1991) in Horwitz, E. P. and Schulz, W. W. (eds.) *New Separation Chemistry, Techniques for Radioactive Waste and other Specific Applications*, Elsevier Applied Science, New York, p. 21.

3. Arnaud-Neu, F., Böhmer, V., Dozol, J.-F., Grüttner, G., Jakobi, R. A., Kraft, D., Mauprivez, O., Rouquette, H., Schwing-Weill, M.-J., Simon, N., and Vogt, W. (1996) *J. Chem. Soc., Perkin Trans. 2* 1175.

4. (a) Boerrigter, H., Verboom, W., and Reinhoudt, D. N. (1997) *J. Org. Chem.* **62**, 7148. (b) Boerrigter, H., Lugtenberg, R. J. W., Egberink, R. J. M., Verboom, W., Reinhoudt, D. N., and Spek, A. L. (1997) *Gazz. Ital. Chim.* **127**, 709. (c) Boerrigter, H., Verboom, W., and Reinhoudt, D. N. (1997) *Liebigs Ann./Recueil* 2247.

5. For reviews on resorcinarenes see: (a) Cram, D. J. and Cram, J. M. (1994) Container Molecules and their Guests, in Stoddart, J. F. (ed.), *Monographs in Supramolecular Chemistry, Vol. 4*, The Royal Society of Chemistry, Cambridge. (b) Timmerman, P., Verboom, W., and Reinhoudt, D. N. (1996) *Tetrahedron* **52**, 2663.

6. (a) Seaborg, G. T. (1993) *Radiochim. Acta* **61**, 115. (b) Choppin, G. R., and Rizkalla, E. N. (1994) in Gschneider, Jr, K. A., Eyring, L., Choppin, G. R., and Lander, G. H. (eds.), *Handbook of the Physics and Chemistry of Rare Earths, Vol. 18*, Elsevier Science, Amsterdam, p. 559.

7. Boerrigter, H.; Verboom, W.; De Jong, F.; Reinhoudt, D. N. (1998) *Radiochim. Acta* accepted for publication.

8. The $\nu_{C=O}$ stretching vibrations of the free ligands **1a** (1671 cm^{-1}) and **1b** (1635 cm^{-1}) shift to a lower wavenumber upon complexation with Eu(III). Approximately, the shifts of **1a** and **1b** are 60 and 25 cm^{-1}, respectively, with an estimated error of ±20 cm^{-1} [4a].

9. Complexation of organic compounds in the cavity of resorcinarene *octols* by $CH_3 \cdots \pi$ interactions has been demonstrated. For example see: (a) Kobayashi, K., Asakawa, Y., Kikuchi, Y., Toi, H., and Aoyama, Y. (1993) *J. Am. Chem. Soc.* **115**, 2648. (b) Leigh, D. A., Linnane, P., Pitchard, R. G., and Jackson, G. (1994) *J. Chem. Soc., Chem. Commun.* 389.
10. Cram, D. J., Karbach, S., Kim, H.-E., Knobler, C. B., Maverick, E. F., Ericson, J. L., and Helgeson, R. C. (1989) *J. Am. Chem. Soc.* **110**, 2229.
11. Tucker, J. A., Knobler, C. B., Trueblood, K. N., and Cram, D. J. (1989) *J. Am. Chem. Soc.* **111**, 3688.
12. Beer, P. D., Tite, E. L., Drew, M. G. B., and Ibbotson, A. (1990) *J. Chem. Soc., Dalton Trans.* 2543.
13. The Gibbs free energies of amide rotations are usually in the range of 78-90 kJ/mol. For a discussion see: Kessler, H. (1970) *Angew. Chem., Int. Ed. Engl.* **9**, 219.
14. Due to the limited number of interesting contacts in the 2D NOESY and ROESY spectra, the exact elucidation of these structures is not viable.
15. It has to be noted that several of the materials contained small amounts of solvent molecules, which can impose large differences in the dynamic behavior including the relaxation behavior. Also, relaxation is strongly dependent on the dynamic flexibility of the system. For several of the materials described more than one conformation exists at room temperature and the interchange between these conformers strongly influences the T_1 times observed. However, in combination with the *NOE* effects observed and the contacts with the pentyl moiety (in particular C^I, C^{III}, and C^{IV}), aggregation behavior is suggested.
16. Aggregation of cavitands in the solid state by bilayer formation with the mutual pentyl chains was observed in the single crystal X-ray structure of a crown[6]cavitand: Higler, I., Boerrigter, H., Verboom, W., Kooiman, H., Spek, A. L., and Reinhoudt, D. N. (1998) *Eur. J. Org. Chem.* accepted for publication.
17. Malonamides are similar to the CMP(O) molecules but contain two amide moieties. For micellar behavior of tetraalkyl-alkyl malonamides see: Nigmond, L., Musikas, C., and Cuillerdier, C. (1994) *Solv. Ex. Ion Exch.* **12**, 261.
18. (a) Jeener, J., Meier, B. H., Bachmann, P., and Ernst, R. R. (1979) *J. Chem. Phys.* **71**, 4546. (b) Neuhaus, D. and Williamson, M. (1989), in *The Nuclear Overhauser Effect in Structural and Conformational Analysis*, VCH Publishers, Cambridge.
19. Bothner-By, A. A., Stephens, R. L., Lee, L., Warren, C. D., and Jeanloz, R.W. (1984) *J. Am. Chem. Soc.* **106**, 811.
20. Bax, A. and Davis, D. (1985) *J. Magn. Reson.* **65**, 355.
21. Bax, A. and Freeman, R. (1981) *J. Magn. Reson.* **44**, 542.
22. States, D. J., Haberkorn, R. A., and Ruben, D. J. (1982) *J. Magn. Reson.* **48**, 286.

NMR STUDIES OF MOLECULAR RECOGNITION BY METALLOPORPHYRINS

NICK BAMPOS, ZÖE CLYDE-WATSON, JOANNE C. HAWLEY,
CHI CHING MAK, ANTON VIDAL-FERRAN, SIMON J. WEBB
AND JEREMY K. M. SANDERS

University Chemical Laboratory, Lensfield Road,
Cambridge, CB2 1EW, UK

1. Introduction

If the past century of chemistry has been primarily concerned with understanding and manipulating the covalent bond, then the next will surely be dominated by non-covalent interactions. We aim to understand how to use a range of non-covalent interactions to create new macrocycles that are capable of recognition and catalysis or that have unusual properties. This chapter will focus on the use of NMR to elucidate the recognition properties of a series of macrocycles that contain metalloporphyrins. Following this brief introduction we will examine a series of systems that display increasing flexibility, and therefore increasing responsiveness to the geometric demands of their ligands. Our work is carried out in non-polar solvents such as $CDCl_3$ or toluene.

Metalloporphyrins are attractive building blocks for supramolecular chemistry [1]: variation of the central metal ion changes the recognition specificity, while the extended aromatic π-system leads to characteristic NMR and UV/visible spectroscopic properties. We will use just two recognition events in this chapter: N-coordination by Zn porphyrins, and carboxylate coordination by Sn porphyrins. The spectroscopic consequences of such binding events are summarised in Figs 1 and 2. As can be seen from the rate constants illustrated in Fig 1(a), Zn–N ligation is generally a fast exchange process. When multiple binding occurs between host and guest the dissociation rates can move into slow exchange; measurement of large binding constants ($> 10^7$ M^{-1}) is then possible *via* simple NMR competition experiments [2]. The UV/visible changes that result from such binding are also summarised in Fig 1: they enable smaller binding constants ($< 10^9$ M^{-1}) to be measured. As Fig 1(b) shows, the spectrum resulting from two porphyrins bound to a single DABCO ligand is highly characteristic, the 6 nm shift from 426 to 420 nm being the result of exciton coupling between the π-systems [3].

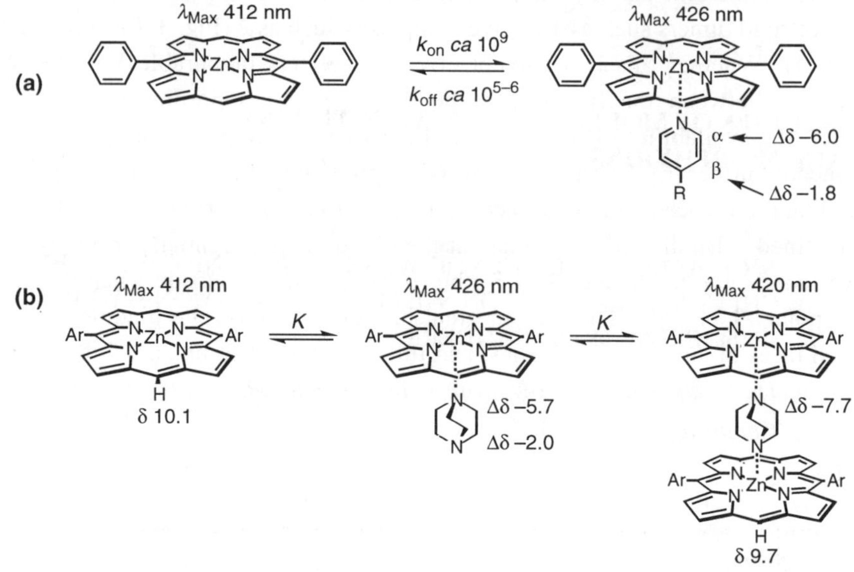

Figure 1. Complexation of Zn porphyrins with (a) pyridines and (b) DABCO, together with the key spectroscopic consequences; Δδ is in ppm.

Binding of carboxylates (and carboxylic acids) by Sn(IV) porphyrins has not been thoroughly studied in the past, but appears very promising: the binding geometry for benzoates is perpendicular to the porphyrin, as shown in Fig 2, but switches to a more coplanar arrangement for 9-anthroate where additional interactions between the porphyrin and anthacene units lead to a stronger complex than expected [4]. Replacement of the bound hydro groups by bound carboxylates does not give any shift in the UV/visible spectrum, and is in slow exchange on all NMR timescales.

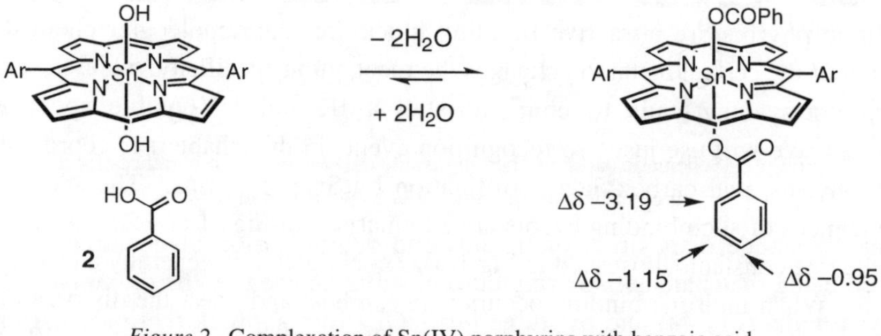

Figure 2. Complexation of Sn(IV) porphyrins with benzoic acid, with the key NMR consequences.

2. Dimers

We showed some time ago [5] that pyridine and its derivatives prefer to bind *inside* the cavities of rigid dimers such as **1** to give complexes such as **2** (Fig. 3; for full structures see Fig. 4) Whether this is a solvent effect or reflects favourable Van der Waals interactions between the ligand and the distal (passive) porphyrin is not clear. However, the availability of mixed metal dimers such as **3** enables us to explore this phenomenon further [6]. Insertion of a $Sn(OH)_2$ group into the passive face (**3**) switches the binding preference at the other face from inside to outside (**4**) *via* an effect that is as yet undefined. Binding a carboxylate at the tin site is preferentially inside when the passive face is non-polar (e.g. **3 → 5**) but is preferentially outside when the passive face is polar (**9 → 10**). There is no great barrier to carboxylate binding at the less favoured faces of any of these species, and fully substituted complexes **7, 8** and **11** are readily prepared. Similar preferences are seen in the corresponding trimeric systems. [6]

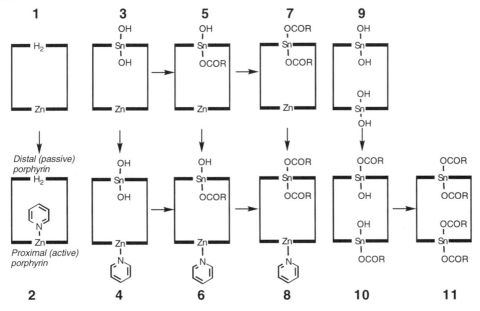

Figure 3 Schematic cartoons showing inside/outside binding preferences for a range of ligands and rigid cofacial porphyrin dimers.

The inside/outside disposition of ligands and determination of geometry is readily achieved using the characteristic ring current shifts induced by the proximal and distal porphyrin units (Fig 4a) and by numerous NOEs (not shown); the inner carboxylate experiences much larger shifts than the outer. In such complexes the internal ligands generally rotate rapidly, but in the tetra(9-anthroate) complex of the rigid Sn_2 dimer

(Fig 4b) the steric crowding is so large that inner anthroate rotation is slow on the chemical shift timescale at room temperature, rendering the two halves of each porphyrin inequivalent. Exchange is rapid on the T_1/NOE timescale, allowing detailed assignments to be made [6].

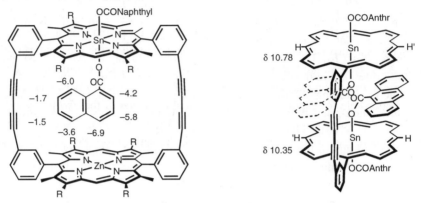

Figure 4 (a) The di(1-naphthoate) complex of a rigid SnZn dimer, showing the ring current induced shifts ($\Delta\delta$) experienced by the inner bound naphthoate. (b) The tetra(9-anthroate) complex of a rigid Sn_2 dimer.

3. Trimers

Our versatile synthesis of cyclic trimers from linear intermediates [7] has enabled us to prepare numerous new mixed metal trimers and to explore their properties. Some of our more unusual examples are illustrated in Fig 5: in **12**, binding of two pyridines of the ligand to the Zn centres spatially forces the third nitrogen to coordinate to an otherwise inert Ni centre, thereby converting it to a rare five-coordinate paramagnetic form. [7] Trimer **13** contains a combination of porphyrins with non-aromatic (and therefore ring current-free) but more strongly binding dioxoporphyrins, allowing us to map binding geometry preferences[8]. **14** is homochiral by virtue of the chiral carboxylates bound to the Sn centres: when these carboxylates themselves possess further functionality such as hydroxyls, one can observe cavity binding preferences by NMR competition experiments [9].

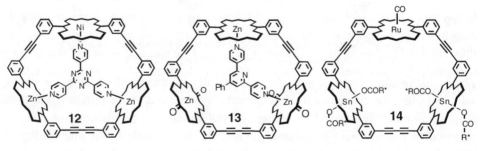

Figure 5. Some asymmetric trimers

However, perhaps the most exciting aspect of these trimers is their ability to influence the outcome of reactions where two finely-balanced pathways compete as summarised in Fig 6 [10]. The reversal of regioselectivity between the two cyclic trimers at 30° C is the result of two separate effects, one predicted and one not: the large (500-fold) *endo*-acceleration induced by the smaller 1,1,2-trimer was expected but the fact that this smaller trimer is ineffective at 30° C in accelerating the *exo*-reaction and binding the *exo*-adduct was a surprise. The key difference appears to lie in the greater flexibility of the larger 2,2,2-system: neither of the trimeric hosts has the ideal equilibrium geometry to bind the *exo*-transition state or adduct, but at 30° C the larger trimer is more flexible and so is better able to respond to the geometrical demands of the *exo*-pathway. One might claim that the stereochemical reversal at 30° C is a major success, but it is important to note that at 60° C the smaller trimer becomes more flexible and loses its regioselectivity.

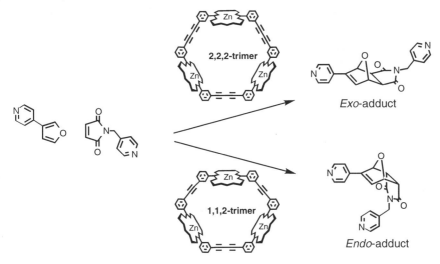

Figure 6 Redirection of a Diels–Alder reaction using the geometrical constraints of a host cavity.

If we create even more conformational freedom in a trimer by hydrogenation of the butadiyne linkers then we obtain the complex mixture of interconverting isomers illustrated in Fig.7. This isomerisation process, and the characterisation of the different recognition selectivities of the various isomers, has been carried out using COSY, NOESY and ROESY as described in detail elsewhere[11].

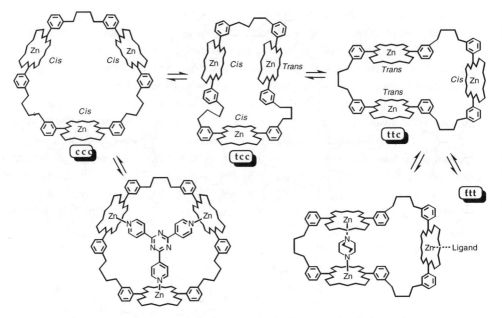

Figure 7 Conformational isomers of a flexible trimer in solution and their major complexes with tripyridyltriazine and DABCO.

4. Dendrimers

We can unleash even more conformational flexibility by creating dendrimers such as **15** (see next page) [12]. This has the intriguing property that in the presence of a bifunctional ligand such as DABCO its arms can be folded in a way that is readily detected by NMR and UV/visible specroscopy (Fig. 8).

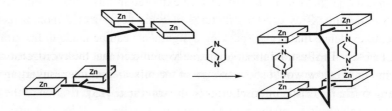

Figure 8 DABCO-induced folding of the arms of dendrimer **15**.

15 R = *n*-hexyl or CH₂CH₂COOMe

5. CPMAS of porphyrins

Finally, given the difficulty of porphyrin crystallography, we consider the question of whether CPMAS spectroscopy of porphyrins can help to elucidate solid state host–guest chemistry [13]. The large intermolecular ring current that one would expect from closely-packed porphyrins is the key feature explored in this work. Comparison of the CPMAS data with X-ray crystal structures of some key selected monomers provides convincing evidence that crystal packing assignments can be made in favourable cases using CPMAS. To illustrate this, we correctly predicted that the coordination polymer shown in Fig. 9 below would show no sign of the offset structures resulting from the π–π packing so characteristic of porphyrins in the solid state [13].

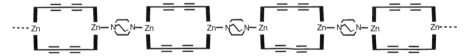

Figure 9. A coordination polymer resulting from extra-cavity binding of DABCO to a rigid porphyrin dimer.

6. Acknowledgments

We thank the EPSRC, the European Union, Ethyl Petroleum, Unilever Research, the Croucher Foundation, and Cambridge Commonwealth Trust for financial support.

7. References

1. Anderson, H. L. and Sanders, J. K. M. (1995) Enzyme mimics based on cyclic porphyrin oligomers: strategy, design and exploratory synthesis, *J. Chem. Soc. Perkin Trans 1*, 2223–2229.
2. Vidal-Ferran, A., Clyde-Watson, Z., Bampos, N. and Sanders, J. K. M. (1997) Ethyne-linked cyclic porphyrin oligomers: synthesis and binding properties, *J. Org. Chem.*, **62**, 240–241.
3. Hunter, C. A., Sanders, J. K. M. and Stone, A. J. (1989) Exciton coupling in porphyrin dimers, *Chemical Physics*, **133**, 395–404.
4. Hawley, J. C., Bampos, N., Abraham, R. J. and Sanders, J. K. M. (1998) Carboxylate and carboxylic acid recognition by tin(IV)porphyrins, *Chem. Commun.*, 661–662.
5. Wylie, R. S. Levy, E. G. and Sanders, J. K. M. (1997) Unexpectedly selective ligand binding within the cavity of a cyclic metalloporphyrin dimer, *Chem. Commun.*, 1997, 1611–1612.
6. Hawley, J. C., Bampos, N. and Sanders, J. K. M. (1997) unpublished results.
7. Vidal-Ferran, A., Bampos, N. and Sanders, J. K. M. (1997) Stepwise approach to bimetallic porphyrin hosts: spatially enforced coordination of a Ni(II) porphyrin, *Inorg. Chem.*, **36**, 6117–6126.
8. Clyde-Watson, Z., Bampos, N. and Sanders, J. K. M. (1998) Mixed cyclic trimers of porphyrins and dioxoporphyrins: geometry vs electronics in ligand recognition, *New. J. Chem.*, Nov 1998, in press.
9. Webb, S. J. and Sanders, J. K. M. (1997) unpublished results.
10. Clyde–Watson, Z., Vidal-Ferran, N., Twyman, L. J., Walter, C. J., McCallien, D. W. J., Fanni, S., Bampos, N., Wylie, R. S. and Sanders, J. K. M. (1998) Reversing the stereochemistry of a Diels–Alder reaction: use of metalloporphyrin oligomers to control transition state stability, *New J. Chem.*, **22**, 493–502.
11. Bampos, N., Marvaud, V and Sanders, J. K. M. (1998) Metalloporphyrin oligomers with collapsible cavities: characterisation and recognition properties of individual atropisomers, *Chem. Eur. J.*, **4**, 335–343.
12. Mak, C. C., Bampos N. and Sanders, J. K. M. (1998) Metalloporphyrin dendrimers with folding arms, *Angew. Chemie*, Nov 1998, in press.
13. Bampos, N., Prinsep, M. R., He, H., Vidal-Ferran, A., Bashall, A., McPartlin, M., Powell H. and Sanders, J. K. M. (1998) ^{13}C CPMAS NMR spectroscopy as a probe for porphyrin-porphyrin and host-guest interactions in the solid state, *J. Chem. Soc. Perkin Trans 2*, 715–723.

REVERSIBLE DIMERIZATION OF TETRAUREAS DERIVED FROM CALIX[4]ARENES

V. BÖHMER[a], O. MOGCK[a], M. PONS[b], E. F. PAULUS[c]

a) Institut für Organische Chemie, Johannes-Gutenberg-Universität,
J.-J.-Becher Weg 34 SB1, D-55099 Mainz, Germany
b) Departament de Química Orgànica, Universitat de Barcelona,
Martí Franquès, 1-11, E-08028 Barcelona, Spain
c) Hoechst-Marion-Roussel Deutschland GmbH,
D-65926 Frankfurt/Main, Germany

Abstract

Calix[4]arene derivatives, substituted at the wider rim by four urea functions are easily available in great variety. In apolar solvents like benzene or chloroform they dimerize via NH···O=C hydrogen bonds encapsulating a single solvent molecule as guest. The exclusive formation of dimers has been proved in solution by NMR, while their shape has been characterized by single crystal X-ray analysis. The kinetic stability and the rate of the guest exchange have been studied by NOESY experiments using a derivative with lower symmetry.

1. Introduction

"Molecular self-assembly is the spontaneous association of molecules under equilibrium conditions into stable, structurally well-defined aggregates joined by noncovalent bonds." [1] If this self-assembly occurs between suitably curved, concave molecules, the formation of molecular containers can be achieved, which, in contrast to the covalently linked carcerands [2], are able to encapsulate smaller guest molecules in a reversible manner. The first example, reported by Rebek and Mendoza [3], has been described as a "molecular tennis ball" due to the shape of the two single molecules which are held together by intermolecular hydrogen bonds in a similar manner as tennis balls are usually made up of two self-complementary pieces. Numerous modifications of this system containing larger internal volumes ("molecular baseball", "soft ball") have been described subsequently [4]. Cavitands, the standard building block of carcerands have been also used to compose a hollow dimer via hydrogen bonds (-O$^{(-)}$···H-O-) [5] or via coordinative binding to metals [6]. Both approaches are appealing due to a comparatively easy access to the monomeric parts. This easy access is true also for calixarenes [7,8] which can be transformed into self-complementary molecules by the attachment of urea functions in their *p*-positions.

2. Synthesis

Tetraurea derivatives of calix[4]arenes fixed in the cone conformation are easily synthesised, in numerous variations, starting from the readily available *t*-butyl calix[4]-arene (**1**) using the reaction sequence shown in Scheme 1. O-Alkylation gives tetraethers **2** fixed in the cone conformation for residues Y larger than ethyl. Similar tetraethers with two different residues Y^1/Y^2 in alternating order are available in two steps. For simple alkyl residues the complete replacement of all *t*-butyl groups by ipso-nitration is possible in excellent yield (CH_2Cl_2, HNO_3, r.t.) [9], while the yield of the tetranitro compounds **3** is somewhat lower with acid sensitive residues Y. Reduction of the nitro groups (by hydrazine or by hydrogenation using Raney-Ni) leads to tetraamines **4** [9] which in the final step are converted to tetraurea derivatives **5** by reaction with isocyanates. The same reaction sequence can be applied also to calix[5]arenes (and probably to larger calixarenes) while thiourea derivatives are available when iso-thiocyanates are used in the last step. Thus, using simple standard reactions of organic chemistry a huge variety of urea (or thiourea) derivatives are easily available even in large quantities.

In solvents like dmso, acetone or methanol the ^{1}H NMR spectra of such C_{4v}-symmetrical tetraurea derivatives show the expected pattern [9], namely a single set of signals for the residues Y and R, one pair of doublets for the protons (axial/equatorial) of the methylene groups, a singlet for the eight aromatic protons of the calixarene and two singlets for the two different NH groups of the urea groups, which in the case of aryl-ureas (see Fig. 1) have quite similar chemical shifts.

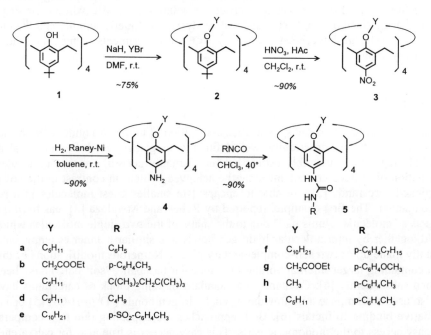

	Y	R		Y	R
a	C_5H_{11}	C_6H_5	f	$C_{10}H_{21}$	p-$C_6H_4C_7H_{15}$
b	CH_2COOEt	p-$C_6H_4CH_3$	g	CH_2COOEt	p-$C_6H_4OCH_3$
c	C_5H_{11}	$C(CH_3)_2CH_2C(CH_3)_3$	h	CH_3	p-$C_6H_4CH_3$
d	C_5H_{11}	C_4H_9	i	C_5H_{11}	p-$C_6H_4CH_3$
e	$C_{10}H_{21}$	p-SO_2-$C_6H_4CH_3$			

Scheme 1 Synthesis of tetraurea derivatives **5**. For compounds used in the following discussion Y and R are indicated.

In apolar solvents such as chloroform or benzene the two singlets for the NH groups are up- and down-field shifted leading to a separation of more than 2 ppm, and the aryl protons appear as two meta-coupled doublets separated also by more than 1.5 ppm. This spectral behaviour, indicating a C_4-symmetrical conformation for the calix[4]arene, cannot be satisfactorily explained by intramolecular hydrogen bonding, neither between adjacent urea functions (which would be unlikely for steric reasons) nor between opposite urea functions (which would be in disagreement with four spectroscopically identical units).

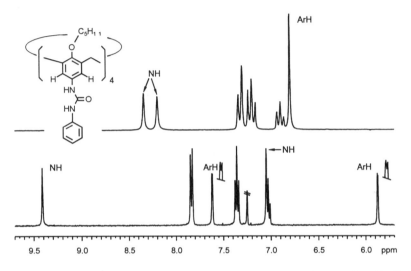

Figure 1 Section of the ^{1}H NMR spectrum of the tetraurea calixarene **5a** in dmso-d_6 and in CDCl$_3$

3. Homo and Hetero Dimers with S_8-Symmetry

A satisfactory explanation for the observed spectra in apolar solvents was first given by Rebek [10], who suggested the formation of dimeric capsules of S_8-symmetry

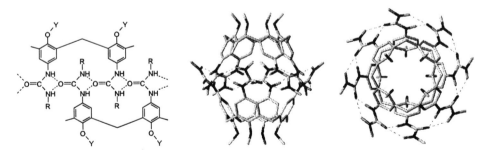

Figure 2 Schematic representation of hydrogen bonding in dimers formed by tetraurea calix[4]arenes and calculated structures of the most simple example (R=Y=Me, MM3)

composed of two enantiomeric C_4-symmetrical calix[4]arenes held together by hydrogen bonds as shown in Fig. 2.

This explanation was, nevertheless, met with some scepticism, at least with respect to its exclusivity (see section 5) and its generality (see below). Additional convincing support for the existence of dimers in solution came mainly from NMR-spectroscopy.

Due to the dimerisation and the directionality of the belt of hydrogen bonded urea functions within these dimers enantiotopic protons or groups attached to the calixarene skeleton via the ether residues Y or the urea residues R become diastereotopic. This is shown in Fig. 3a for **5b** ($Y=CH_2COOEt$) where the CH_2 group shows a singlet in benzene containing 5% methanol, but appears as a doublet with geminal coupling in pure benzene-d_6. Similarly the two methyl groups of $R=C(CH_3)_2CH_2C(CH_3)_3$ in **5c** give rise to two singlets in benzene-d_6 (Fig. 3b), while they show one singlet in dmso-d_6. Again this reduction of the symmetry from C_{4v} to C_4 or S_8 respectively can also be seen for the CH_2 protons.

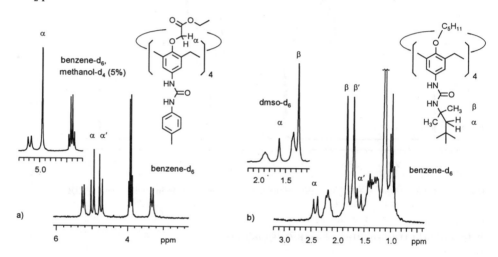

Figure 3 Sections of the NMR-spectra of tetraureas **5b** (left) and **5c** (right) demonstrating the dimerisation in benzene-d_6 by the transformation of enantiotopic atoms or groups into diastereotopic atoms/groups.

If tetraurea derivatives **5** exist as dimers, and if this dimerisation is a (more or less) general phenomenon, a mixture of two different tetraureas A and B should contain not only the homodimers AA and BB but also the heterodimer AB. This should be seen by additional peaks in the NMR spectrum of such a mixture. In fact this is the case! [11] While the ^{1}H-NMR-spectrum of a dmso-d_6 solution of **5a** ($Y=C_5H_{11}$, $R=C_6H_5$) and **5d** ($Y=C_5H_{11}$, $R=C_4H_9$) corresponds exactly to the superimposition of the spectra of **5a** and **5d**, additional peaks with the correct multiplicity and intensity corresponding to the heterodimer are found in $CDCl_3$. This is most easily seen in the region of the Ar-CH_2-Ar protons (Fig. 4) where four pairs of doublets for the diastereotopic protons in axial and equatorial position can be distinguished, one pair for each of the homodimers **5a/5a** and **5d/5d** and two pairs for the heterodimer **5a/5d**.

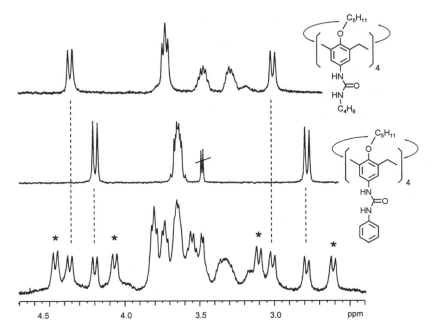

Figure 4 Section of the NMR spectra of **5a**, **5d**, and a mixture of **5a** and **5d** (200 MHz, CDCl$_3$).

In all examples studied by our group, these heterodimers formed more or less in statistical amounts. A mixture of Z tetraureas should therefore result in the formation of a "library" containing $0.5 \cdot Z \cdot (Z+1)$ dimers [12]. Very recently, Rebek et al. reported that the tosyl urea derivative **5e** (R=SO$_2$C$_6$H$_4$-p-CH$_3$) forms only heterodimers in mixtures with an aryl urea like **5f** (R=C$_6$H$_4$-p-C$_7$H$_{15}$). While the exact reason for this preference is as yet unclear, this observation opens the way to a controlled formation of larger assemblies [13].

4. Single Crystal X-Ray Analysis

Single crystals suitable for an X-ray analysis were obtained for **5b** by recrystallisation from benzene [14]. Although the NMR-evidence for the formation of dimers in solution was already overwhelming, this structure gave the first experimental description of their shape (compare Fig. 5). The dimer is composed of two crystallographically independent molecules which lie on a fourfold crystallographic axis. They are turned around this axis by 43° with respect to each other in entire agreement with the (time averaged) S$_8$-symmetry in solution, which requires 45°.

The individual bisaryl urea structures are nearly planar (interplanar angle between the phenolic rings of the calixarene and the tolyl residues of the urea 12.4°/22.0°) and the interlocking N-CO-N elements of both parts are nearly perpendicular to each other (dihedral angles 73.8°/83.7°).

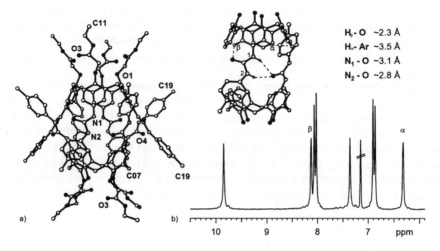

Figure 5 Molecular conformation of the dimers **5b/5b** in the crystalline state: a) "ball and stick" representation; b) explanation of the chemical shifts observed for the Ar-H and NH protons.

The N····O=C distances are quite different, 2.85/2.84 Å for the tolyl-NH and 3.16/3.10 Å for the calix-NH, indicating hydrogen bonds of quite different strengths. In combination with shielding/deshielding effects this explains the strong splitting of the NH singlets observed in solution. For the two calixarene aryl protons one is at a distance of 2.3 Å from the carbonyl oxygen (deshielding) while the other is 3.5 Å from the centre of the tolyl residue (shielding) which again confirms their down- and upfield shifts (compare Fig. 5).

TABLE 1 Diagonal distances (Å) of selected atoms in both calix[4]arenes of the dimer **5b/5b** in the crystal and in the energy minimized structure [14]. For the numbering see Fig. 5.

	X-ray analysis		calculated structure		
atom	calix 1	calix 2	empty	1 benzene*	2 benzene*
C 11	13.217	11.187	5.68	5.68	5.68
O 3	8.159	7.698	5.92	5.93	5.89
O 1	4.375	4.469	4.30	4.29	4.28
C 07	7.223	7.212	7.20	7.20	7.20
N 1	9.479	9.300	9.08	9.05	9.22
O 4	11.212	11.641	10.43	10.43	10.35
N 2	12.011	11.934	11.31	11.29	11.32
C 19	19.447	20.141	17.82	17.83	17.54

* included as guest

The shape of the dimer found in the crystalline state is in remarkable agreement with the calculated structure (Table 1). Larger deviations are found only for the flexible pendant ether residues. This is not surprising, since in the crystal their conformation is strongly determined by packing effects while in the calculation many local minima with similar energy exist.

Fig. 6 shows the crystal packing from two different perspectives. Dimers with the same direction of the carbonyl groups are arranged in infinite columns and columns with the opposite direction are packed in a tetragonal arrangement. The crystal contains completely disordered benzene (not shown in Fig. 6) for which only maxima of electron density are found. The following quantities (per dimer) can be estimated: 1.06 benzene molecules in the cavity formed by the dimerization, 0.81 benzene molecules in the similar sized cavity formed between the estergroups of adjacent dimers along the columns, and 5.8 molecules of benzene in further voids of the crystal lattice between the columns.

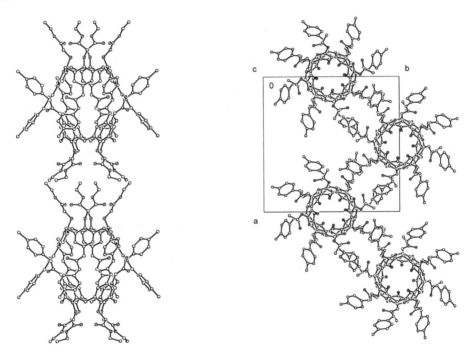

Figure 6 Crystal packing of dimers **5b/5b** seen perpendicular and along the columns extending in the direction of the c-axis.

Strictly speaking the existence of dimers in the crystalline state does not prove their existence in solution (while it gives in addition to the solution results a fairly accurate description of their shape). However, a further argument for the existence of dimers in apolar solvents comes from crystals obtained for **5g** from dmso/water/acetonitrile (Fig. 7) [15]. Although not yet entirely refined the following statements are possible: In contrast to the crystals obtained for the very similar **5b** from benzene the single molecules of **5g** assume a "pinched" cone conformation in which two opposite phenolic units are nearly parallel (interplanar angle 22.8°) and two are bent outwards (interplanar angle 110.5°). There are exclusively intermolecular hydrogen bonds between urea groups, not leading to the formation of dimers. Instead. the molecules are arranged in a complicated way in infinite chains via two H-bonding sequences of the type $(CH_3)_2S=O\cdots(HNR)_2C=O\cdots(HNR)_2C=O$ at each side of every molecule, as shown in Fig. 7.

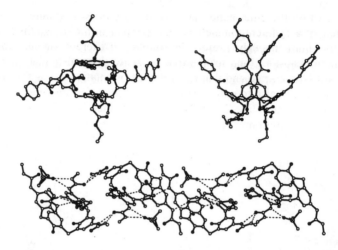

Figure 7 Molecular conformation (above, two different perspectives) and arrangement in infinite chains (below) in crystals of **5g** obtained from dmso/water/acetonitrile. (Substituents on the oxygen are omitted and the urea residues are "shortened" to one carbon for clarity in the lower part.)

5. Guest Inclusion and Thermodynamic Stability of the Dimers

The inclusion of appropriate guest molecules in solution can be shown in different ways.

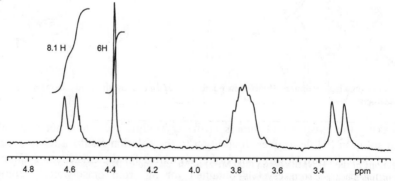

Figure 8 Section of the ^{1}H NMR spectrum (200 MHz) of **5b** in C_6H_6

If C_6H_6 is added in increasing amounts to a solution of **5b** in C_6D_6 a peak of increasing intensity appears at 4.38 ppm. Its relative intensity decreases again upon addition of benzene-d_6. This peak must be attributed therefore to benzene included in the cavity of the dimers which experiences a high field shift of about 2.8 ppm in comparison to free benzene. Fig. 8 shows the corresponding section of an ^{1}H-NMR spectrum measured in pure benzene-H_6, obviously the largest excess that can be applied. Its integration reveals, that within the experimental accuracy **one** molecule of benzene is included per dimer.

Fig. 9 demonstrates on the other hand that in a mixture of two tetraureas the included benzene experiences a different magnetical environment in the heterodimer and in the two homodimers simultaneously present in solution. Different singlets for benzene included in different capsules (and for the free benzene) show also, that their exchange is slow on the NMR-time scale and that no preferred orientation exists for the included molecule. It is interesting also to note, that different signals for the dimeric capsule can be found in a solvent mixture like $C_6D_6/CDCl_3$ showing, that the different guests cause a magnetically different environment for parts of the surrounding host (compare also [10]).

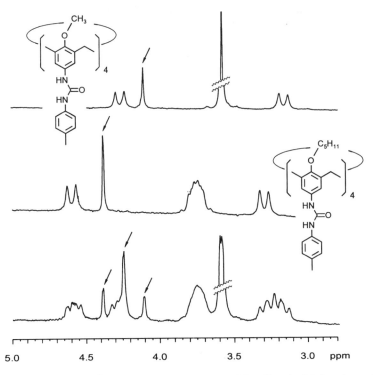

Figure 9 Sections of the ^{1}H NMR spectra (200 MHz) of **5h**, **5i** and of their mixture in C_6D_6/C_6H_6. The peaks for the included benzene are marked by an arrow.

Capsules are formed also in p-xylene-d_{10}, a solvent which is obviously less readily included than other solvents. Thus, Rebek at al. could establish a relative inclusion tendency for different guests from competitive measurements [16] by comparing the ratio of included guests to the ratio of free (offered) guests. They found for instance the following order for monosubstituted benzenes C_6H_5R and different substituents R: F>H>OH>Cl≈NH$_2$>CH$_3$. From a $\Delta\delta > 2$ ppm for the ortho and para hydrogens of included fluorobenzene the authors also conclude, that this guest (although obviously rotating fast around its axis) assumes a preferred orientation within the capsule. This is not surprising for a molecule having a dipole moment within an anisotropic closed cavity. The volume of the dimers is also large enough (or the capsule is flexible enough) not to increase the ring-inversion barrier of included cyclohexane [17].

Up to now nothing is known about the thermodynamic stability of the dimers in solution, characterised for instance by an equilibrium constant $A + A \rightleftharpoons A_2$. However, no conditions have been found under which dimers and the single tetraurea molecules can be observed simultaneously.

For instance no fundamental change in the ^{1}H NMR spectrum can be observed for a solution of **5i** in $C_2D_2Cl_4$ when the temperature is raised from 20°C to 110°C, at which temperature all signals are still in agreement with a single species, the tetraurea dimer. Fig. 10 shows for the same tetraurea **5i** a dilution experiment in CDCl$_3$. While for c = 24 mmol/l only the dimer is observed, additional peaks appear at c = 1.8 mmol/l, where, however, the NH-singlet at 9.3 ppm still indicates dimers as the predominant species. At c = 0.6 mmol/l this signal has disappeared, but no definite species can be seen.

Obviously on dilution the dimers are replaced by various species partly associated via hydrogen bonds and exchanging with rates similar to the NMR-time scale. Similar observations are made for the stepwise addition of small amounts of solvents like methanol-d_4 or dmso-d_6 until at concentrations of >1-2% only the monomer is present,

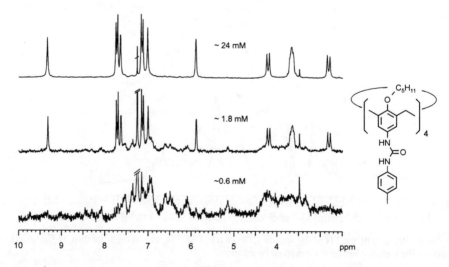

Figure 10 ^{1}H-NMR spectrum of **5i** (CDCl$_3$, 200 MHz) as a function of the concentration. A concentration of 1-1.5 mmol/l may be deduced as "critical dimer concentration".

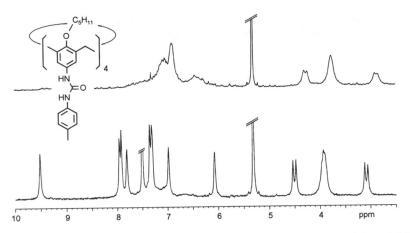

Figure 11 ¹H-NMR spectrum (200 MHz) of **5i** in CD$_2$Cl$_2$ (above) and after addition of 5% benzene-d$_6$ (below).

in which (most likely) the urea functions are solvated by this polar solvent. Similarly a destruction of the capsules has been reported by addition of simple dialkyl- or diaryl ureas [10].

As with the dimerisation of cavitands [5] the presence of a suitable guest is necessary for the formation of tetraurea dimers as demonstrated in Fig. 11. The rather broad and in the aromatic region unstructured spectrum of **5i** in CD$_2$Cl$_2$ changes into the clear spectrum of the dimer upon addition of 5% benzene-d$_6$, obviously a favoured guest. Thus, it seems still possible, that the dimerisation of pentaurea calix[5]arenes **6** for which also very broad spectra are observed in solvents like CDCl$_3$ or benzene-d$_6$ can be induced by suitable guests, although inclusion of various potential guest molecules such as adamantane, naphthalene, tetraline etc. has been unsuccessful.

Molecular mechanic calculations lead to analogous energies ($\Delta E = E_{dimer} - 2\, E_{monomer}$) for the dimerisation of calix[4]- and calix[5]arenes [18]. For [N'-methyl-ureido]-methyl ethers (R=Y=CH$_3$) these values are $\Delta E_4 = -87.8$ kcal/mol and $\Delta E_5 = -111.5$ kcal/mol, which means -21.95 and -22.3 kcal/mol per unit. Although these calculations do not account for the "inner" and "outer" solvation of the dimer, the identical values per unit show that the interaction via H-bonding is analogous and that no additional steric strain would be present in a calix[5]arene dimer in comparison to the calix[4]arene dimers.

6. Kinetic Stability and Guest Exchange

Mixing of two solutions containing different homodimers leads immediately to the observation of heterodimers, showing that the equilibrium AA + BB ⇌ 2 AB is reached within less than one minute. In principle now the exchange rate of A (or B) between AA (or BB) and AB can be determined by EXSY experiments. An even more detailed information on the kinetic stability of the dimers was obtained with the tetraurea derivative **7** [19], which is available from the corresponding 1,3-dimethyl-2,4-dipentyl ether of *t*-butylcalix[4]arene exactly as described in Scheme 1.

By its constitution tetraurea **7** is C_{2v}-symmetrical, as expressed by its ^{1}H-NMR spectrum under conditions where no dimerisation takes place (Fig. 12). This observed C_{2v}-symmetry (in combination with the usual pair of doublets for the protons of the methylene bridges and the strong chemical shift difference for the two singlets of the aromatic protons) also proves that already the monomer assumes a cone conformation and that this cone conformation is not caused by the dimerisation. The same is true in our hands for the tetramethoxy tetrurea derivative **5h**, where also no "conformational control through self-assembly" [20] could be observed.

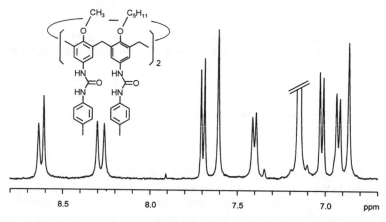

Figure 12 Section of the ^{1}H-NMR-spectrum (400 MHz) of **7** in benzene-d_6 containing 2% dmso-d_6.

In comparison to the monomeric form, the NMR-spectrum of the dimer **7/7** is very complex (Fig. 13) but entirely interpretable on the basis of the following symmetry considerations. While the C_{4v}-symmetry of tetraureas **5** is reduced to C_4 by the directionality of the carbonyl groups, the symmetry of **7** is reduced from C_{2v} to C_2. Again a pair of enantiomeric molecules is combined in the dimer, but due to the lower symmetry this combination does not result in a meso dimer with a S_n axis. The C_2-axis

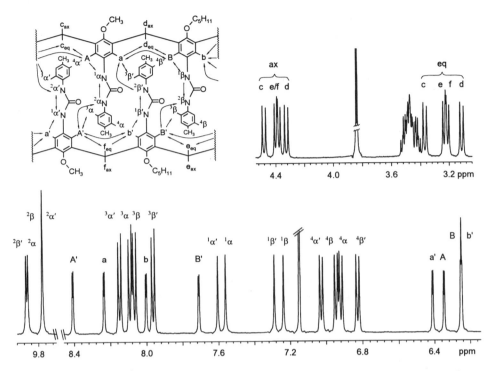

Figure 13 Sections of the ^{1}H-NMR spectrum (500 MHz) of 7 in benzene-d_6 including the labelling scheme. The observed NOEs/ROEs used for the peak assignment are indicated by arrows in the schematic formula of the dimer.

remains the only symmetry element of the dimer and the two calixarene parts are non-equivalent within the dimer. Thus we have four different phenolic units in the dimer (two anisol and two pentyl ether units) and each has two different aromatic protons. Consequently eight doublets with meta coupling can be distinguished for the aromatic protons of the calixarene skeleton (Fig. 13). We find four singlets at higher field for the NH groups attached to the calixarene and three (ratio 1:1:2) at lower field for the NH-groups attached to the tosyl residues. The four different tosyl residues give four pairs of doublets with ortho-coupling, indicating rapid rotation around the Ar-N bonds. Four pairs of doublets four the Ar-CH$_2$-Ar groups can be distinguished in the methylene region, as well as two singlets for the methoxy groups in the two calixarenes.

An unambiguous assignment of all protons was possible using TOCSY experiments (to correlate protons of the same spin system, e.g. A, a or c_{ax}, c_{eq}), HMBC experiments (to correlate through three bond couplings protons in the calixarene aryl ring with ether substituents via the carbon atom bearing the ether oxygens) and the various NOEs/ROEs indicated in Fig 13. Based on this assignment exchange processes could be studied kinetically by NOESY experiments as discussed subsequently.

For the subsequent discussion we label the NMR distinguishable sites as follows (see also Fig. 14):
  (i) Protons of anisole rings are assigned A/a, those of the pentyl ether rings by B/b
  (ii) Protons pointing towards the same (the different) type of ring in the second calixarene are assigned capital letters A,B (small letters a,b).
  (iii) A dash (A', a') is used to characterise protons of rings where the urea carbonyl group is pointing to a ring of the same type.

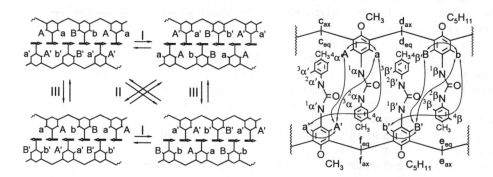

Figure 14 Schematic representation of the possible homomerization pathways of the dimer 7/7 (left side). The corresponding observed exchange cross-peaks in NOESY experiments are indicated in the schematic formula on the right side.

Having four different environments in the dimer for each proton in a given aromatic ring (e.g. A, a, A', a' for the aromatic protons of the anisol rings) offers the possibility to study three different exchange processes. This is the striking advantage in comparison to S_8-symmetrical dimers resulting from C_{4v}-symmetrical tetraureas, where two different environments allow to study only one exchange process. This exchange may occur by different mechanisms, between which under these circumstances no distinction is possible.

These three exchange processes, schematically represented in Fig. 14, can be realised by the following pathways:
  (I) A change of the direction of the C=O···HN hydrogen bonds would convert a → a', A → A' etc. This could occur **either** *within a dimer* by a more or less concerted rotation of the urea segments around the aryl-NH bonds **or** via *dissociation* of the dimer followed by *recombination*.
  (II) The *dissociation/recombination* could occur without a change in the direction of the carbonyl groups, but with the two calixarenes being turned by 90° with respect to each other. This would convert a → A', A → a', etc.
  (III) The *dissociation/recombination* could occur with rotation by 90° accompanied by a change in the direction of the C=O groups (formally a combination of the processes I and II). This would convert a → A, A' → a', etc.

Of course these exchanges can be recorded by the signals of various protons, and they were studied in various experiments with different mixing times. As an average the following rate constants for the three different processes were obtained:

$k_I = 0.066 \pm 0.015 \text{ s}^{-1}$ $k_{II} = 0.069 \pm 0.015 \text{ s}^{-1}$ $k_{III} = 0.055 \pm 0.015 \text{ s}^{-1}$

Apparently these values are identical within the experimental error, and therefore it is most reasonable to assume, that all three exchange processes occur via the same mechanism. And this can be only via *dissociation/recombination*. (It should be mentioned that in the presence of traces of water, process I is significantly faster, suggesting an additional reorientation of the C=O groups *within* a dimer.)

Considering the fact, that a dissociation can be followed by a recombination in exactly the same orientation, an overall rate constant for the dissociation/recombination process can be calculated as

$k_{d/r} = 4 \times k_{I-III} = 0.26 \pm 0.06 \text{ s}^{-1}$.

An exchange cross-peak can also be observed for "free" and "included" benzene (compare Fig. 8) which allows the estimation of the exchange rate from a NOESY experiment. In a mixture of 77% C_6H_6 and 23% C_6D_6 a value of $k_e = 0.47 \pm 0.1 \text{ s}^{-1}$ was found for this exchange rate, which is in reasonable agreement with the rate of the dissociation/recombination process.

Rebek et al. report a half-life time of about 8 min for the uptake of benzene (5 μl) in tetraurea capsules in *p*-xylene-d_{10} [16]. Although not clearly stated, the process recorded under their conditions seems to be a bimolecular reaction between free benzene and capsules with *p*-xylene as guest (or without benzene as guest), and most probably this half-life time would decrease, if the concentration of benzene were increased.

7. Concluding Remarks

The "exclusive" formation of dimers observed for a large variety of calix[4]arene tetraurea derivatives represents an excellent example for artificial systems showing "self-assembly". These dimers are able to encapsulate various medium sized molecules as guests [21] This principle of self-organisation may be used also to construct larger assemblies (see ref. [13]) and further developments in various directions can be easily predicted for the near future. Although it was possible to demonstrate the existence of dimers also by mass spectrometry [10] the proof of their existence was mainly based on NMR-studies. The elucidation of the "structure and dynamics" of these dimers by NMR-spectroscopy may be regarded even as a "model case" to demonstrate the application of NMR to supramolecular systems.

8. Acknowledgement

Financial support by the Deutsche Forschungsgemeinschaft, the Direccion General de Enseqanza Superior and the European Community is gratefully acknowledged.

9. References

1. Whitesides, G. M., Mathias, J. P., and Seto, C. T. (1991) Molecular Self-Assembly and Nanochemistry: A Chemical Strategy for the Synthesis of Nanostructures, *Science* **254**, 1312-1319.
2. Cram, D. J. and Cram, J. M. (1994) *Container Molecules and Their Guests*, Royal Society of Chemistry, Cambridge.
3. Wyler, R., de Mendoza, J., and Rebek, Jr., J. (1993) A Synthetic Cavity Assembles through Self-Complementary Hydrogen-Bonds, *Angew. Chem. Int. Ed. Engl.* **32**, 1699.
4. Rebek, Jr., J. (1996) Assembly and Encapsulation with Self-complementary Molecules, *Chem. Soc. Rev.* 255-264.
5. Chapman, R. G. and Sherman, J. C. (1995) Study of Templation and Molecular Encapsulation Using Highly Stable and Guest-Selective Self-Assembling Structures, *J. Am. Chem. Soc.* **117**, 9081-9082.
6. Jacopozzi, P. and Dalcanale, E., (1997) Metal-Induced Self-Assembly of Cavitand-Based Cage Molecules, *Angew. Chem. Int. Ed. Engl.* **36**, 613-615.
7. Vicens, J. and Böhmer, V. (1991) *Calixarenes: A Versatile Class of Macrocyclic Compounds*, Kluwer Academic Publishers, Dordrecht.
8. Böhmer, V. (1995) Calixarenes, Macrocycles with (Almost) Unlimited Possibilities, *Angew. Chem. Int. Ed. Engl.* **34**, 713-745.
9. Jakobi, R. A., Böhmer, V., Grüttner, C., Kraft, D., and Vogt, W. (1996) Long-chain alkyl ethers of p-nitro and p-amino calixarenes, *New J. Chem.* **20**, 493-501.
10. Shimizu, K. D. and Rebek, Jr., J. (1995) Synthesis and assembly of self-complementary calix[4]arenes, *Proc. Natl. Acad. Sci.* **92**, 12403-12407.
11. Mogck, O., Böhmer, V., and Vogt, W. (1996) Hydrogen Bonded Homo- and Heterodimers of Tetra Urea Derivatives of Calix[4]arenes, *Tetrahedron*, **52**, 8489-8496.
12. Compare Crego Calama, M., Hulst, R., Fokkens, R., Nibbering, N. M. M., Timmerman, P., and Reinhoudt, D. N. (1998) Libraries of non-covalent hydrogen-bonded assemblies; combinatorial synthesis of supramolecular systems, *Chem. Commun.* 1021-1022.
13. Castellano, R. K. and Rebek, Jr., J. (1998) Formation of Discrete, Functional Assemblies and Informational Polymers through the Hydrogen-Bonding Preferences of Calixarene Aryl and Sulfonyl Tetraureas, *J. Am. Chem. Soc.* **120**, 3657-3663.
14. Mogck, O., Paulus, E. F., Böhmer, V., Thondorf, I., and Vogt, W. (1996) Hydrogen-bonded dimers of tetraurea calix[4]arenes: unambiguous proof by single crystal X-ray analysis, *Chem. Commun.* 2533-2534.
15. Böhmer, V., Mogck, O., Paulus, E. F., and Shivanyuk, A. unpublished results.
16. Hamann, B. C., Shimizu, K. D. and Rebek, Jr., J. (1996) Reversible Encapsulation of Guest Molecules in a Calixarene Dimer, *Angew. Chem. Int. Ed. Engl.* **35**, 1326-1329.
17. O'Leary, B. M., Grotzfeld, R. M., and Rebek, Jr., J. (1997) Ring Inversion Dynamics of Encapsulated Cyclohexane, *J. Am. Chem. Soc.* **119**, 11701-11702.
18. Thondorf, I. and Böhmer, V. unpublished results.
19. Mogck, O., Pons, M., Böhmer, V., and Vogt, W. (1997) NMR Studies of the Reversible Dimerization and Guest Exchange Processes of Tetra Urea Calix[4]arenes Using a Derivative with Lower Symmetry, *J. Am. Chem. Soc.* **119**, 5706-5712.
20. Castellano, R. K., Rudkevich, D. M., Rebek, Jr., J. (1996) Tetramethoxy Calix[4]arenes Revisited: Conformational Control through Self-Assembly, *J. Am. Chem. Soc.* **118**, 10002-10003.
21. See also Castellano, R. K., Kim, B. H., and Rebek, Jr. J. (1997) Chiral Capsules: Asymmetric Binding in Calixarene-Based Dimers, *J. Am. Chem. Soc.* **119**, 12671-12672.

SELF-ASSEMBLING PEPTIDE NANOTUBES

JUAN R. GRANJA[1], M. REZA GHADIRI[2]

[1]*Departamento de Química Orgánica y Unidad Asociada al CSIC, Universidad de Santiago, 15706 Santiago de Compostela, Spain.*
[2]*Department of Chemistry & Molecular Biology and The Skaggs Institute for Chemical Biology, The Scripps Research Institute, La Jolla, California 92037, USA.*

1. Introduction

There is great excitement about the possible application of nanosized structures since they may have novel material properties owing to their finite small size [1]. Innovative material-processing methods based on nanophysical or nanochemical techniques have revolutionized the design and manufacture of microscopic and submicroscopic devices [2]. For example, atomic-force and scanning tunnelling microscopy techniques, or self-organizing and self-assembling processes are being used for atomic- and molecular-scale manipulations. The goal is to construct structurally and functionally predetermined nanoscale objects by chemical processes from simple components. There is particular interest in the design of molecular scaffolding and container devices with well-defined nanoscopic cavities [3]. Much of this interest has fuelled research on nanotubes because of their potential utility in applications as diverse as molecular inclusion and separation technologies, catalysis, preparation of nanocomposites, construction of optical and electronic devices; and as novel therapeutic agents, transmembrane channels, and drug delivery vehicles. This has resulted in numerous reports of nanosized tube-shaped objects from different fields, including all-inorganic zeolites, all-carbon graphite nanotubes, lipid-based tubular assemblies, cyclodextrin-based materials, and a number of crown-ether based systems.

2. Design Principles

The utility of nanotubes depends in part on the ability to construct them with specified, uniform internal diameter, a task that has remained largely unfulfilled. To this end we have devised a new strategy for the preparation of tububular structures by self-assembly of small ring-shaped molecules (Fig. 1). This strategy should allow construction of open-ended hollow objects (tubes) whose internal diameter depends only on the ring-size of the monomer. The basic subunit envisaged for this task was a cyclopeptide made up of an even number of alternating *D* and *L* amino acids [4]. Such peptides could adopt a flat ring-shaped conformation with carbonyl and amide protons in the backbone oriented in opposite directions and approximately perpendicular to the plane of the ring structure, and with the side-chain pointing outwards.

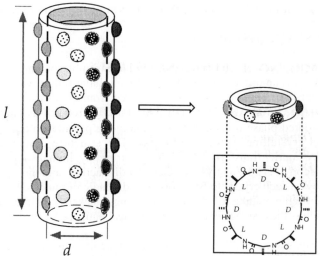

Figure 1. Strategy for the construction of self-assembled nanotubes. The ring size of the subunit determines the internal diameter of the ensemble. The chemical structure of the proposed cyclic peptide is also shown (*D* or *L* refers to the chirality of each amino acid).

These peptides thus have hydrogen bond donor and acceptor sites on either side of their ring structure, and in favourable conditions they could stack to form uniformly-shaped and continuously hydrogen-bonded β-sheet like tubular ensembles (Fig. 2). These tubular ensembles would have the desired hollow core and the characteristics of their outer surface could be varied by substituting the peptide side-chain.

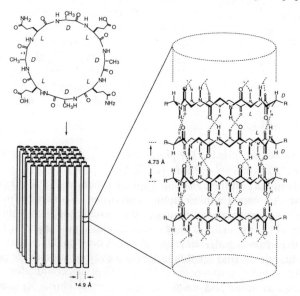

Figure 2. In suitable conditions the cyclic peptide subunits stack to form hydrogen-bonded tubular ensembles. Model of crystal packing of the self-assembled peptide nanotubes (for clarity, most side chains have been omitted), as deduced from TEM, FTIR and electron diffraction patterns studies. Meridional spacing of 14.9 Å was observed, corresponding to the center-to-center spacing for the closely packed nanotubes, and axial periodicity of 4.73 Å, corresponding to the intersubunit distance (the distance between two hydrogen-bonded peptides).

3. Synthesis and Characterization of Nanotube Structures

3.1. SOLID-STATE ENSEMBLES

The first example of a well-characterized tubular ensemble prepared by the strategy described above was an ordered array of crystalline tubular object with a uniform 7.5 Å internal diameter [5]. This ensemble was prepared from eight-residue cyclic peptides $Cyclo[-(Glu-D-Ala-Gln-D-Ala)_2]$ by a novel proton-triggered self-assembly process that promotes an ordered phase-transition from monomeric peptides in aqueous solution to highly ordered crystalline nanotube objects. The structure of the nanotubes was characterized by Fourier transform IR (FTIR) spectroscopy, high resolution transmission electron microscopy (TEM), analysis of electron diffraction, and molecular modelling. The results showed each crystal to be an organized bundle of hundreds of tightly packed nanotubes, and each nanotube to be made up of rings stacked with intersubunit distance of 4.73 Å, which is the distance expected for an ideal antiparallel β-sheet structure (Fig. 2).

More recently, several other examples of solid-state nanotubular structures have been prepared from C_4-symmetric peptides composed of L-glutamine alternating with a hydrophobic D-amino acid [6, 7]. These peptides assembled in aqueous trifluoroacetic acid to form needle-shape crystalline objects, which were characterized as described previously. Unfortunately, the crystals obtained were not large enough for X-ray analysis to unequivocally establish the proposed structure. It is noteworthy that these ensembles had different properties from the previous one, due to the type of amino acid used in the basic subunits and to the highly cooperative nature of the self-assembly process which simultaneously reinforces multiple noncovalent interactions throughout the lattice. This is a unique design feature of this class of nanotubes, the choice of amino acid side-chain functionalities determines the properties of the assembled structures such as the mode of initiation of the self-assembly process, the stability of the nanotube, etc.

The proton-triggered self-assembly process was extended to the preparation of a larger tubular ensemble with an internal pore diameter of 13 Å, simply by using a larger dodecameric cyclic peptide as subunit [8]. This confirm that the internal diameter of the peptide nanotubes can be adjusted simply by altering the number of amino acids residues in the subunit.

3.2. SOLUTION-PHASE ENSEMBLES

The design principles have been further extended to the preparation of solution-phase and solid-state cyclindrical ensembles from $cyclo[-(Phe-D-N-MeAla)_4-]$ [9]. This cyclic peptide has N-methylated amino acid residues on one face of its ring and it self-assembles in an antiparallel fashion to form a dimeric cylindrical structure (fig. 3). The N-methyl groups prevent formation of an extended network of hydrogen bonds, which limits self-assembly to formation of the dimer. Solution-phase studies using NMR and FTIR spectroscopy have shown that, in non-polar solvents, this octapeptide takes a flat, ring-shaped conformation, thus favouring ring-stacking and intermolecular hydrogen-bond formation.

Solution-phase studies of this and other N-methylated peptides have allowed the thermodynamics of the self-assembly process to be examined, including determination of association constants (Ka), and of the energy difference between dimers hydrogen-bonded as in parallel β-sheets (formed from mixtures of cyclo[-(L-Phe-D-N-MeAla)4-] and cyclo[-(D-Phe-L-N-MeAla)4-]) and antiparallel dimers [10]. It was also established that the side-chain does not appreciably affect the stability of the β-sheet structure in these dimeric systems [10].

The crystals of cyclo[-(Phe-D-N-MeAla)4-] were suitable for X-ray analysis. The solid-state structure was a cylindrical dimeric ensemble analogous to the structure deduced from solution studies, corroborating the structure proposed for the nanotubes. Furthermore, the interior of the cylinder contained partially disordered water molecules, confirming the hydrophilic nature of the cavity in these nanotubes.

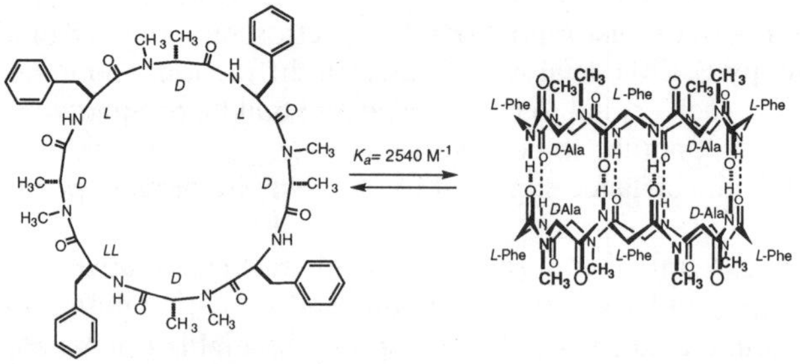

Figure 3. Chemical structure of the peptide subunit, and two-dimensional sketch of the corresponding cylindrical ensemble (for clarity, side chains are not shown). The association constant (K_a) was measured in CDCl$_3$ at 293 K.

4. Practical Applications: Artificial Transmembrane Channels

The unique properties of this class of nanotubes, such as the hydrophilic character of their internal cavity and the possibility of tailoring their external surface characteristics, suggested that they might have applications as mimics of transmembrane pores or ion channels. Cyclic peptides with appropriate external surface characteristics were designed and shown to partition into phospholipid bilayers and spontaneously assemble into hydrogen-bonded tubular structures, which could be structurally characterized by FTIR [11, 12]. The first such channel-forming nanotubes were formed by assembly of cyclo[Gln-(D-Leu-Trp)3-D-Leu] (Fig. 4a), and allowed efficient passage of metal ions at rates in excess of 10^{-7} ions s^{-1} [13], which is roughly three times faster than in the naturally occurring gramidicin A channel. Subsequently, a larger ten-residue cyclic peptide was used to prepare a transmembrane nanotube with an internal diameter of 10 Å (Fig. 4b) [14]. This was found to transport glucose very efficiently, whereas the octapeptide nanotubes could not do so because passage of glucose requires a pore of diameter greater than 9 Å. The transport rate was dependent on glucose concentration, which provides strong evidence that transport was by channel-mediated diffusion rather than carrier-mediated. These results confirm that although these transmembrane nanotubes exhibit only weak charge-selectivity they exhibit selectivity based on size and so could be exploited to effect size and shape-selective molecular transport across lipid membranes [15].

To sum up, these self-assembling peptide nanotubes constitute a new class of nanoscale tubular biomaterials whose design offers great flexibility as regards structure, and whose potential applications in biology, chemistry and materials science cover a wide range. The transmembrane channels described here could be adapted for use as novel cytotoxic agents themselves [16], or for the transport of biologically active molecules across lipid bilayers e.g. for drug delivery into living cells or in gene therapy. Also, it may be possible to modify nanotubes covalently [17], and exploit their remarkable stability to use them in biosensors and optical or electronic devices [13b].

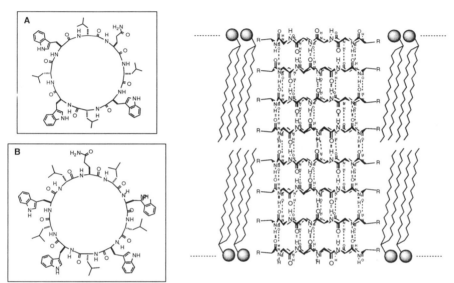

Figure 4. Schematic representation of a nanotube self-assembled in a lipid membrane. The chemical structures of peptide subunits used in transport experiments are shown on the left.

Acknowledgements

We thank the US Office of Naval Research for financial support.

5. References

1. Special section (1991) "Engineering a Small World" *Science* **254**, 1300-1335
2. Ozin, G. A. (1992) Nanochemistry synthesis in diminishing dimensions, *Adv. Mater.* **4**, 612-649.
3. Cram, D. J. (1992) *Nature* **356**, 29-36; Wyler, R., de Mendoza, J., Rebek, J. (1993) A synthetic cavity assembles through self-complementary hydrogen bonds, *Angew Chem. Int. Ed. Engl.* **354**, 1699-1701; T., Rebek, J. (1996) Assembly and encapsulation with self-complementary molecules, *Chem. Soc. Rev.*, **25**, 255-263.
4. Early theoretical analysis predicted the formation of this type of cyclindrical structures: DeSantis, P., Morosetti, S., and Rizzo, R. (1974) Conformational analysis of regular enantiomeric sequences, *Macromolecules* **7**, 52-58
5. Ghadiri, M. R., Granja, J. R., Milligan, R. A., McRee, D. E., and Khazanovich, N. (1993) Self-assembling organic nanotubes based on a novel cyclic peptide architecture *Nature* **366**, 324-3276
6. Hartgerink, J., Granja, J. R., McRee, D. E., Milligan, and R. A., Ghadiri, M. R. (1996) Self-assembling peptide nanotubes, *J. Am. Chem. Soc.*, **117**, 43-50.
7. Some other nanotubes have been recently prepared using Lys or Glu instead of Gln; Hartgerink, J., Ghadiri, M. R. unpublished results.
8. Khazanovich, N., Granja, J. R., Milligan, R. A., McRee, D. E., and Ghadiri, M. R. (1994) Nanoscopic tubular ensembles with specified internal diameters. Design of a self-assembled nanotube with a 13-Å pore, *J. Am. Chem. Soc.*, **1994**, **116**, 6011-6012
9. a) Ghadiri, M. R., Kobayashi, K., Granja, J. R., Chadha, R. K., and McRee D. E. (1995) The structural and thermodynamic basis for the formation of self-assembled peptide nanotubes, *Angew. Chem. Int. Ed. Engl.*, **34**, 93-95. b) Clark, T.D., Buriak, J.M., Kobayashi, K., Isler, M.P., McRee, D.E., Ghadiri, M.R. (1998) Cylindrical β–sheet peptide assemblies, *J. Am. Chem. Soc.*, **120**, 8949-8962.
10. Kobayashi, K., Granja, J. R., and Ghadiri, M. R. (1995) β-sheet peptide architecture: measuring the relative stability of parallel vs. antiparallel β-sheets, *Angew. Chem. Int. Ed. Engl.*, **34**, 95-98.
11. Ghadiri, M. R., Granja, J. R., and Buehler, L. K. (1994) Artificial transmembrane ion channels from self-assembling peptide nanotubes, *Nature*, **369**, 301-304.
12. Recently these and other nanotubes formed across phosphatidylcholine (PC) bilayers have been shown to align parallel to the PC hydrocarbon chains. See: Kim H. S., Hartgerink, J. D., Ghadiri, M. R. (1998) Oriented self-assembly of cyclic peptides nanotubes in lipid membranes, *J. Am. Chem. Soc.*, **120**, 4417-4424.
13. a) These transmembrane channels can very efficiently transport Li, Na, K, Cs and Cl ions, Ghadiri, M. R.; Granja, J. R.; and Buehler, L. K., unpublished results. b) For a study of a transport of different ions see: Motesharei, K., Ghadiri, M. R. (1997) Diffusion-limited size selective ion sensing based on SAM-supported peptide nanotubes, *J. Am. Chem. Soc.*, **119**, 11306-11312.
14. Granja, J. R., and Ghadiri, M. R. (1994) Channel-mediated transport of glucose across lipid bilayers, *J. Am. Chem. Soc.*, **116**, 10785-10786.
15. Recently it has been shown that cyclic tetramers of β-amino acids can form similar hollow tubular stuctures, and that hydrophobic decorated cyclopeptides can form transmembrane ion channels with a central hole 2.6-2.7 Å in diameter. Clark, T.D., Buehler, L. K., and Ghadiri, M. R. (1998) Self-assembling cyclic β^3-peptide nanotubes as artificial transmembrane ion channel, *J. Am. Chem. Soc.*, **120**, 651-656.
16. These channels are antibacterially active against gram-positive strains and cytotoxic in a test with human kidney cells. Ghadiri, M. R. unpublished results.
17. Clark, T.D. and Ghadiri, M. R. (1995) Supramolecular design by covalent capture. Design of a peptide cylinder via hydrogen-bond-promoted intermolecular olefin metathesis, *J. Am. Chem. Soc.*, **117**, 12364-12365.

SYMMETRY: FRIEND OR FOE?

M. PONS[1]*, M.A. MOLINS[1], O. MILLET[1], V. BÖHMER[2],
P. PRADOS[3], J. de MENDOZA[3], J. VECIANA[4], J. SEDÓ[4]

[1]*Departament de Química Orgànica. Universitat de Barcelona,
Martí i Franquès, 1 , E-08028 Barcelona, Spain.*
[2]*Institut für Organische Chemie, Johannes Guttenberg Universität
Mainz, J.J. Becher-Weg 18-20, D-55099 Mainz, Germany*
[3]*Departamento de Química Orgánica. Universidad Autónoma de Madrid.
Cantoblanco, E-28049 Madrid, Spain.*
[4]*Institut de Ciència de Materials de Barcelona. CSIC,
Campus de la Univ. Autònoma de Barcelona, E-08193 Bellaterra, Spain.*

Abstract

While symmetry in molecules or supramolecular complexes is desirable from the design point of view and affords simpler NMR spectra, the study of symmetrical species by NMR is often subject to ambiguities resulting from signal degeneracy. In this paper we shall discuss different ways of breaking symmetry in order to obtain additional structural information by NMR.

1. Symmetry in the design of supramolecular complexes

Supramolecular structures are designed by introducing into the structure of the constituent units the information needed to direct the spontaneous formation of a well defined larger entity usually held together by non-covalent bonds. Optimization of the design requires that the desired structure is formed with high association constants, that competing structures are destabilized and that the complexity of the starting materials is as low as possible. Self-association of a single precursor molecule provides a starting situation of the lowest complexity and often leads to symmetrical structures.
In order to obtain large association constants using only weak non-covalent bonds, cooperative effects are exploited. Multiple non-covalent interactions involving symmetry related atoms to form a symmetrical supramolecular structure provide an additional entropic stabilization due to the formation of several degenerate forms (microstates) of the same species. Simultaneously, the requirement for multiple interactions provides the desired selectivity.
DNA binding proteins are often dimeric and recognize palindromic sequences of matching C_2 symmetry. They provide a natural example of these basic "design" principles. The same concepts have been applied to the de novo design of peptides and

proteins that self assemble through hydrophobic interactions or by the formation of multiple disulfide bonds. In an early design by our group a parallel cyclic dimer of a tetrapeptide inspired by the highly symmetric structure of the potassium complex of valinomycin (a depsipeptide of twelve residues) was found to form very stable ion complexes with a very high selectivity for barium [1]. In a very recent example a palindromic peptide with two cysteine residues has been found to form a cyclic trimeric structure with quasi octahedral symmetry spontaneously (Figure 1) [2].

Ac-CBKLHAELSSLEAHLKBCG-NH$_2$
B = α-aminoisobutyric acid

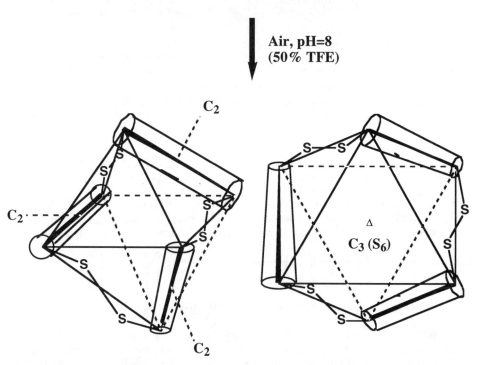

Figure 1. Schematic representation of two views of the trimer formed spontaneously by a palindromic peptide with two cysteine residues. Each cylinder represents a peptide chanin in α–helical conformation. The symmetry elements relate only the peptide side-chains and the S_6 symmetry would appear only in an achiral system with the same architecture and topology.

Self assembled α-helical peptide oligomers of cylindrical symmetry -four helix bundles- with and without disulfide bond stabilization have been extensively studied by de Grado [3], Hodges [4] and others. Non peptidic, self assembled hydrogen bonded structures approaching spherical symmetry and containing solvent filled cavities have been presented, especially by the group of Rebeck [5] using the directional properties of hydrogen bonds included in relatively rigid structures to direct self-assembly and to avoid polymerization.

2. Isotopic desymmetrization: quenching of NOEs

Isotopic labeling has been used extensively to aid in the structure determination of dimeric proteins by NMR. In a mixture of labeled and unlabeled protein three different dimers can be formed by combination of two labeled, two unlabeled or one labeled and one unlabeled protein. In these mixed dimers symmetry is lost and it is possible to distinguish between intramolecular and intermolecular NOEs. This is achieved by filtering the signal from protons that are bound to a particular isotopic form (^{12}C / ^{13}C or ^{14}N / ^{15}N) before and after the mixing time during which the NOE effect develops [6]. While ^{13}C or ^{15}N isotope labeling provides a fairly general method for solving ambiguities in symmetrical dimers that can be obtained by expression in a suitable organism, it may require a considerable synthetic effort in other instances. Deuterium exchange of labile protons can offer an easier alternative method, although with a much more restricted range of application.

A partially exchange-deuterated sample of a dimeric molecule containing labile protons will contain mixed dimers in which only one of the symmetry related sites has been exchanged. The difference between the two sites can be propagated to other protons that are coupled to the exchangeable one by a TOCSY type magnetization transfer (Figure 2).

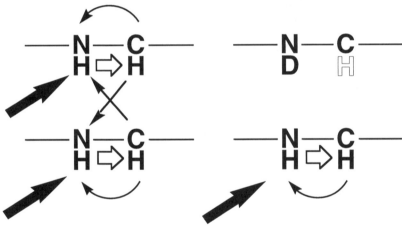

Figure 2. The Sping-pong experiment. Magnetization-transfer pathways in a semi-selective TOCSY-NOESY experiment in a partially exchanged sample of a peptide allows the differentiation of inter- and intramolecular NOEs in symmetrical oligomers.

After selective excitation of the exchangeable protons (e.g. amide protons in a peptide) and TOCSY transfer (e.g to the alpha protons), only the alpha protons in non deuterated residues are excited.

If the scalar coupled protons are also close in space, cross-relaxation causes the appearence of NOEs irrespective of wheter the two protons are located in a dimer that is partially deuterated or fully protonated. If samples with different degrees of deuteration are compared, the cross-peak intensity will be directly proportional to the intensity of the signal from the residual, non-deuterated, amide proton.

On the other hand, NOEs involving an exchangeable proton and a second one non scalar coupled to it (i.e. that belongs to a different residue or molecule) are not present in mixed dimers. Therefore the intensity of these cross-peak will be further attenuated with respect to the intensity of the amide proton signal. If we call x the fraction of amide groups that are deuterated, the intensity of the cross-peak arising from an intermolecular NOE is expected to be proportional to $(1-x)^2$ while that arising from an intra-residue NOE will be proportional to $(1-x)$.

This experiment, that we dub SPING-PONG because magnetization is transferred back and forth between the exchangeable and coupled non exchangeable proton [7], is especially suitable for the detection of parallel β-sheets involving identical peptide chains as the cross-strand NOEs characteristic of these supramolecular structures have to be distinguished from trivial NOEs involving directly coupled protons belonging to the same residue. Figure 3 shows the observed cross-peak intensities in a NOESY experiment with a partially deuterated sample of peptide T, an octapeptide derived from the external envelope glycoprotein gp120 of HIV, dissolved in d_6-DMSO

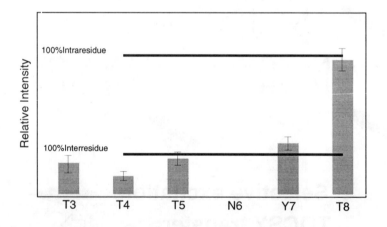

Figure 3. Cross-peak intensities of NH(i)CH(i) NOEs in a partially exchanged sample of peptide T in d_6-DMSO. Intensities are corrected for the degree of deuteration and normalized for the intensity observed in a non-deuterated sample. The expected intensities for intra- and intermolecular NOEs are also shown. Some cross-peaks could not be measured because of spectral overlap.

All the measured NH(i)CH(i) except that for the C-terminal residue are shown to arise exclusively from inter-strand NOEs, i.e. they are in fact NOEs from the CH_α proton of residue i in one molecule to the NH proton of residue i of <u>a different molecule</u>. It is interesting that IR experiments show that the β-sheet concentration is not more than 10% of the total peptide concentration even in very concentrated solutions. The almost exclusive observation of intermolecular NOEs can be explained by the different correlation times of the aggregated and non-aggregated forms of the peptide that cause very week NOEs for a single molecule as $\omega\tau_c \approx 1$ but gives rise to strong negative NOEs for the higher molecular weight form. The NOEs observed in that particular example can be considered as a special case of transferred NOE involving not a small and a large molecule but a single molecule in two different aggregation forms.

3. Isotopic desymmetrization: isotopic shifts

A different consequence of partial deuterium exchange that can be useful for breaking symmetry in NMR spectra is the presence of isotopic effects. This approach has been used for the study of the hydrogen bond array present in the cone conformation of unsubstituted calix[4]arenes The C_4 symmetry of the completely protonated form gives a single peak for the four phenol protons of **1**. However, addition of microliter amounts of deuterated methanol so that the calixarene has an average deuteration level of ca. 50% causes the appearance of five additional upfield shifted hydroxyl signals (Figure 4).

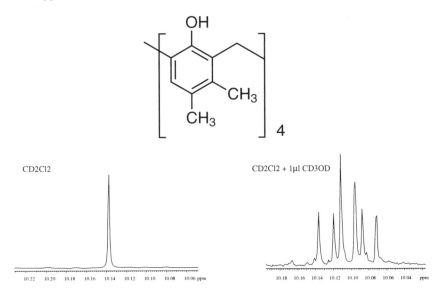

Figure 4. ^{1}H-NMR spectra of the phenol region of **1** before and after the addition of 1 μL of CD$_3$OD.

Assuming fast reversal of the hydrogen bond direction in this system, a given proton can be isotopically shifted by deuterium atoms located either two- or four bonds away in the array. The observed isotopic multiplet can be explained by a statistical deuteration pattern and additive effects for isotopics shifts arising from different deuterium atoms present in the array. This is illustrated in figure 5.

This interpretation is supported by the apparent T_1 values measured for the different isotopically shifted peaks given in Table 1.

Dipolar relaxation by deuterium atoms is much less efficient than by protons due to the different gyromagnetic ratios of the two isotopes. On the other hand, larger isotopic shifts correspond to the presence of deuterium atoms in the neighboring positions of the observed proton. Therefore, the more isotopically shifted protons should show longer relaxation times which is what is observed experimentally.

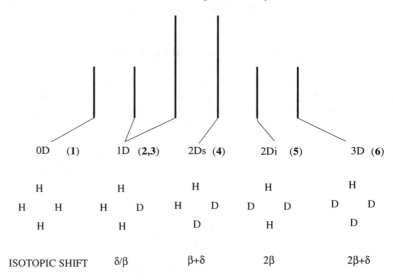

Figure 5. Shifts and intensities of the observed multiplet can be explained by the analysis of individual isotopic shifts and symmetry properties of the complete set of isotopomers.

TABLE 1. ^{1}H Relaxation times of isotopically shifted peaks in ca. 50% deuterated **1**

Proton site	Number of neighbouring deuterium atoms		Apparent T_1 (s)
	2-bond	4-bond	
1	0	0	0.76
2	0	1	0.79
3	1	0	1.02
4	1	1	1.04
5	2	0	1.41
6	2	1	1.57

Partially deuterated species have reduced symmetry and interactions that were not detectable because of the degeneracy of the affected nuclei now become observable. Figure 6 shows an expansion of a NOESY/EXSY experiment on 50% deuterated **1**. Lines 2 and 3 of the isotopic multiplet correspond to two different proton sites situated two- or four bonds away from the point of deuteration in the singly deuterated isotopomer. The negative cross-peak between these lines arises from a NOE effect between these two protons within a hydrogen bond array.

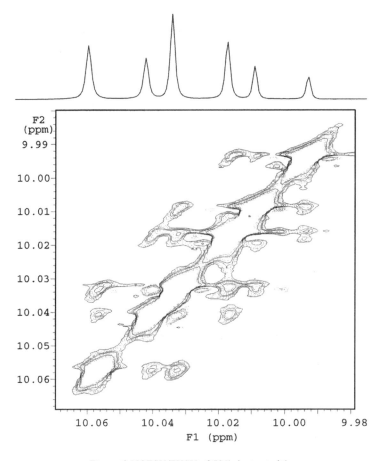

Figure 6. NOESY/EXSY of 50% deuterated **1**.

Exchange between different sites within the hydrogen bond array can also be studied in the isotopically resolved system. Exchange cross-peaks can originate from i) proton-deuterium exchange with the solvent resulting in the conversion between isotopomers containing different numbers of deuterium atoms, ii) proton-proton exchange between non-equivalent positions in a particular isotopomer and iii) proton-deuterium exchange within the array preserving the number of deuterium atoms. Processes i and iii are illustrated in Figure 7.

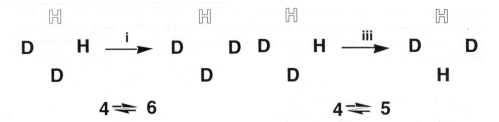

Figure 7. Two types of exchange processes observed in the EXSY spectra of **1**.

Analysis of the exchange pattern in the first two cases is straightforward. The first type of exchange can be easily visualized as the interconversion between different species (two isotopomers). In the second type, the deuterium atom is not involved in the exchange and therefore the problem is reduced to the classical chemical exchange problem between two non-equivalent sites. The analysis of the third type of exchange requires considering that when a deuterium and a proton exchange their positions the isotopic shifts and therefore the chemical shifts of other protons not directly involved in the exchange may also change.

In the particular example considered, only two types of exchange could be observed. Exchange between sites 2 and 3 may be obscured by the presence of a strong NOE cross-peak between these protons Obviously two consecutive exchanges with the solvent can result in an apparent intramolecular exchange. However, the absence of exchange cross-peaks between sites belonging to isotopomers that differ by more than one deuterium atom indicate that multiple exchange processes are slow and that direct proton deuterium exchange between two calixarene molecules does not take place in the time scale of the experiment. This suggests that intra-molecular exchange within the hydrogen bond array is actually taking place.

4. Symmetry reduction through dimerization

Although dimerization processes should apparently give rise to species of higher symmetry, as the final product contains two identical units, this is not always the case. One possible cause is a decreased mobility of the dimer that destroys average symmetry elements present in the monomer.

This effect is illustrated by the decreased symmetry observed in dimers formed by calix[4]arenes containing urea substituents [8] and is discussed in detail in a separate article by Prof. Böhmer in this book.

Figure 8 shows the symmetry properties of a dimerizable calix[4]arene with two different substituents.

The monomer has effective C_{2v} symmetry because the urea substituents can rotate freely. However, in the dimer the urea groups are hydrogen bonded to form a cyclic array and the rate at which it can change direction is slow on the NMR time scale. This reduces the symmetry of each monomer within a dimer to C_2. The fact that two different types of substituent are present in each monomer and that the interaction between the monomers involves interdigitation of the substituents prevents the appearance of new symmetry elements in the dimer. As a consequence, its global symmetry is only C_2. In this particular example, the low symmetry of the dimer allowed the observation of signals belonging to the two calixarene units and therefore of dynamic processes, e.g. dissociation-reassociation, which exchange the identity of protons in both units.

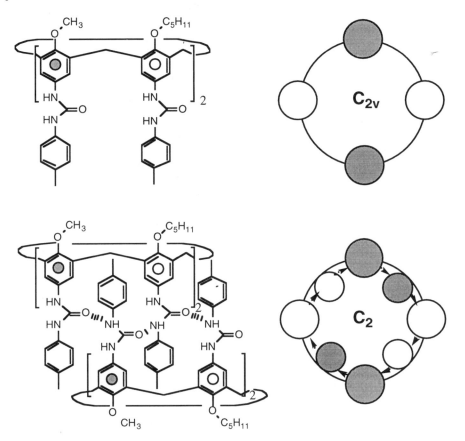

Figure 8. Dimerization of a calix[4]arene with alternating substituents reduces the symmetry of each monomer and allows the separate observation of each of them by NMR

5. Symmetry-breaking and -restoring exchange pathways

The apparent symmetry observed at room temperature in many flexible molecules such as calix[6]arenes is the result of fast exchange between a limited number of non symmetrical forms. At low temperature the individual forms can be observed and exchange between the different conformations can be followed at some intermediate temperature. When the molecule has constitutional symmetry, the different conformations have to be symmetry related and only one set of resonances is observed. The exchange between different conformations then appears in the NMR spectra as exchange between different sites in a single molecule. The observed exchange pattern maps the pathway between an average symmetrical species and a frozen asymmetric conformation. While the decrease in the conformational exchange rate that takes place at low temperature reveals the inherent asymmetry of the low energy conformations and allows the separation of previously degenerate NMR sites, the observation of the actual exchange pathway offers a new source of evidence for the structure of both the low temperature non-symmetric and the average symmetric conformation.

Calix[6]arenes have been extensively used as building blocks for the design of receptors due to their larger cavities as compared to calix[4]arenes. However, in contrast to calix[4]arenes, calix[6]arenes display complex conformational equilibria that lead to average structures at room temperature. The room temperature NMR spectrum of 5,17,29-tri-*tert*-butyl-11,23,35-trichlorocalix[6]arene (**2**), a molecule with constitutional three-fold symmetry, in CDCl$_3$ show the expected number of signals for the average species: e.g. two different phenol protons. At 223K the NMR spectrum reveals a completely asymmetrical conformation with six different phenolic protons (Figure 9).

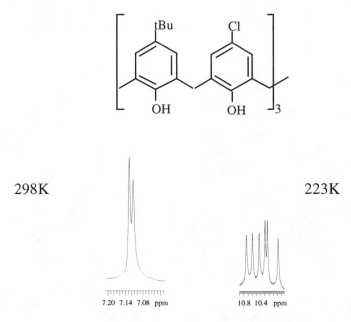

Figure 9. Expansion of the phenol region in the ^{1}H-NMR spectra of **2** at room temperature and 223K.

After complete assignment of the low temperature spectrum, the conformation of **2** can be tentatively deduced from a number of NOEs. However, at least two different conformations are compatible with the observed NOEs. The conformation initially suggested by us [9] is a winged cone with four aryl groups in an "up" alignment and two aromatic rings in opposite positions across the ring bent outside. The six methylene groups have their protons pointing outside from the macrocyclic cavity (Figure 10).

Figure 10. Schematic representation of the winged cone conformation of a calix[6]arene and the macrocycle inversion process. The methylene protons located between the two "up" or "down" rings, indicated by arrows, are not exchanged by this conformational interconversion

An alternative conformation, that is in better agreement with the conformation observed in crystals and which has a lower energy *in vacuo*, has been suggested [10]. In this conformation two of the methylene groups are "pinched" and the bisector of the two C-H bonds is pointing toward the interior of the cavity. The two suggested structures can undergo conformational exchange processes that could explain the completely symmetrical spectra observed at room temperature and therefore can be considered as alternative desymmetrized modes of the average structure. In both models the C_{3v} symmetry of a perfect cone is reduced initially to C_s symmetry and finally to C_1 by additional distortions or by the presence of a circular hydrogen bond array. At room temperature one has to consider additionally the inversion of the cone. The conformational interconversion suggested for the winged cone consists of a pseudorotation process that restores the C_{3v} symmetry and that is independent of the inversion of the cone. The lowest energy interconversion pathway calculated for the pinched cone consists of a concerted pseudorotation and inversion process (Figure 11).

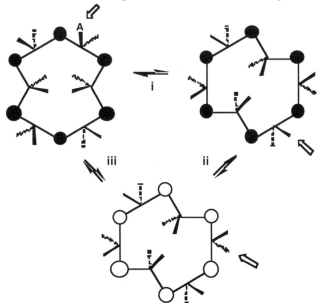

Figure 11. Schematic representation of the pseudorotation (i), inversion (ii) and concerted pseudorotation-inversion (iii) for a pinched cone conformation of a calix[6]arene. Filled and empty circles represent the up and down orientations of the aromatic rings. The arrows indicate the proton with the chemical shift of A after the conformational interconversions

The two models predict different patterns in EXSY spectra. In particular, the pinched cone model predicts exchange cross-peaks of the same intensity between all methylene protons while the winged cone model can explain the absence of a cross-peak between the high-field and low-field protons of the methylene group located between the two "up" aryl groups (Figure 12).

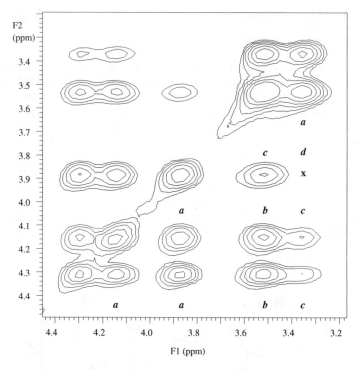

Figure 12. Exansion of the methylene exchange region of a ROESY spectrum (200 ms mixing time) of **2** dissolved in CDCl$_3$ at 223K.. All cross peaks originate from exchange processes. Peaks labeled *a* have been interpreted as the result of pseudorotation while peaks labeled *b* have been assigned to the macrocycle ring inversion. Note the absence of a cross-peak at position *d* consistent with the lack of exchange between the methylene protons located between the two "up" or "down" rings during the ring inversion of the winged cone conformation. Peaks labeled *c* are the result of two exchange processes. At longer mixing times a cross peak is also observed at position *d* probably also as a result of a combination of multiple exchange processes.

The winged cone or a related conformation seem therefore to be in better agreement with the exchange pattern observed in CDCl$_3$. Although there is no clear explanation for the observation of a different conformation in solution and in the solid state, the pinched conformation is a "closed" conformation with the methylene groups filling the macrocycle cavity. On the other hand the winged cone has an "empty" cavity that may include CDCl$_3$ which has the correct size. It is interesting to notice that at the same temperature **2** shows much higher mobility in CD$_2$Cl$_2$ than in CDCl$_3$.

6. Symmetry and NMR of tri-triarylmethanes

The group of Prof. Veciana in the Material's Science Institute of the CSIC in Bellaterra (Spain) has been studying systems formed by several fused triarylmethane systems with chlorine atoms in all the aromatic positions. Although in their planar representations these molecules are highly symmetric, steric hindrance forces each triarylmethane unit to adopt a propeller shape. In addition, the tetrahedral carbon atoms may also act as stereogenic centers depending on the symmetry of the molecule [11]. The three-helix compound **3** appears as a mixture of six diastereoisomers (five enantiomeric pairs of C_1 symmetry and one of C_3 symmetry) interconverting slowly on the NMR time scale. The proton NMR spectrum consists of 16 singlets. Three for each C_1 symmetrical compound and one for the C_3 symmetrical one (Figure 13).

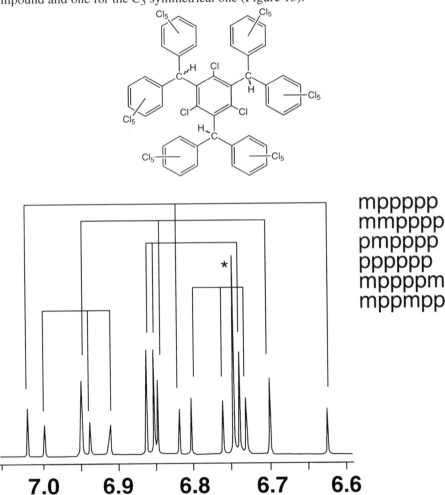

Figure 13 ^{1}H-NMR spectrum of the equilibrium isomer mixture of **3**. Stereochemical labels refer to the helicity of each triarylmethane unit and the configuration of the sp^3 centers

Complete assignment of the NMR spectra could be accomplished by initially grouping the signals belonging to the same conformation using HMBC spectra as schematically illustrated in figure 14. Each proton (labeled withe letters) shows long range couplings to two different carbon atoms in the center aromatic ring and each one of these carbons is coupled to two different protons giving a characteristic pattern in the HMBC spectrum.

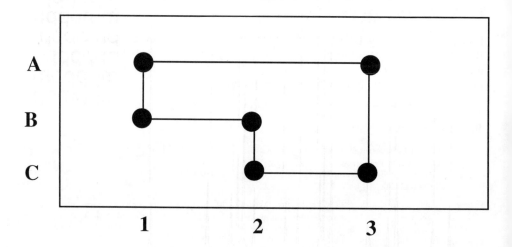

Figure 14. Schematic representation of the strategy for the assignment of the mixture of stereoisomers of **3** using HMBC spectra.

Assignment of each group to a particular isomer was done by calculating the shifts induced by the aromatic rings. While the precision with which each individual chemical shift can be computed is comparable to the differences between different signals in the spectrum and therefore alternative assignments are possible, the relative shifts of protons belonging to the same molecule match the calculations unambiguously.

7. Symmetry breaking in anisotropic environments

The symmetry displayed by a molecule or a supramolecular complex depends also on its environment. Very recently there has been a renewed interest for the use of partially oriented samples to help in the structure determination of macromolecules [12].
In an oriented system, dipolar couplings that average to zero in isotropically reorienting molecules, can be measured and their structural dependency exploited. Additionally, the new couplings may allow the creation of additional coherent states that may increase the amount of information that could be eventually encoded in a single molecule used, for example, in storage devices.
Orientation can be obtained directly in the spectrometer magnetic field if the anisotropy of the magnetic susceptibility is high enough (e.g. in heme containing proteins or in nucleic acids) or by using liquid crystals [13].
The complexity of the resulting spectra may be extremely high in molecules with a large number of coupled nuclei. However in special cases the use of liquid crystals may be a reasonable alternative. This is the case of **3** with only three weakly coupled protons per molecule and such studies are currently under way in our group.

8. Acknowledgements

Financial support by DGES (PB97-0933) and Generalitat de Catalunya (Centre de Referència en Biotecnologia i Grup Consolidat) is acknowledged. The peptides mentioned in Section 1 were synthesized by Dr. C. García-Echeverría and Dr. M. Royo in the group of Profs. F. Albericio and E. Giralt.

9. References

1. García-Echeverría, C., Albericio, F., Giralt, E., and Pons, M. (1993) Design, Synthesis, and Complexing Properties of (^{1}Cys-$^{1'}$Cys, ^{4}Cys-$^{4'}$Cys)-dithiobis(Ac-L-^{1}Cys-L-Pro-D-Val-L-^{4}Cys-NH$_2$). The First Example of a New Family of Ion-Binding Peptides, *J. Am. Chem. Soc.* **115**, 11663-11670.

2. Royo, M., Contreras, M.A., Giralt, E., Albericio, F., and Pons, M.(1998) An easy entry to a new high symmetry, large molecular framework for molecular recognition and de novo protein design. Solvent modulation of the spontaneous formation of a cyclic monomer, dimer or trimer from a bis-cysteine peptide, *J. Am. Chem. Soc.* **120**, 6639-6650.

3. Betz, S.F., Raleigh, D.P. and DeGrado, W.F. (1993) De novo protein design: from molten globules to native-like states, *Curr. Opin. Struct. Biol.*, **5**, 475-463.

4. Monera, O.D., Zhou, N.F., Kay, C.M., and Hodges, R.S. (1993) Comparison of antiparallel and parallel two starnded a-helical coiled-coils. Design, synthesis, and characterization, *J. Biol. Chem.* **268**, 19218-19227.

5. Rebeck, Jr., J. (1996) Assembly and encapsulation with self-complementary molecules, *Chem. Soc. Rev.* **1996**, 255-264.

6. Wörgötter, E., Wagner, G. and Wüthrich, K. (1986) Simpliification of two-dimensional 1H-NMR spectra using an X-filter, *J. Am. Chem. Soc..* **108**, 6162-6167. Zwahlen, C., Legauult, P., Vincent, S. J. F., Greenblatt, J., Konrat, R. and Kay, L. (1997) Methods for measurement of intermolecular NOEs by multinuclear NMR spectroscopy: application to a bacteriophage λ N-peptide/boxB RNA complex, *J. Am. Chem. Soc.*, **119**, 6711-6721.

7. Pons, M. and Giralt E. (1991) Determination of Interchain NOEs in symmetrical dimer peptides, *J. Am. Chem. Soc.* **113**, 5049-5050.. Ruiz-Gayo, M., Royo, M., Fernandez, I., Albericio, F., Giralt , E., and Pons, M. (1993).Uniequivocal synthesis and characterization of a parallel and an antiparallel bis-cystine peptide, *J. Org. Chem.* **58**, 6319-6328.

8. Mogck, O., Pons, M., Böhmer, V., and Vogt, W., (1997) NMR-Studies of the reversible dimerization and guest exchange processes of tetra urea calix[4]arenes using a derivative with lower symmetry, *J. Am. Chem. Soc.* **119**, 5706-5712.

9. Molins, M.A., Nieto, P.M., Sanchez, C., Prados, P., Mendoza, J., and Pons, M. (1992) Solution structure and conformational equilibria of a symmetrical calix[6]arene. Complete sequential and cyclostereospecific assignment of the low temperature NMR spectra of a cycloasymmetric molecule,.*J. Org. Chem.* **57**, 6924-6931.

10. van Hoorn, W.P., van Veggel, F.C.J.M., Reinhoudt, D.N. (1996) Conformation of hexahydroxycalix[6]arene, *J. Org. Chem.*, **61**, 7180-7184.

11. Gust, D. and Mislow, K. (1973) Analysis of isomerization of compounds displaying restricted rotation of aryl groups, *J. Am. Chem. Soc.* **95**, 1535-1547.

12. Tjandra, N., Omichinski, J.G., Gronenborn, A.M., Clore, G.M., and Bax, A. (1997) Use of dipolear 1H-15N and 1H-13C couplings in the structure determination of magnetically oriented macromolecules in solution *Nat. Struc. Biol.* **9**, 732-738.

13. Gayathri, C., Bothner-by, A.A., van Zijl, P.C.M., and Maclean, C. (1982) Dipolar magnetic field effects in NMR spectra of liquids, *Chem Phys. Lett.*, **87**, 192-196. Tolman, J.R., Flanagan, J.M., Kennedy, M.A.,and Prestegard, J.H. (1995) Nuclear magnetic dipole interactions in field-oriented proteins: information for structure determination in solution. *Proc. Natl. Acad. Sci.* USA **92**, 9279-9283.

QUENCHING SPIN DIFFUSION IN TRANSFERRED OVERHAUSER STUDIES AND DETECTION OF OCTUPOLAR ORDER IN GASEOUS XENON-131

MICHAËL DESCHAMPS AND GEOFFREY BODENHAUSEN
*Département de chimie, Ecole Normale Supérieure,
24 rue Lhomond, 75231 Paris Cedex 05, France.*

1. Introduction

In so-called "transferred NOE" studies, one seeks to determine the conformation of a small molecule (drug, substrate, coenzyme, inhibitor, etc.) which is bound to a larger molecule. If the exchange between bound and free forms is sufficiently rapid, and if the free form is in excess, the spectrum features narrow lines typical of small molecules, while Overhauser effects between spins belonging to the small molecule are predominantly determined by the conformation in the bound form, where the cross-relaxation rates are far larger than those of the free form. Thus we can have the best of both worlds: narrow lines characteristic of small molecules, and rapid cross-relaxation typical of large molecules. This explains the popularity of exchange-transferred NOE studies in enzymology, drug design, and many areas of supramolecular chemistry. Problems may occur however if spin diffusion processes are not properly taken into account [1].

Exchange-transferred NOE studies bear analogies with other exchange processes. In particular, quadrupolar spins such as ^{23}Na ($I = 3/2$) can be used to study exchange between free and bound forms, for example in proteins. It is also possible to study the interface between solid surfaces and cavities filled with gaseous ^{131}Xe ($I = 3/2$). Physisorbed ^{131}Xe is usually not directly observable by NMR spectroscopy, but information about the quadrupolar interactions experienced at the surface can be transferred to the gas phase through desorption. The exchange process may lead to a characteristic distribution of populations of the energy-levels known as octupolar order, and to single-quantum coherences of second and third rank, which are related to antiphase magnetization in spin $I = 1/2$ systems. These can be discriminated by multiple-quantum filtration from the background of xenon gas that has not been in contact with the surface. These methods might open new avenues for the characterization of surfaces [2].

2. Spin diffusion

Spin diffusion occurs when magnetization migrates through a set of nuclei in several consecutive steps. For example, the transfer of longitudinal magnetization from one proton to another, e.g. $I_{Az} \rightarrow I_{Xz}$, is often overshadowed by two-step processes such as $I_{Az} \rightarrow I_{Kz} \rightarrow I_{Xz}$. Spin diffusion can in principle be taken into account by the "total relaxation matrix" method [3-7], provided all cross-relaxation pathways can be identified. Spin diffusion pathways can be blocked by a variety of methods [8-13]. We shall focus attention on the principle of Quenching Undesirable Indirect External Trouble in Nuclear Overhauser Effect SpectroscopY (QUIET-NOESY) [14-16]. This method involves a doubly-selective inversion (Fig. 1a) of the longitudinal magnetization components of a source spin A and a target spin X to measure the cross-relaxation rate (Overhauser effect) between A and X without significant perturbation by spin diffusion. An inversion pulse with the envelope of a Q^3 Gaussian cascade [17] and a duration of, say, 12 ms can be made doubly-selective by placing the carrier frequency midway between the two chemical shifts and by modulating the amplitude with a function $\cos \omega_a t$, where ω_a is equal to half the difference between the chemical shifts. During the first half of the mixing time, the magnetization of the source spin migrates not only to the target spin, but also to other protons in the vicinity, including protons that belong to the binding pocket of the enzyme. The doubly-selective inversion pulse changes the sign of the longitudinal magnetization components of both the source and target spins, so that their mutual Overhauser effect continues to build up during the second half of the mixing time. However, barring accidental degeneracy, the other (clandestine) spins are not inverted by the doubly-selective inversion pulse, and the flow of magnetization from the source to the clandestine spin in the first half of the mixing time is reversed in the second half. Likewise, the flow from the clandestine spin to the target spin is largely cancelled. The net effect is that, to a good approximation, only direct cross-relaxation between the source and target spins will give rise to a cross-peak in (ET-)NOESY, and that contributions due to clandestine protons will be suppressed, provided the latters' magnetization components are not accidentally inverted by the doubly-selective pulse. The effect bears some similarities, but is not identical, to what can be obtained by replacing all protons belonging to the binding pocket by deuterium nuclei [18]. In this case, one would not only expect to suppress the cross-relaxation rates σ_{ij} between cofactor protons and enzyme deuterons, but also to reduce the diagonal elements ρ_{ii} that describe the longitudinal self-relaxation of the cofactor protons in the Solomon equations. The attenuation of self-relaxation represents an additional benefit of deuteration that cannot be obtained by simple manipulations of the magnetization.

In nitrogen-15 or carbon-13 enriched proteins, a doubly-selective inversion can also be achieved by using a BIlinear Rotation Decoupling (BIRD) sequence for the selective inversion of all amide protons [15].

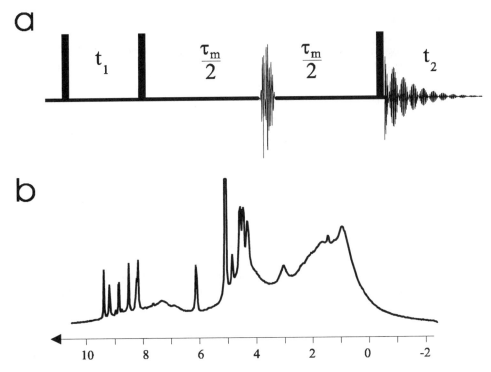

Figure 1. (a) Pulse sequence used for QUIET-ET-NOESY where spin diffusion pathways are quenched by doubly-selective inversion of selected source and target protons. (b) One-dimensional spectrum of the LDH·NAD$^+$ complex in D$_2$O, with [LDH] ≈ 0.5 mM, [NAD$^+$] ≈ 5 mM, recorded at 8°C on a Bruker DMX 300 spectrometer, showing sharp resonances due to the free cofactor and broad resonances due to the enzyme. (Reproduced from ref. (1) with permission.)

3. Exchange transferred nuclear Overhauser effect spectroscopy

In systems where the exchange between bound and free forms is sufficiently rapid, information from X-ray diffraction can be supplemented by two-dimensional exchange transferred nuclear Overhauser effect spectroscopy (ET-NOESY) [19-20]. The transferred NOE method uses excess cofactor, so that the lineshapes resemble those of the cofactor in free solution, while cross-relaxation or nuclear Overhauser effects are predominantly determined by internuclear distances within the cofactor in the bound form. In applications to the cofactor nicotinamide adenine dinucleotide (NAD$^+$) in rapid exchange with dogfish lactate dehydrogenase (LDH) [1, 21-23], it is usually assumed that ET-NOESY data can be interpreted on the assumption that, at least for short mixing times, spin-diffusion effects can be ignored.

In actual fact, spin diffusion turns out to be of particular importance for exchange-transferred NOE studies. Thus an intermolecular two-step process such as

$I_{Az}(H^{cofactor}) \to I_{Kz}(H^{enzyme}) \to I_{Xz}(H^{cofactor})$ could "contaminate" an intramolecular one-step process $I_{Az}(H^{cofactor}) \to I_{Xz}(H^{cofactor})$. The manifestations of such a two-step process might not be obvious at first sight, since the resonance of H^{enzyme} would be very broad, and, if at all recognizable, difficult to assign. The danger of an intermolecular two-step process is exacerbated by the fact that the longitudinal relaxation of the "clandestine" spin can be very slow, since it belongs to a large molecule, so that the Zeeman order $I_{Kz}(H^{enzyme})$ tends to have a long memory. On the other hand, this spin is likely to be dipolar-coupled to many other spins within the large molecule, which may act as "heat sinks". The Zeeman order $I_{Kz}(H^{enzyme})$ will therefore tend to dissipate over many spins, thus diminishing the likelihood of the misleading second step $I_{Kz}(H^{enzyme}) \to I_{Xz}(H^{cofactor})$.

In the conformation of NAD^+ bound to LDH determined by QUIET-ET-NOESY [1], the structure of the adenosine moiety is in agreement with X-ray diffraction in the solid state, but the structure of the nicotinamide deviates significantly from the structure found in the solid. The QUIET data indicate conformational averaging about the glycosidic bond between nicotinamide and nucleotide. We have determined that spin-diffusion effects in the LDH·NAD^+ complex involve protons belonging to the binding pocket of the enzyme. Such indirect processes cannot be analyzed properly by full relaxation matrix methods, because it is impossible to identify all protons in the enzyme pocket and their dipolar coupling partners.

In order to identify protons belonging to the binding pocket that could mediate spin diffusion, we used the X-ray coordinates of a ternary complex of dogfish LDH with NADH and oxamate determined from diffraction data at 2.1 Å resolution (Protein Data Bank file 1ldm.) Oxamate was excluded, and the positions of the protons were determined based on geometry and optimal hydrogen bonding using the program CHARMM [24]. The energy of the resulting structure was minimized to remove poor nonbonded interactions using parameters described previously [25], and sixty steps of the steepest descent algorithm with main-chain atoms harmonically constrained, followed by one hundred steps of the conjugate gradient algorithm without constraints. All protons within a radius of 4.5 Å around the relevant NAD^+ protons were identified and taken into account.

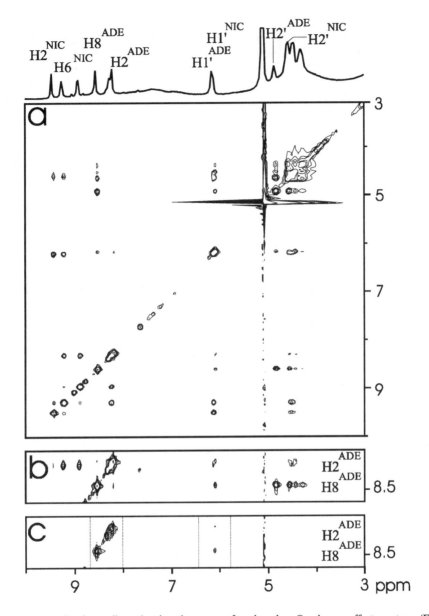

Figure 2. (a) Conventional two-dimensional exchange transferred nuclear Overhauser effect spectrum (ET-NOESY) of the LDH·NAD$^+$ complex, recorded with a mixing time τ_m = 150 ms. Relevant NAD$^+$ protons are labeled with "ADE" for adenosine and "NIC" for nicotinamide. (b) Strip from the NOESY spectrum of Fig. 2a shown with lower contours. (c) Corresponding strip from a QUIET-ET-NOESY spectrum recorded under identical conditions. Doubly-selective inversion was achieved by applying a 12 ms cosine-modulated Gaussian cascade Q^3 in the middle of the mixing time to invert two frequency bands (represented by dashed lines), each of ~1.1 ppm width, centered around 6.07 ppm (to invert H1'^{ADE} of the adenosine) and 8.36 ppm (to invert both adenosine protons H2^{ADE} at 8.13 ppm and H8^{ADE} at 8.47 ppm). No water suppression methods were required. (Reproduced from ref. (1) with permission.)

A naive interpretation of the ET-NOESY intensities in Fig. 2 would indicate that, in the bound form, the aromatic H_2^{ADE} and H_8^{ADE} protons appear to be at similar distances from the $H_{1'}^{ADE}$ proton. This is an unreasonable conclusion given the structural constraints of adenosine nucleoside. The energy-minimized structure derived from X-ray diffraction indicates an extended conformation of the cofactor within the binding pocket, shown in Fig. 3, with distances of 4.6 Å for H_2^{ADE} - $H_{1'}^{ADE}$ and 3.9 Å for H_8^{ADE} - $H_{1'}^{ADE}$.

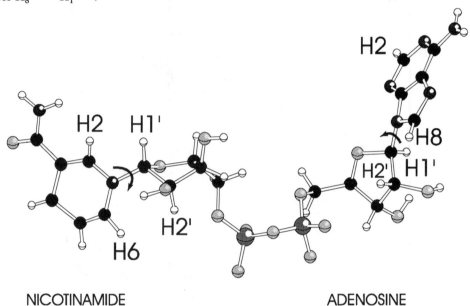

Figure 3. Structure of the cofactor NAD^+ in the binding pocket of LDH as derived from X-ray diffraction. (Reproduced from ref. (1) with permission.)

The problem inherent to conventional NOE studies stems from the possibility that the magnetization is not transmitted directly from H_2^{ADE} (or H_8^{ADE}) to $H_{1'}^{ADE}$, but first from H_2^{ADE} (or H_8^{ADE}) to one or several unknown "clandestine" protons that belong to the binding site of LDH, and from there to the $H_{1'}^{ADE}$ proton. In QUIET-ET-NOESY, the $H_{1'}^{ADE}$ proton of the adenosine moiety was taken to be the source spin and the H_2^{ADE} and H_8^{ADE} protons of the adenosine ring were both target spins. In Fig. 4, the build-up behaviour of the cross peaks $H_{1'}^{ADE} \rightarrow H_8^{ADE}$ and $H_{1'}^{ADE} \rightarrow H_2^{ADE}$ show a different behaviour when spin diffusion is quenched: the former is essentially unaffected, while the latter is strongly attenuated in the modified experiment, thus indicating that the corresponding peak in conventional ET-NOESY must be due largely to spin diffusion. Thus, the cross-relaxation rate between the protons H_2^{ADE} and $H_{1'}^{ADE}$ tends to be overestimated in conventional ET-NOESY, so that the distance between these protons will be underestimated.

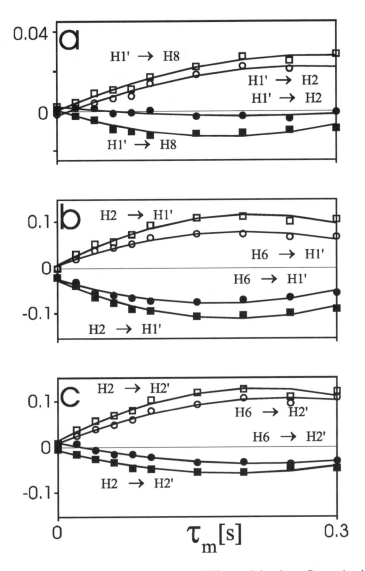

Figure 4. Build-up curves obtained by recording spectra at different mixing times. Conventional ET-NOESY amplitudes, which are positive (i.e. which have the same sign as the diagonal peaks), are represented by empty circles and squares, while QUIET-ET-NOESY amplitudes, which are negative because of the inversion in the middle of the mixing time, are represented by filled symbols. The cross-peak amplitudes were normalized with respect to the diagonal peaks. (a) Cross-relaxation from $H_{1'}^{ADE}$ to H_8^{ADE} (squares) and from $H_{1'}^{ADE}$ to H_2^{ADE} (circles). The latter build-up is strongly attenuated when spin diffusion is suppressed. (b) Cross-relaxation from nicotinamide H_2^{NIC} to ribose $H_{1'}^{NIC}$ (squares) and from H_6^{NIC} to $H_{1'}^{NIC}$ (circles). (c) Cross-relaxation from H_2^{NIC} to $H_{2'}^{NIC}$ (squares) and from H_6^{NIC} to $H_{2'}^{NIC}$ (circles). (Reproduced from ref. (1) with permission.)

For cross-relaxation between $H_{1'}^{ADE}$ and H_2^{ADE}, there are several protons belonging to the LDH binding pocket that may be involved in the flow of magnetization. The energy-minimized proton coordinates derived from diffraction data indicate that the protons H^α(Gly27), H^α(Asp52), and the methyl protons of Val53 and Ala96 should all lie in the proximity of H_2^{ADE} and $H_{1'}^{ADE}$ of the adenosine moiety (Fig. 5a). Although their identity can be inferred from the X-ray structure, the protons belonging to the binding pocket of a large enzyme cannot be probed by magnetic resonance because of their linewidths, and therefore cannot be taken into account in a full relaxation matrix analysis. Nonetheless, their contributions to spin diffusion are suppressed in QUIET-ET-NOESY, since their chemical shifts do not lie in one of the inverted regions. The intensity of the cross peak between H_8^{ADE} and $H_{1'}^{ADE}$ is hardly attenuated when spin diffusion is suppressed (Fig. 4a), which indicates a direct nuclear Overhauser effect between these two spins. In the bound form, the H_8^{ADE} and $H_{1'}^{ADE}$ protons are indeed at a distance greater that 3.3 Å from any other proton except $H_{2'}^{ADE}$.

Further experiments were developed to attempt to identify the protons belonging to the binding pocket that could be responsible for spin diffusion. In principle, contributions to ET-NOESY cross-peaks involving aliphatic protons of the enzyme can be suppressed in an experiment using a band-selective pulse instead of the doubly-selective pulse of Fig. 1a. In this context, "aliphatic" stands for all protons with resonances below 4 ppm, which can be inverted in the middle of the mixing time by a band-selective inversion pulse applied at −0.3 ppm with a bandwidth of about 2400 Hz (8 ppm at 300 MHz). This can be achieved by using a Q^3 pulse [17] with a duration of 2 ms, which should be reasonably effective even if the T_2's of the LDH protons are very short. In this experiment (data not shown), the peak intensities remained the same as in conventional ET-NOESY. We may therefore conclude tentatively that the protons belonging to the LDH binding pocket that mediate spin diffusion between NAD^+ protons do not resonate below 4 ppm. This excludes a participation of the H^γ(Val53) and H^β(Ala96), although the X-ray distances would seem to make these protons possible candidates. It is possible that H^γ(Val53) and H^β(Ala96) do not participate actively because they cross-relax rapidly with other spins within the enzyme. Similarly, $H^{\beta'}$(Asn138), H^γ(Val31) and $H^{\beta'}$(Arg99) can be excluded from playing a role in spin diffusion. These conclusions appear to be compatible with the X-ray structure. On the other hand, the negative outcome of the experiment with band-selective inversion of all aliphatic protons does not allow one to exclude H^α(Gly27) and H^α(Asp52), which are likely to resonate above 4 ppm, from playing a role in spin diffusion between $H_{1'}^{ADE}$ and H_2^{ADE} (Fig. 5a), nor H^α(Ser137) and H^α(Asn138) from being involved in spin diffusion between $H_{1'}^{NIC}$ and H_2^{NIC} (Fig. 5b). In general, the band-selective inversion experiment does not allow one to exclude any backbone H^α, any aromatic protons, or, if the studies are carried out in light water, any amide protons.

91

a ADENOSINE

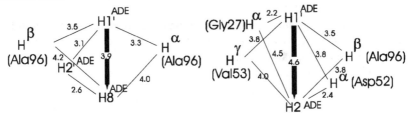

b NICOTINAMIDE

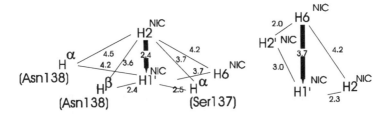

c NICOTINAMIDE

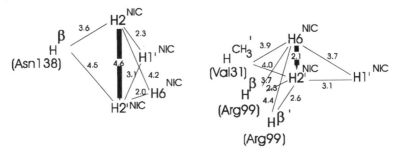

Figure 5. Representation of the X-ray distances between spins that are relevant to some of the cases studied. (Reproduced from ref. (1) with permission.)

The interpretation of QUIET-ET-NOESY with suppression of spin diffusion is much more straightforward than conventional ET-NOESY. A simplified analysis is legitimate and allows one to extract accurate relative internuclear distances.

4. Multiple-Quantum Filtered Xenon-131 NMR as a Surface Probe.

Xenon is relatively inert but highly polarizable, making it a sensitive probe for porous materials because it can provide information about the structure and symmetry of internal cavities [26]. Chemical shifts and relaxation effects of xenon-129 (I = 1/2) are indicative of the properties of host materials. Xenon-131 (I = 3/2) has a small gyromagnetic ratio (8.25% of the proton Larmor frequency) and a low natural abundance (21.18%). Compared to xenon-129 (27.81% with respect to protons and 26.44% natural abundance), it seems less attractive for surface studies, unless its quadrupolar moment is exploited. Physisorbed ^{131}Xe is usually not directly observable by NMR spectroscopy, but information can be transferred into the gas phase through desorption. The exchange process may lead to higher-rank order of the nuclear spins in gaseous xenon which can be detected by multiple-quantum filtered NMR.

Simpson [27] has shown that the free induction decay of optically pumped ^{201}Hg (spin I = 3/2) exhibits a beating pattern which depends not only on the symmetry of the cell that contains the Hg vapor but also on the orientation of this cell with respect to the external magnetic field. Following the work of Volk and coworkers [28, 29] on spin polarized ^{83}Kr and ^{131}Xe, Happer [30, 31], Pines [32] and Mehring [33] and their respective coworkers have shown that gaseous xenon-131 in very low fields (about 10^{-5} T) features small quadrupolar splittings (between 0.05 and 0.5 Hz) when the gas is contained in cells with suitable geometry and orientation.

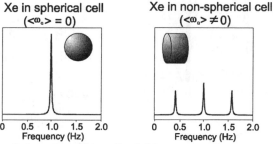

Figure 6. Quadrupolar splittings observed in cells of different geometries. $<\omega_Q>$ represents the averaged quadrupolar interaction with the surface. Adapted from ref (33).

Meersmann et al. [2] have demonstrated that multiple-quantum filtered NMR signals of xenon-131 can be also observed in the gas phase at high fields. Xenon-131 is a quadrupolar nucleus (I = 3/2), and its quadrupolar moment can interact with an electric field gradient (EFG) that is generated by the surrounding electron shell. The coupling with the anisotropic part of the EFG may lead to a splitting of the NMR transition. The quadrupolar interaction is believed to be quite large (20 to 30 MHz) when xenon is physisorbed to the surface. If a signal of surface-bound xenon-131 could be observed, one would expect three non-degenerate transitions $|-3/2> \leftrightarrow |-1/2>$, $|-1/2> \leftrightarrow |+1/2>$, and $|+1/2> \leftrightarrow |+3/2>$ separated by the quadrupolar splitting ω_Q. For a gas-phase NMR spectrum of xenon-131, no quadrupolar coupling is expected since the spherical symmetry of the electron shell should not create any anisotropic EFG.

Although this high symmetry may be disturbed by gas-phase collisions between xenon atoms, random rapid motion in the gas-phase should lead to a zero average of all quadrupolar interactions created by such processes.

It is convenient to represent the evolution of quadrupolar spins in terms of irreducible tensor operators [34]. Thermal equilibrium at high fields is described by a tensor $T_{1,0}$ of first rank and zeroth order. This can be transformed by a 90° radio-frequency pulse into single-quantum coherence described by a tensor $T_{1,\pm1}$ of first rank and first order. Only the order of the coherence can be changed by an rf pulse. However, the rank of a coherence can be altered through evolution under a quadrupolar interaction or through relaxation. Coherent evolution of xenon-131 under the quadrupolar coupling can lead to the transformation of $T_{1,\pm1}$ coherence into second and third rank terms $T_{2,\pm1}$ and $T_{3,\pm1}$.

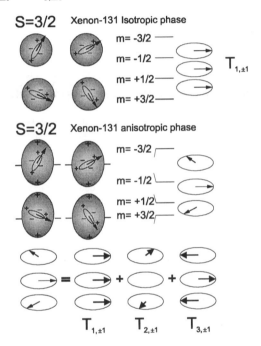

Figure 7. Energy levels of a quadrupolar nucleus (S = 3/2) with surrounding electron shell of (top) spherical symmetry and (below) of non-spherical symmetry causing an anisotropic EFG.

The splitting observed in the gas phase originates from collisions and adsorption of the xenon atoms on the surface. Such process reduces the symmetry of the electron shell, and the resulting quadrupolar couplings do not average to zero due to the fixed orientation of the surface with respect to the magnetic field. The quadrupolar evolution on the surface is transferred by exchange into the gas phase, where it is detected. This evolution leads to the transformation of $T_{1,\pm1}$ coherence into second and third rank terms $T_{2,\pm1}$ and $T_{3,\pm1}$. It is possible to track these terms by using multiple quantum-filters. Figure 8 summarizes the surface-induced coherent evolution and its detection with multiple quantum filters.

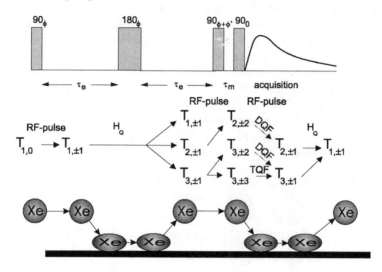

Figure 8. The initial 90° pulse generates first-rank single quantum coherence $T_{1,\pm 1}$. During the interval $2\tau_e$, second- and third-rank single-quantum coherences $T_{2,\pm 1}$ and $T_{3,\pm 1}$ can be created due to exchange between gaseous and physisorbed xenon. The 180° pulse allows one to refocus offset effects. In the brief interval τ_m (typically a few microseconds to allow the transmitter phase to be shifted), the transverse magnetization is temporarily converted into second- and third-rank double- and triple-quantum coherences $T_{2,\pm 2}$, $T_{3,\pm 2}$, and $T_{3,\pm 3}$. After cancellation of all terms other than the desired double- or triple-quantum coherences by an appropriate phase cycle, they are converted back into $T_{2,\pm 1}$ and $T_{3,\pm 1}$ by the last pulse. Although the latter terms are not directly observable, they can be reconverted through further physisorbtion processes into $T_{1,\pm 1}$ which can be detected by the receiver. (Adapted from ref. (2) with permission.)

Starting from a $T_{1,\pm 1}$ term, one can calculate the evolution of coherences of different ranks in the density operator under the quadrupolar coupling. This leads to:

$$\sigma(t) = \frac{i}{\sqrt{10}}\left(3\cos\left(\frac{eQV_{zz}t}{2\hbar}\right)+2\right)\{T_{1,+1}^{3/2}+T_{1,-1}^{3/2}\}+\sqrt{\frac{3}{2}}\sin\left(\frac{eQV_{zz}t}{2\hbar}\right)\{T_{2,+1}^{3/2}-T_{2,-1}^{3/2}\}$$
$$-i\sqrt{\frac{3}{5}}\left(\cos\left(\frac{eQV_{zz}t}{2\hbar}\right)-1\right)\{T_{3,+1}^{3/2}+T_{3,-1}^{3/2}\}$$

The double- and triple-quantum filters allow one to measure selectively the intensities of each term in this equation as a function of the evolution time.

The $T_{3,\pm 1}$ terms can also be generated by quadrupolar relaxation, as shown for lithium-7, sodium-23, and magnesium-25 [34-37]. For the case where the extreme narrowing condition is fulfilled, longitudinal and transverse relaxation of quadrupolar nuclei may be described by single exponential decays. This should also be true for xenon-131 in the gas phase, since it undergoes rapid isotropic motion. If the nuclei under investigation are associated with macromolecules, clusters, slowly tumbling complexes or rigid surfaces, the relaxation is multiexponential. Relaxation studies can be interesting because they provide additional information that cannot be revealed by coherent evolution alone. Figure 9 illustrates the effect of multiexponential relaxation on the evolution of the density operator.

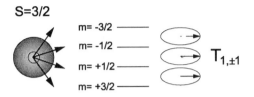

a) Monoexponential transverse relaxation

b) Biexponential transverse relaxation

c) Without quadrupolar coupling :

Figure 9. Relaxation of spin S = 3/2 atoms in isotropic phase where the energy levels are equally split. After excitation, first rank tensor elements are created. (a) All coherences relax at the same rate if the extreme narrowing condition is fulfilled ($\tau_c\omega_0 \ll 1$). The resulting coherence can be described again by attenuated first-rank tensor elements. (b) The situation changes if the fast motion limit is violated ($\tau_c\omega_0 \approx 1$) since the outer transitions relax at a higher rate than the central coherence. The resulting term can be described as a superposition of first and third rank tensor elements. (c) Second rank tensor elements cannot be created through relaxation alone. (Adapted from ref. (2) with permission.)

The $T_{3,\pm1}$ terms can be generated either through quadrupolar relaxation or through a quadrupolar splitting. The $T_{2,\pm1}$ term characterizes the surface-induced coherent evolution only, without relaxation effects. The DQF (double quantum filtered) sequence allows us to observe $T_{2,\pm1}$ and $T_{3,\pm1}$ together. The $T_{2,\pm1}$ term can be isolated by a modified DQF sequence introduced by Navon et al. [37, 38], shown in Fig. 10.

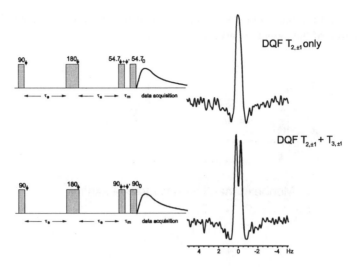

Figure 10. A conventionnal DQF measurement makes use of two 90° pulses and allows both, $T_{2,\pm1}$ and $T_{3,\pm1}$ elements to pass the filter. By using 54.7° pulses, one selects the $T_{2,\pm1}$ term.

It is possible to probe the multiexponential character of longitudinal ("T_1") relaxation by conducting double- or triple-quantum filtered inversion-recovery experiments (Fig. 11.). Thus it is possible to observe selectively the effects of relaxation and coherent evolution.

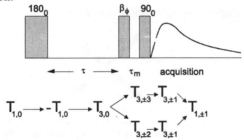

Figure 11. Multiple-quantum filtered inversion-recovery experiment. Quadrupolar relaxation in the τ interval leads to a partial conversion of inverted Zeeman order $-T_{1,0}$ into third-rank octupolar order $T_{3,0}$, which may be temporarily converted by the second pulse into third-rank double- and triple-quantum coherences $T_{3,\pm2}$ and $T_{3,\pm3}$, and subsequently into observable coherence $T_{1,\pm1}$ after reconversion by the last pulse and physisorption during the detection interval. Coherent evolution under a quadrupolar Hamiltonian cannot give rise to any signal in this experiment. The flip angle β should be 54.7° for double-quantum filters and 90° for triple-quantum filters.

The simulations of Figure 12 show a strong dependence on many parameters such as exchange rate, surface coverage, surface angle with respect to the magnetic field, and quadrupolar coupling constant. Factors representing relaxation and diffusion have not been incorporated into these calculations, so that the fits are merely qualitative at this point. These experiments can be realised with microporous materials like aerogels. The local asymmetry of the pores give rise to a non-zero average quadrupolar coherent interaction. The complexity of the system make simulations difficult, unless more independent measurements are obtained.

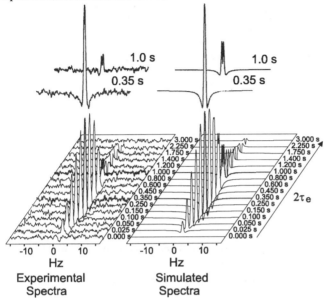

Figure 12. Stacked plot of DQF xenon-131 NMR spectra obtained with the sequence of Fig. 8 (increments of 2τ not linear). Left: Experimental spectra recorded at 24 MHz (7 T or 300 MHz for protons). Xenon gas at natural abundance at 1.3 MPa and 303 K was contained in a Pyrex tube with 6 mm i.d. and 8 mm o.d., inserted in a standard 10 mm NMR tube. The annulus between the tubes was filled with D_2O for field-frequency locking. Right: Simulated spectra. The physisorbed xenon was assumed to have an axially symmetric quadrupolar tensor with its symmetry axis normal to the surface and a coupling constant of 23.2 MHz [33]. The average residence time of the xenon on the surface was assumed to be 3.3 ns and the surface/gas population ratio was estimated to be 2×10^{-7}. The calculations were performed in composite Liouville space. The experimental spectrum at $2\tau = 0.35$ s exhibits a typical lineshape. For 2τ in the vicinity of 1 s, a doublet is observed due to interference of signals with opposite signs stemming from second- and third-rank tensors. For longer intervals, the DQF signal undergoes an inversion. (Reproduced from ref. (2) with permission.)

5. Conclusions

This contribution has highlighted some of the analogies and differences between exchange-transferred NOE studies and surface studies with xenon-131.

6. References

[1] Vincent, S. J. F., Zwahlen, C., Post, C. B., Burgner, J. W. & Bodenhausen, G. (1997) *Proc. Nat. Acad.* **94**, 4283-4388
[2] Meersmann, T., Smith, S.A. & Bodenhausen, G., (1998) *Phys. Rev. Lett.* **80**, 1398.
[3] Boelens, R., Koning, T. M. G. & Kaptein, R. (1988) *J. Mol. Struct.* **173**, 299-311.

[4] Borgias, B. A. & James, T. L. (1988) *J. Magn. Reson.* **79**, 493-512.
[5] Post, C. B., Meadows, R. P. & Gorenstein, D. G. (1990) *J. Am. Chem. Soc.* **112**, 6796-6803.
[6] Zheng, J. & Post, C. B. (1993) *J. Magn. Reson. Ser.* B **101**, 262-270.
[7] Ni, F. (1994) *Progr. NMR Spectroc.* **26**, 517-606.
[8] Massefski Jr., W. & Redfield, A. G. (1988) *J. Magn. Reson.* **78**, 150-155.
[9] Fejzo, J., Krezel, A. M., Westler, W. M., Macura, S. & Markley, J. L. (1991) *J. Magn. Reson.* **92**, 651-657.
[10] Burghardt, I., Konrat, R., Boulat, B., Vincent, S. J. F. & Bodenhausen, G. (1993) *J. Chem. Phys.* **98**, 1721-1736.
[11] Boulat, B., Burghardt, I. & Bodenhausen, G. (1992) *J. Am. Chem. Soc.* **114**, 10679.b
[12] Hoogstraten, C. G., Westler, W. M., Macura, S. & Markley, J. L. (1993) *J. Magn. Reson. Ser.* B **102**, 232-235.
[13] Zolnai, Z., Juranic, N., Markley, J. L. & Macura, S. (1995) *Chem. Phys.* **200**, 161-179.
[14] Zwahlen, C., Vincent, S. J. F., Di Bari, L., Levitt, M. H. & Bodenhausen, G. (1994) *J. Am. Chem. Soc.* **116**, 362-368.
[15] Vincent, S. J. F., Zwahlen, C., Bolton, P. H., Logan, T. M. & Bodenhausen, G. (1996) *J. Am. Chem. Soc.* **118**, 3531-3532.
[16] Vincent, S. J. F., Zwahlen, C. & Bodenhausen, G. (1996) *J. Biomol. NMR* **7**, 169-172.
[17] Emsley, L. & Bodenhausen, G. (1992) *J. Magn. Reson.* **97**, 135-148..
[18] Roberts, G. C. K. (1996) *13th European Experimental NMR Conference*, Paris, France.
[19] Bothner-By, A. A. & Gassend, R. (1972) *Annals N. Y. Acad. Sci.* **222**, 668-676.
[20] Rosevaer, P. R. & Mildvan, A. S. (1989) in: *Methods in Enzymology*, Oppenheimer, N. J. & James, T. L. Eds. (Academic Press, New York), Vol. 177, pp. 333-357.
[21] Holbrook, J. J., Liljas, A., Steindel, S. J. & Rossmann, M. G. (1975) in: *The Enzymes*, Boyer, P. D. Ed. (Academic Press, New York), Vol. XI, pp. 191-292.
[22] Stinton, R. A.& Holbrook, J. J. (1973) *Biochem. J.* **131**, 719-728.
[23] Heck, H. D. (1969), *J. Biol. Chem.* **244**, 4375-4381.
[24] Brooks, B. R., Brucoleri, R. E., Olafson, B. D., Slater, D. J., Swaminathan, S. & Karplus, M. (1983) *J. Comput. Chem.* **4**, 187-217.
[25] Young, L. & Post, C. B. (1993) *J. Am. Chem. Soc.* **115**, 1964-1970.
[26] Raftery, D. & Chmelka, B.F.(1994), *NMR Basic Princ. Prog.* **30**, 111.
[27] Simpson, J.H. (1978), *Bull. Am. Phys. Soc.* **23**, 394.
[28] Volk, C.H., Mark, J.G. & Grover, B. (1979), *Phys. Rev. A* **20**, 2381.
[29] Kwon, T.M., Mark, J.G. & Volk, C.H. (1981), *Phys. Rev. A* **24**, 1894.
[30] Wu, Z., Happer, W. & Daniels, J.M. (1987), *Phys. Rev. Lett.* **59**, 1480.
[31] Wu, Z.,. Schaefer, S , Cates, G.D. & Happer, W. (1988), *Phys. Rev. A* **37**, 1161.
[32] Raftery, D., Long, H., Shykind, D., Grandinetti, P.J. & Pines, A. (1994), *Phys. Rev. A* **50**, 567.
[33] Butscher, R., Wäckerle, G. & Mehring, M. (1994), *J. Chem. Phys.* **100**, 6923.
[34] Jaccard, G., Wimperis, S. & Bodenhausen, G. (1986), *J. Chem. Phys.* **85**, 6282.
[35] Müller, N., Bodenhausen, G. & Ernst, R.R. (1987), *J. Magn. Reson.* **75**, 297.
[36] Chung, C.-W. & Wimperis, S. (1990), *J. Magn. Reson.* **88**, 440.
[37] Eliav, U., Shinar, H. & Navon, G. (1992), *J. Magn. Reson.* **98**, 223.
[38] Eliav, U. & Navon, G. (1994), *J. Magn. Reson. B* **103**, 19.

APPLICATIONS OF NMR SPECTROSCOPY TO THE STUDY OF THE BOUND CONFORMATION OF O- AND C-GLYCOSIDES TO LECTINS AND ENZYMES

J. JIMÉNEZ-BARBERO[1], J. CAÑADA[1], J. L. ASENSIO[1], J. F. ESPINOSA[1], M. MARTÍN-PASTOR[1], E. MONTERO[1], AND A. POVEDA[2]
[1]*Instituto Química Orgánica, CSIC, Juan de la Cierva 3, 28006 Madrid, Spain.*
[2]*RMN-SIdI, Universidad Autónoma de Madrid, Cantoblanco-28047, Madrid, Spain.*

Abstract

TR-NOE experiments have been used to characterise the bound conformation of oligosaccharide and synthetic analogues to different carbohydrate binding proteins. In particular, the bioactive conformations of lactose, C-lactose, and N-acetyl glucosamine-containing oligosaccharides when bound to Ricin, *E. coli* β-galactosidase and hevein is presented.

1. Introduction

The recognition events among biomolecules are strongly dependent on the three dimensional structure of the interacting species. Among biomolecules, carbohydrates differ from the other classes in that their constituting moieties (monosaccharides) may be connected to one another by a great variety of linkage types. In addition, they can be highly branched, thus allowing oligosaccharides to provide an almost infinite array of structural variations. The decoding process of the existing information in oligosaccharide structures involve their recognition by other biomolecules. Thus, they are most often specifically recognized by proteins (so called lectins) and these interactions may mediate a particular biological response, such as host-parasite interactions, fertilization, autoinmune disorders, cellular differentiation,..., etc [1]. Therefore, the study about how oligosaccharides are recognised by the binding sites of enzymes, antibodies, and lectins is a topic of major interest. It is today evident that the knowledge of the three dimensional structure of these biomolecules (proteins and carbohydrates) has the potential to assist in the design of new carbohydrate-based therapeutic agents. Current technical facilities and biophysical techniques, mainly X-ray crystallography analysis, have permitted the access to detailed information on the three dimensional structure of protein-carbohydrate complexes [2]. This data, complemented mainly with those obtained through titration microcalorimetry, have allowed to postulate which are the major factors involved in these interactions [3]. These are usually hydrogen bonds and van der Waals forces, often including packing of a hydrophobic sugar face against aromatic amino acid side chains.

Until very recently, the large size of carbohydrate-binding proteins (lectins, antibodies and enzymes) have prevented their direct studies by means of nuclear magnetic resonance spectroscopy (NMR). However, in the last few years, researchers in the field of NMR have applied this spectroscopy to the study of these molecular recognition processes at different levels of complexity. Moreover, the carbohydrate also exhibits greater dynamic fluctuations than the protein and, therefore, NMR measurements may well offer new insights on the conformation of bound oligosaccharides and/or on the corresponding motional timescales. Thus, using NMR nowadays it is possible to deduce the specificity and affinity of binding, including measurements of association constants and equilibrium thermodynamic parameters (titration experiments). It may also be possible to deduce which aminoacid residues are involved in binding (titration, NOESY), even the three dimensional structures of the bound carbohydrate and/or the protein (NOESY experiments).

2. TR-NOE experiments

Different examples on the study of the structural events that mediate the molecular recognition processes between proteins and carbohydrates have been recently presented, using NMR spectroscopy, examining interactions involving lectin-, antibody-, and enzyme-type receptors [4]. Specific comparisons of the differential binding of natural and modified analogues have also been reported [4, 5]. In some cases, protein-induced conformational changes of the oligosaccharide ligand have been observed. Nevertheless, in other cases the lectin selects one out of the conformers present in solution or a structure close to the major one existing in solution. Finally, there are also cases in which the ligand retains at least part of the flexibility that was present in the isolated state. It is obvious that the knowledge of the recognised conformation of a biologically active carbohydrate presents considerable implications for rational drug design. From the three dimensional point of view, the transferred nuclear Overhauser enhancement (TR-NOE) may permit the assessment of the conformation of protein-bound oligosaccharides [6]. NOESY experiments provide information about which protons are close in space and, therefore, they may be used to deduce conformational information. The TR-NOESY is a regular NOESY experiment, but it is applied to a protein/ligand system in dynamical exchange in which the ligand is present in excess. For ligands which are not bound tightly and exchange with the free form at reasonably fast rate, as usually observed for carbohydrates, the transferred nuclear Overhauser enhancement (TR-NOE) experiment provides an adequate means to determine their bound conformation (Figure 1). In complexes involving large molecules, cross relaxation rates of the bound compound (σ^B) are opposite in sign to those of the free one (σ^F) and produce negative NOEs. Therefore, the existence of binding may be easily deduced from visual inspection, since NOEs for small molecules are positive (Figure 1). The conditions for the applicability of this approach are well established, considering the well known equilibrium and the molar fractions of free and protein bound sugar,

$$\text{PROTEIN} + \text{SUGAR}_{(excess)} \leftrightarrow \text{PROTEIN} \cdot \text{SUGAR}$$

$$K_a = \frac{[\text{PROTEIN} \cdot \text{SUGAR}]}{[\text{PROTEIN}] \times [\text{SUGAR}]}$$

$$p_b \sigma^B > p_f \sigma^F$$
$$k_{-1} >> \sigma^B$$

where p_b and p_f are the fractions of bound and free sugar ligand and σ^B and σ^F the cross relaxation rates for the bound and the free ligand, respectively. k_{-1} is the off-rate constant.

Under these conditions, it can be considered that

$$\sigma^{obs} = p_b \sigma^B + p_f \sigma^F$$

This approach has been recently applied to several studies of lectin- and antibody-bound oligosaccharides [4, 5].

TR-NOESY experiments are usually performed at different mixing times and ligand/protein ratios and produce strong and negative NOEs upon existence of ligand binding. However, one of the major drawbacks of this experiment is the possible existence of spin diffusion effects, which are typical for large molecules. In this case, apart of direct enhancements between protons close in space, other spins may mediate the exchange of magnetization, thus producing negative cross peaks between protons far apart in the macromolecule. Thus, protein-mediated, indirect TR-NOE effects may lead to interpretation errors in the analysis of the ligand bound conformation. In TR-ROESY [7], spin -diffusion (three spin) effects appear as positive cross peaks and therefore, the application of this experiment permits to distinguish direct from indirect enhancements, and thus complements those measured under regular conditions, providing conformational information which is less contaminated by artifacts (Figure 1).

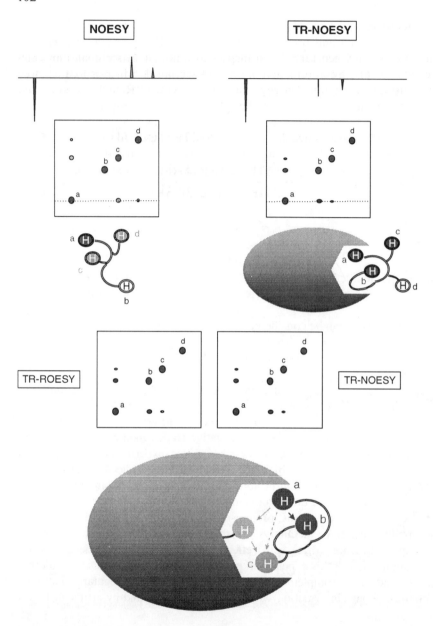

Figure 1. Top Left. Schematic representation of a NOESY spectrum for a free sugar. Cross peaks and diagonal peaks have different signs. Top right. Schematic representation of a TR-NOESY spectrum recorded for an exchanging sugar/protein system. Cross peaks and diagonal peaks have the same signs, as expected for a large molecule, thus indicating binding to the protein. The relative sizes of the peaks and the apparison of new ones may be used to detect conformational variations. Bottom left. Schematic representation of TR-ROESY spectra for an exchanging sugar/protein system. In TR-ROESY spin diffussion (three spin) cross peaks (i.e. a/c) and diagonal peaks have the same signs. On the other hand, direct cross peaks (a/b) show different sign to diagonal peaks (a).

3. The ricin/lactose case

Although there is not any general rule, in many cases, protein binding sites are well preorganized to recognise a conformation of the oligosaccharide which is located close to the its global minimum energy region. Several TR-NOE studies on protein/carbohydrate interactions have studied the complexes between Ricin-B chain and different oligosaccharides [8]. Ricin is a dimeric (A and B chains) galactose-binding lectin which has been shown to be 10-100 fold more toxic to some transformed cell lines than to normal cells and has therefore been considered as a potential antitumoral agent. The structures of several sugars used are presented in figure 2.

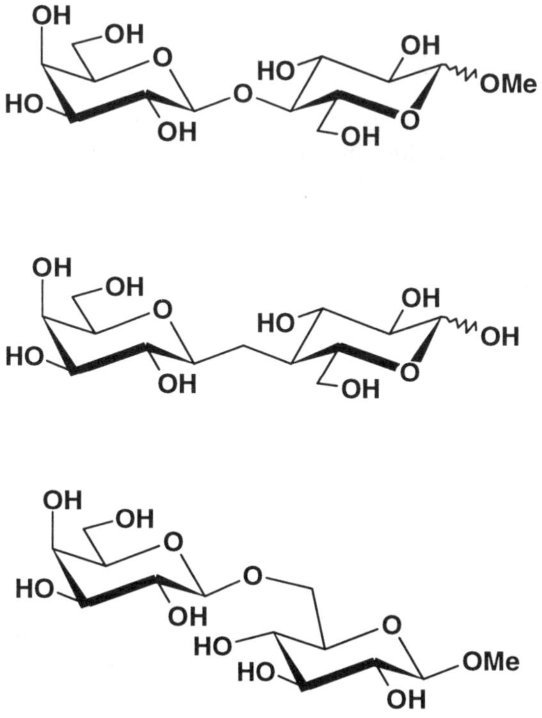

Figure 2 Structure of different galactose-containing disaccharides and analogues. From top to bottom, lactose, , C-lactose, methyl β-allolactose

The conformational changes that occur when methyl α-lactoside is bound to Ricin-B chain in aqueous solution have been studied [8]. The observed data indicate that the protein causes a slight conformational variation in the glycosidic torsion angles of methyl α-lactoside, although the recognized conformer was still within the lowest energy region. The results were in agreement with a pioneer NMR study of Ricin-

B/disaccharide complexes that used monodimensional transferred NOE (1D-TR-NOE) experiments to study the binding of Ricin-B by methyl ß-lactoside [9]. In this example, it was demonstrated, by using a selectively deuterated substrate that there were minor changes in the conformation of free methyl ß-lactoside (Figure 2) upon binding to Ricin-B. Molecular modeling using molecular dynamics, minimization, and docking of the disaccharide within the binding site of Ricin B strongly suggested that, apart from the expected contacts between the galactose moiety and different aminoacid residues, there were also van der Waals contacts between the protein and the remote glucose moiety, as previously deduced from binding studies using modified lactoside derivatives. Thus, both van der Waals contacts and hydrogen bonding contribute to the stability of the complex [8].

We have found examples in which the protein does not select a single conformer. For instance, in the complex between methyl ß-allolactoside (Galß(1→6)Glcß-OMe) and Ricin B [8]. In this case, and contrary to the observations for lactose, different conformations around the ϕ, ψ, and ω glycosidic bonds of methyl ß-allolactoside were recognized by the lectin. In fact, for this complex, only the TR-NOESY cross-peaks corresponding to the protons of the galactose residue were negative, as expected for a molecule in the slow motion regime. In contrast, the corresponding cross peaks for the glucose residue were *ca.* zero, as expected for a molecule whose motion is practically independent of the protein.

4. The ricin/C-lactose case.

Obviously, not only natural ligands, but also structurally-modified oligosaccharides may be used as lectin ligands or as inhibitors of carbohydrate-processing enzymes, and thus, these analogues may be employed to deduce enzymatic mechanisms. Moreover, conformational differences between free and protein-bound natural carbohydrates and synthetic analogues may also be assessed by TR-NOE experiments.

Ricin B has also been used as a model to study the bound conformation of potential glycosidase inhibitors such as C-glycosides [10, 11]. Although many reports have usually assumed that the conformation of free C-glycosides was the same as that of the corresponding O-analogues, it has recently been reported that, at least for O- and C-lactoses, this is not the case [10]. The conformational study of C-lactose in the free state showed that the exo-anomeric conformation around the C-glycosidic bond was adopted. However, the conformation around the aglyconic bond was rather different to that of the natural compound. For O-lactose, ca. 90% of the population was located around the so called minimum syn, ϕ/ψ: 54/18 and ca. 10% of population around minimum anti, ϕ/ψ: 36/180. However, C-lactose was shown to exhibit much higher flexibility than its O-analogue and three conformational regions (syn, anti and gauche-gauche) were significantly populated in solution (Figure 3). NMR and modeling studies on free C-lactose demonstrated that three different conformational families coexist in solution. Approximately, 55% of population adopts the *anti* conformation (ϕ/ψ 36/180, where the glycosidic torsion angles ϕ and ψ are defined as H1'-C1'-Cα-

C4 and C1'-Cα-C4-H4 respectively) while about 40% presents the *syn* conformation (ϕ/ψ 54/18), which is the conformation displayed by the previously reported structures for different β(1->4) equatorial linked disaccharides. The experimental data also indicate that a high energy minimum (ϕ/ψ, 180/0) is also slightly populated (*ca.* 5%). This last conformation has been called *gauche-gauche* since displays the Cα-C4 linkage gauche with respect to both C1'-O5' and C1'-C2' bonds, and represents a conformation never observed for free β(1->4) natural disaccharides.

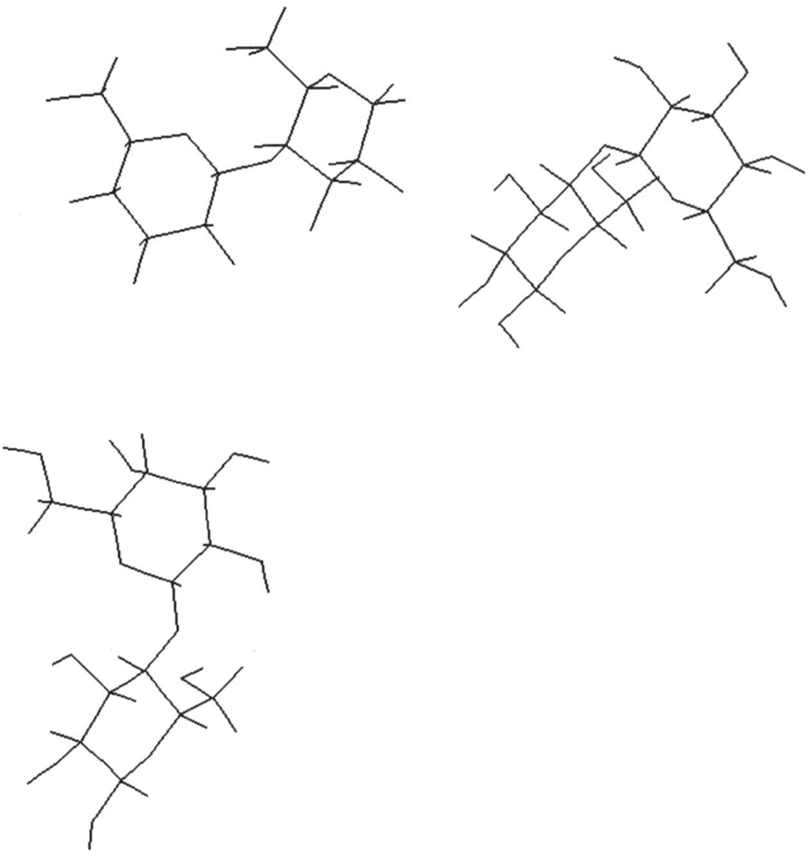

Figure 3. Schematic view of putative conformational changes around the glycosidic torsion angles of a disaccharide. Three different conformations are shown: *syn (bottom left), anti (top right), gauche-gauche (top left)*.

The existence of these three conformational families could be detected by the presence of NOEs that unequivocally characterize the different regions of the conformational map. These NOEs have been dubbed exclusive NOEs. For C-lactose, H1'-H4, H1'-H3 and H4-H2' are exclusive NOEs for the *syn*, *anti* and *gauche-gauche* conformations, respectively. Consequently, the corresponding NOE intensities will be sensitive to their respective populations. Therefore, and at least qualitatively, a first indication of the bound conformation can be obtained by focusing on these key NOEs. Thus, 2D transferred NOESY experiments were recorded to study the complexation of C-lactose by Ricin B. The comparison between the NOESY and ROESY spectra of C-lactose, recorded in the absence and in the presence of the lectin, indicated that conformers syn and gauche-gauche were not bound. Therefore, the experimental results indicated that Ricin-B selects different conformers of C-lactose, (anti), and its O-analogue, (syn) [10, 11]. In order to estimate the relative binding affinities of the C- and O-glycosides, competitive TR-NOEs, with different O-lactose/C-lactose ratios, were also performed. It was demonstrated that both ligands compete for the same binding sites of the lectin and that the affinity constant of C-lactose is smaller than that of its O-analogue. Although merely speculative, and since the flexibility of C-lactose in the free state is much higher than that of O-lactose, the cause of the recognition of different conformations could be of entropic origin.

5. The β-galactosidase/C-lactose case.

Since ricin-B is not an enzyme, these conclusions are not general. Thus, binding of C-lactose to *E. coli* β-galactosidase [12, 13] was explored. At present moment there is not much information on how oligosaccharide substrates are recognized by their appropiate glycosidase. *E. coli* β-galactosidase is a retaining enzyme with tetrameric structure and one active site per monomer whose X-ray structure is known [14]. Each monomer consists of 1023 amino acid residues and its natural substrate is lactose. The mechanism of action has not been firmly established, but it is accepted that involves a double displacement reaction in which the enzyme first forms (Figure 4) and then hydrolyses a glycosyl-enzyme intermediate via oxocarbonium ion-like transition state [12-18].

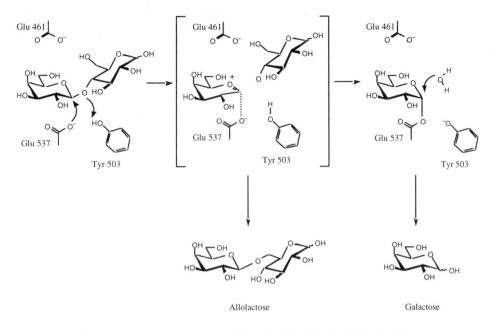

Figure 4. Schematic representation of lactose hydrolysis by *E. coli* β-galactosidase.

Several studies have allowed the identification of the residues that are responsible for catalysis. Specifically, Glu-461, Glu-537 and Tyr-503 have been confirmed to be essential for enzymatic activity [19-22]. Glu-537 is the active nucleophile. With respect to Tyr-503, inactivation has suggested that this residue is probably acting as the Broensted catalyst in the mechanism. A recent report has also suggested the involvement of His-540 in the stabilization of the transition state by forming a hydrogen bond with galactose [23].

These Glu-461, Tyr-503 and Glu-537 aminoacids are found together within a deep pocket within a distorted "TIM" barrel in the X-ray structure. Only the structure of the free enzyme is available, and therefore the orientation, position and conformation of lactose within the binding site is uncertain. This information may be essential for a fully understanding of the catalytic mechanism, as well as for the knowledge of the role that each aminoacid plays in the reaction. As the huge size of β-galactosidase precludes direct ^{1}H-NMR observations using current strategies [24], information about the conformation of complexed oligosaccharides can only be derived from transferred NOE studies. Obviously, the hydrolysis of the natural substrate does not allow the use of lactose for this sort of experiments but C-lactose could be used instead. Indeed, the C-analogue is an inhibitor of β-galactosidase at millimolar concentrations, in the same range of concentrations described for lactose. Therefore, the determination of the conformation and orientation of this inhibitor within the binding pocket was pursued..

TR-NOESY experiments were performed at different mixing times giving rise to strong and negative NOEs, as expected for ligand binding. The comparison between the NOESY spectra of C-lactose (Fig. 5) recorded in the absence and in the presence of the enzyme shows important and clear differences.

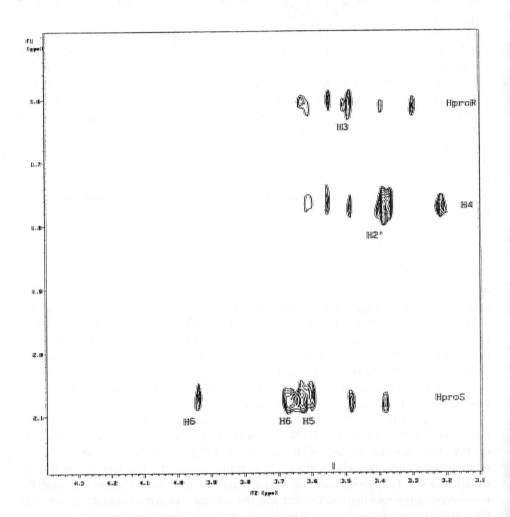

Figure 5. NOESY ^{1}H-NMR spectrum recorded for C-lactose in the presence of *E. coli* β-galactosidase.

Some of the cross peaks in the NOESY spectrum of the free ligand are no longer present in the TR-NOESY spectrum of the complex. The disappearance of both H1'-H3 and H1'-H4 NOEs provided evidence that neither the *anti* conformation (see above) nor the *syn* conformation are recognized by the enzyme. By contrast, the H4/H2' NOE that is very weak in the free ligand is now the strongest one in the spectrum. These findings clearly indicate that the bound conformation belongs to the local minimum *gauche-gauche* family. Fortunately, the presence of two additional methylene protons for C-lactose, in contrast to the natural glycosides, permits detection of more experimental constraints. Indeed, since HproR/H3, HproR/H5, HproS/H3, HproS/H5, and HproS/H6 cross peaks (Fig. 5) are also observed in the TR-NOESY spectra series, indicating the recognition of the *gauche-gauche* conformation (Fig. 3). TR-ROESY experiments confirmed that the above mentioned cross peaks showed different sign to the diagonal peaks thus excluding the possibility of protein-relayed or spin difussion mediated correlations.

It is noteworthy to point out that these experimental results indicate that β-galactosidase recognizes a high energy local minimum of the inhibitor. This conformation has a small population for free **1** (<5%), while it is not present for natural lactose in water solution. Taking into account the energetic differences between the high energy and the global minimum conformer of C-lactose in solution, which amounts to about 2.5 kcal/mol, it may be postulated that the enzyme shows a intrinsic binding energy around this size. A competitive TR-NOE experiment was performed by adding increasing amounts of IPTG to a NMR tube containing C-lactose and the enzyme. It was observed that the TR-NOE signals of C-lactose disappear at equimolecular concentrations of IPTG with the appearance of strong NOEs for this molecule. This results indicate that both molecules compete for the same binding site.

Although the natural glycoside and the C-glycoside are indeed fairly structurally related analogues, the recognized conformation of the synthetic analogue does not necessarily represents that of the natural compound. Nevertheless, in this case, direct comparison of both compounds is not possible, since obviously, lactose, the natural substrate, after being recognized by the protein, is fastly transformed into products. There are cases in which a protein bind oligosaccharides near their global minimum conformation, although there are also examples of major conformational variations upon binding [24, 25].

The following step was to dock C-lactose within the enzyme binding site to determine the 3D structure of the complex, using the structure deduced from the X-ray crystallographic analysis. The galactose moiety was superimposed onto the geometry of a previously optimized enzyme/monosaccharide complex. Then, a systematic search of the possible conformations around the glycosidic linkages was performed by systematic rotations of φ and ψ angles. Since the binding site is located very deeply in the structure, indeed only one conformational family was allowed, corresponding to the *gauche-gauche* conformation. Neither the *syn* nor the *anti* conformations can exist in the binding site without severe steric interactions. As discussed above, the obtained geometry is in fair agreement with the experimental NMR results.

If it is assumed that the experimentally and theoretically deduced *E.coli* β-galactosidase-bound conformation of C-lactose is indeed significant with respect to the initial binding mode of lactose, the recognition of the *gauche-gauche* conformer could present important implications on the reaction mechanism. The preferential recognition of a high energy structure of the substrate could decrease the activation energy of the reaction, thus accelerating the catalytic process [25-27]. Besides, it is possible to especulate about the role that each key amino acid residue plays in the mechanism, taking into account its orientation and distance with respect to the pseudodisaccharide (Fig. 4). Indeed, the geometry of the substrate-enzyme complex, at least qualitatively, is consistent with multiple biochemical data described in the literature for *E.coli* β-galactosidase, and with the proposed roles for the different amino acid residues located around the binding site [28].

The observed bound *gauche-gauche* conformation of C-lactose could also explain the transglycosidation properties of *E. coli* β-galactosidase. Retaining mechanisms go through a glycosyl-enzyme reaction intermediate ,which allows the transfer of glycosyl residues to other nucleophiles different than water.. *E. coli* β-galactosidase in the presence of lactose synthesises allolactose (Fig. 4). It has been demonstrated that this reaction occurs without the need for the aglycon to leave the enzyme active site [29, 30]. Interestingly, only the *gauche-gauche* conformation of the disaccharide places glucose OH-6 group on the β-face of galactose, close to the C-1 anomeric carbon (<4 Å). Thus, it is fair to think that the binding of the *gauche-gauche* conformation would make easier the internal transfer of the galactosyl residue from O-4 to O-6 of glucose with minor movements of the glucose aglycon inside the active site.

There is not a direct way to demonstrate that the conformations of C-lactose and lactose itself when bound to the enzyme are the same. However, this possibility deserves some comments, since structural, biochemical, and modeling studies point to this possibility. C-lactose and lactose display minor geometric and structural differences according to molecular mechanics calculations (C1'-Cα, 1.538 *versus* C'1-O1', 1.426 Å and O5'-C1'-Cα 107.4° *versus* O5'-C1'-O1', 108.3°). In addition, the inhibition constants determined for β-methyl-C-lactoside and C-lactose are of the same order of the Ki for lactose itself. The observed conformation of C-lactose is reminiscent of the deformations in the reaction pathway. Glycosyl oxocarbonium-like transition state structures have been proposed for glycosidases. For disaccharides, the variation from a regular chair conformation towards the distorted conformations proposed for a glycosyl oxocarbonium, produces a significant upwards shift of the aglycon moiety with respect to the plane defined by C-5, O-5, C-2 and C-3 in the initial $^4C_1$ chair of the non reducing end. Several studies have recently addressed this topic in structural terms [31-34]. In all of them, the conformational changes of the residues attached to the scissile bond place the aglycon moiety of the substrate in a pseudoaxial orientation, as preferred for a leaving group. For β-galactosidase/C-lactose, if the galactose moiety would adopt a half chair or sofa conformations, H-1' proton should move away from H-3' and H-5' atoms with the consequent decrease in the intensities of the corresponding H-1'/H-3' or H-1'/H-5' intraresidue NOEs. However, since the experimental intraresidue NOEs are strong, it can be excluded that the galactopyranose ring of C-lactose suffers a significant conformational change.

This strategy represents a novel and different approach to the understanding of the properties of this important enzyme, which, in turn, could be extended to related systems. The TRNOE technique has is applicability window limited by the kinetics underneath the receptor-ligand equilibrium. However, it seems that the study of carbohydrate-enzyme interactions is an area where this methodology can have a remarkable applicability [35-41]. The reason of this favourable situation probably lays in different facts: These interactions are not extremely strong, there is fast exchange between the free and the bound states of the ligand, and the perturbations of the conformational equilibrium of a given oligosaccharide upon binding to a protein are accessible to observation by TRNOE. We have applied this approach for the first time to study a glycosidase enzyme. In order to know if this methodology can be of general use, studies with other glycosidase enzymes and similar nonhydrolizable substrates, together with the reverse approach of studying the conformation of natural substrates in the presence of mutated and nonactive enzymes are now in progress.

6. The hevein case.

In a few favourable cases, dealing with protein receptors small enough to be amenable to direct analysis to NMR methods, ^{1}H-NMR techniques have been used to deduce the three dimensional structure of protein-carbohydrate complexes.

Hevein [42, 43] is a protein of 43 amino acids, whose structure has independently been solved by X-ray at 0.28 nm resolution, and by NMR methods. Interestingly, although the structure of hevein in water-dioxane and water solutions differs significantly from that observed in the crystal, it closely resembles the solid state structures of the domains of wheat germ agglutinin (WGA). We [43, 44] have recently reported on the determination of the binding site of hevein by using NMR spectroscopy and different N-acetyl glucosamine-derived (GlcNAc) ligands. The GlcNAc, chitobiose, and chitotriose (Figure 6) specific binding constants were also determined by 1D NMR spectroscopy. These constants increase in one order of magnitude when passing from the mono- to the di- and to the tri-saccharide. In addition, the thermodynamic parameters for chitotriose-hevein and chitobiose-hevein interactions were obtained from a van't Hoff analysis, indicating that the association process is enthalpy driven, while entropy opposes binding. This behaviour is usually observed for protein-carbohydrate interactions [1-5]. The deduced negative signs indicated that hydrogen bonding and/or van der Waals forces are the major interactions stabilizing the complex. The differences in binding constants were explained in terms of the three-dimensional structure of the complexes, also obtained from NOESY NMR spectroscopy.

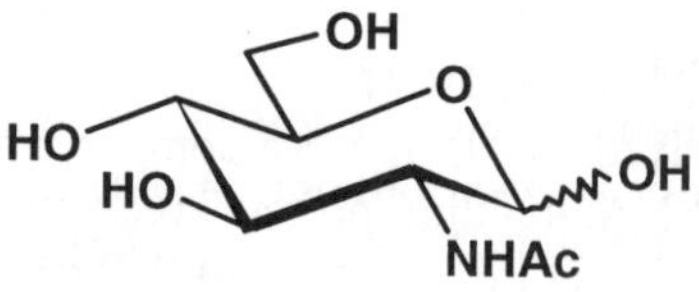

N-Acetyl glucosamine

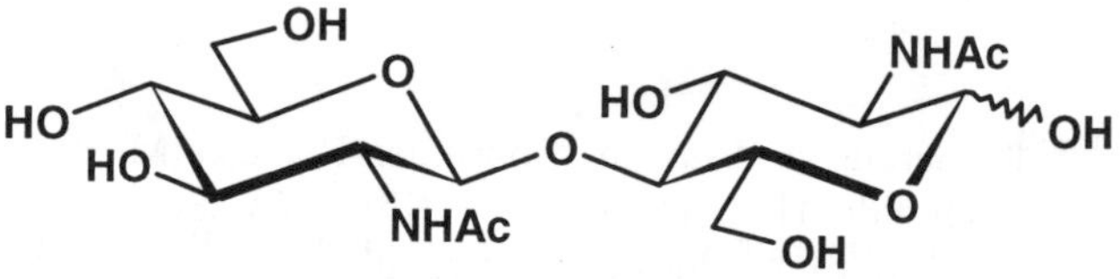

Chitobiose

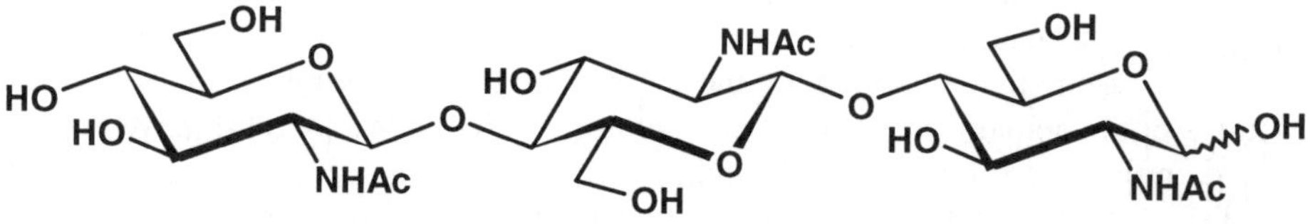

Chitotriose

Figure 6 Structure of N-acetyl glucosamine (GlcNAc) containing sugars. From top to bottom : GlcNAc, chitobiose, chitotriose.

Besides we also have determined of the structure of the complex of hevein with chitobiose, by using NMR spectroscopy. Using NOESY spectroscopy and restrained molecular dynamics, they also presented a refined NMR structure of free hevein in water. The structure of the complex of hevein with methyl ß-chitobiose has also been derived recently [44]. Protein-carbohydrate nuclear Overhauser enhancements measured for the hevein-chitobiose and hevein/methyl ß-chitobiose complexes allowed to deduce the conformation of these complexes. Obviously, the presence of NOESY cross peaks between certain protons of the sugar and the protein permit to infer that these atoms are close in space and therefore, to derive the three dimensional structure of the complex. No important changes in the protein NOE's were observed, indicating that carbohydrate-induced conformational changes in the protein are small. The N-acetyl methyl signal of the non-reducing GlcNAc moiety of ß-chitobiose displayed NOE contacts with Tyr30 and Trp21 residues and appeared strongly shielded. From the inspection of the model, a hydrogen bond between Ser19 and the non reducing N-acetyl carbonyl group was suggested as well as between Tyr30 and HO-3 of the same sugar residue. The previously mentioned N-acetyl methyl group of the non-reducing GlcNAc displayed non polar contacts to the aromatic Tyr30 and Trp21 residues. Moreover, the higher affinities deduced for the ß-linked oligosaccharides with respect to GlcNAc and GlcNAcα-(1→6)-Man could be explained by favorable stacking of the second ß-linked GlcNAc moiety and Trp21.

The final 3D structures derived by NMR were compared to those of WGA [45], Ac-AMP II (which is also a GlcNAc-binding protein) recently solved by NMR, Martins *et al.* [46] and to the crystal structure of hevein [47]. The corresponding average rmsd are 0.060 nm (B domain of WGA, residues 16-32), 0.100 nm, (Ac-AMP2, residues 12-32), and 0.269 nm (crystal of hevein, residues 16-41).

7. Acknowledgments

We thank Drs. J. L. García, A. Vian, A. Rodríguez-Romero, M. Bruix, C. González, A. Imberty, R. R. Schmidt, H. J. Dietrich, N. Khiar, M. Bernabé, and M. Martín-Lomas for helpful discussions and for their contributions. We also thank SIdI-UAM for the facilities provided and DGICYT (PB96-0833) as well as the Mizutani Glycoscience Foundation for financial support.

8. References

1. Dwek, R.A. (1996) *Chem. Rev. Glycobiology*, **96**, 683-710. Gabius H. J. and Gabius S. (1996) *Glycosciences, status and perspectives*, Chapman and Hall, GmbH Weinheim.
2. Weis W.I. and Drickamer, K. (1996) Structural basis of lectin-carbohydrate recognition, *Ann. Rev. Biochem.* **65**, 441.
3. Toone, E. (1994) Structure and energetics of protein-carbohydrate complexes, *Curr. Opin. Struct. Biol.*, **4**, 719-725, and references therein.
4. Poveda, A. and Jiménez-Barbero, J. (1998) NMR studies of carbohydrate-protein interactions in solution, *Chem. Soc. Rev.* **27**, 133-143. and references therein
5. Peters T. and Pinto, B. M. (1996) Structure and dynamics of oligosaccharides: NMR and modeling studies, *Curr. Opin. Struct. Biol.* **6**, 710-716, and references therein.
6. Ni, F. (1994) Recent developments in transferred NOE experiments, *Prog. NMR Spectroscopy* **26**, 517-606.
7. Arepalli, S.R. Glaudemans, C.P.J. Davis, D.G. Kovac, P., and Bax, A. (1995) Identification of protein-mediated indirect NOE effects in a disaccharide-Fab complex by TR-ROESY, *J. Magn. Reson. B*, **106**, 195-198.
8. Asensio, J.L., Cañada, F.J., and Jimenez-Barbero, J. (1995) Studies of the bound conformation of methyl α-lactoside and methyl β-allolactoside to ricin-B chain using transferred NOE experiment in the laboratory and the rotating frames, assisted by molecular mechanics and dynamics calculations, *Eur. J. Biochem.*, **233**, 618-630, and references therein.
9. Bevilacqua, V.L :, Thomson, D.S., Prestegard, J.H. (1990) Conformation of methyl beta lactoside bound to the ricin B chain : interpretation of TR-NOE effects

facilitated by sin simulation and selective deuteration, *Biochemistry*, **29**, 5529-5537.
10. Espinosa, J.F., Cañada, J., Asensio, J.L., Martín-Pastor, M., Dietrich, H.J., Schmidt, R.R., Martín-Lomas, M., and Jiménez-Barbero, J. (1996) Experimental evidence of conformational differences between C-glycosides and O-glycosides in solution and in the Protein-bound state: the C-lactose/O-lactose case, *J. Am. Chem. Soc.*, **118**, 10682-10691, and references therein.
11. Espinosa, J.F. , Cañada, J. , Asensio, J.L., Dietrich, H.J., Schmidt, R. R., Martín-Lomas, M., and Jiménez-Barbero, J. (1996) Conformational differences of O- and C-glycosides in the protein bound state: different conformation of C-lactose and its O-analogue are recognized by Ricin-B, a galactose-binding protein, *Angew. Chem. Int. Ed. Engl.*, **35**, 303-306.
12. Sinnott, M.L. (1990) Catalytic mechanisms of enzymic glycosyl transfer, *Chem. Rev.* **90** , 1171-1202.
13. Huber, R.E., Gupta, M.N., and Khare, S.K. (1994) The active site and mechanism of the β-galactosidase from *Escherichia coli*. *Int. J. Biochem.* **26**, 309-318.
14. Jacobson, R.H., Zhang, X.-J., DuBose, R.F., and Matthews, B.W (1994) Three-dimensional structure of β-galactosidase from *E. coli*, *Nature* **369,** 761-766.
15. Phillips, D.C. (1967) The hen egg-white lysozyme molecule, *Proc. Natl. Acad. Sci. USA* **57**, 484-495.
16. Legler, G. (1990) Glycoside hydrolases: Mechanistic information from studies with reversible and irreversible inhibitors, *Adv. Carb. Chem. Biochem.* **48,** 319-385.
17. McCarter, J.D. and Withers, S.G. (1994) Mechanism of enzymatic glycoside hydrolisis, *Curr. Op. Str. Biol.* **4**, 885-892.
18. Davies, G. and Henrissat, B.H. (1995) Structures and mechanisms of glycosyl hydrolases, *Structure* **3**, 853-859.
19. Huber, R.E. and Chivers, P.T. (1993) beta-Galactosidases of *Escherichia coli* with substitutions for Glu-461 can be activated by nucleophiles and can form beta-D-galactosyl adducts, *Carbohydr Res* **250,** 9-18.
20. Gebler, J.C., Aebersold, R., and Withers, S.G. (1992) Glu-537, not Glu-461 is the nucleophile in the active site of (*lac Z*) β-galactosidase from *Escherichia col, J. Biol. Chem* **267**, 11126-11130.
21. Ring, M. and Huber, R.E. (1990) Multiple replacements establish the importance of tyrosine-503 in beta- galactosidase (*Escherichia coli*), *Arch Biochem Biophys* **283,** 342-350.
22. Richard, J.P., Huber, R.E., Heo, C., Amyes, T.L., and Lin, S. (1996) Structure-reactivity relationships for beta-galactosidase (*Escherichia coli, lac Z*). 4. Mechanism for reaction of nucleophiles with the galactosyl-enzyme intermediates of E461G and E461Q beta-galactosidases, *Biochemistry* **35**, 12387-12401.
23. Roth, N.J. and Huber, R.E. (1996) The β-galactosidase (*Escherichia coli*) reaction is partly facilitated by interaction of His-540 with the C6 hydroxyl of galactose, *J. Biol. Chem.* **271,** 14296-14301.

24. James, T.L. and Oppenheimer, N.J. (1994)   Nuclear magnetic resonance. Part C. *Methods in Enzymology* **239**, 1-813.
25. Jencks, W.P. (1975) Binding energy, specificity, and enzymic catalysis: the circe effect. *Adv. Enzymol.* **43**, 219-410.
26. Lightstone, F.C. and Bruice, T.C. (1996) Ground state conformations and entropic and enthalpic factors in the efficiency of intramolecular and enzymatic reactions. 1. cyclic anhydride formation by substituted glutarates, succinate, and 3,6-endoxo-Δ4-tetrahydrophthalate monophenyl esters, *J. Am. Chem. Soc* **118**, 2595-2605.
27. Cannon, W.R., Singleton, S.F., and Benkovic, S.J. (1996) A perspective on biological catalysis, *Nature Struct. Biol.* **3**, 821-833.
28. Espinosa, J.F., Montero, E., Vian, A., García, J.L., Imberty, A., Dietrich, H.J., Schmidt, R.R., Martín-Lomas, M., Cañada, J., and Jiménez-Barbero, J. (1998) E. coli galactoside recognizes a high energy conformation of C-lactose, a non hydrolizable substrate analogue. NMR investigations of the molecular complex, *J. Am. Chem. Soc.*, **120**, 1309-1316.
29. Huber, R.E., Kurz, G., and Wallenfels, K. (1976) A quantitation of the factors which affect the hydrolase and transgalactosylase activities of beta-galactosidase (*E. coli*) on lactose, *Biochemistry* **15**, 1994-2001.
30. Adelhorst, K. and Bock, K. (1992) Derivatives of methyl β-lactoside as substrates for and inhibitors of β-D-galactosidase from *E. coli*, *Acta Chem. Scand.* **46**, 1114-1121.
31. Hadfield, A.T., *et al.* (1994) Crystal structure of the mutant D52S Hen Egg White lysozyme with an Oligosaccharide product, *J. Mol. Biol.* **243**, 856-872
32. Kuroki, R., Weaver, L.H., and Matthews, B.W. (1993) A covalent enzyme-substrate intermediate with saccharide distortion in a mutant T4 lysozyme, *Science* **262**, 2030-2033.
33. Tews, I., *et al.* (1996) Bacterial chitobiase structure provides insight into catalytic mechanism and the basis of Tay-Sachs disease, *Nature Struct. Biol.* **3**, 638-648.
34. Sulzenbacher, G., Driguez, H., Henrissat, B., Schülein, M., and Davies, G.J. (1996) Structure of the *Fusarium oxysporum* endoglucanasa I with a nonhydrolyzable substrate analogue: Substrate distortion gives rise to the preferred axial orientation for the leaving group, *Biochemistry* **35**, 15280-15287
35. Casset, F., Peters, T. Etzler, M., Korchagina, E., Nifant'ev, N., Perez, S., and Imberty, A. (1996) Conformational analysis in blood group A trisaccharide in solution and in the binding site of Dolichos biflorus lectin using transient and transferred NOE and rotating-frame NOE experiments, *Eur. J. Biochem.*, **239**, 710-719.
36. Siebert, H. C., Guilleron, M., Kaltner, H., Von der Lieth, C. W., Kozar, T., Bovin, N., Korchagina, E. Y, Vliegenthart, J. F. G., and Gabius, H. J. (1996) NMR-based, molecular dynamics and random walk molecular mechanics-supported study of confoemational aspects of a carbohydrate ligand (Gal beta 1-2Gal beta 1-R) for an animal galectin in the free and in the bound state, *Biochem. Biophys. Res. Comm.*, **219**, 205-212.

37. Weimar, T. and Peters, T. (1994) Aleuria aurantia agglutinin recognizes multiple conformations of Fuc-(1->6)-GlcNAc-OMe, *Angew. Chem. Int. Ed. Engl.*, **33**, 88-91.
38. Poppe, L., Brown, G.S., Philo, J.S., Nikrad, P.V., and Shah, B.H. (1997) Conformation of sLeX tetrasaccharide, free in solution and bound to E-, P-, and L-selectin, *J. Am. Chem. Soc.*, **119**, 1727-1736, and references therein.
39. Bundle, D.R., Baumann, H., Brisson, J. R., Gagne, S., Zdanov, A., and Cygler, M. (1994) Structure of a trisaccharide-antibody complex : comparison of NMR measurements with a crystal structure, *Biochemistry,* **33**, 5183-5192.
40. Casset, F., Imberty, A. , Perez, S. , Etzler, M., Paulsen, H. , Peters, T. (1997) Transfer NOEs and ROEs reflect the size and amino acid composition of the binding pocket of a lectin, *Eur. J. Biochem* **244**, 242-251.
41. Meyer, B., Weimar, T., and Peters, T. (1997) Screening mixtures for biological activity by NMR, *Eur. J. Biochem.*, **246**, 705-712.
42. Andersen, N.H., Cao, B., Rodríguez-Romero, A., and Arreguin, B. (1993) Hevein: NMR assignment and assessment of solution-state folding for the agglutinin-toxin motif, *Biochemistry.* **32**, 1407-1422.
43. Asensio, J.L., Cañada, F.J., Bruix, M., Rodriguez-Romero, A., and Jimenez-Barbero, J. (1995) The interaction of Hevein with N-acetylglucosamine-containing oligosaccharides. Solution structure of hevein complexed to chitobiose, *Eur. J. Biochem.* **230**, 621-633.
44. Asensio, J.L., Cañada, F.J., Bruix, M., Rodriguez-Romero, A., Gonzalez, C., Khiar, N., and Jimenez-Barbero, J. (1998) The interaction of Hevein with N-acetylglucosamine-containing oligosaccharides. Solution structure of hevein complexed to chitobiose, *Glycobiology.* **8**, 569-577.
45. (a) Wright, C. S. (1984) Structural comparison of the two distinct sugar binding sites in wheat germ agglutinin isolectin II, *J. Mol. Biol.* **178** , 91-104. (b) Wright, C. S. (1990) 2.2 Å resolution Structure Analysis of Two Refined N-acetyl-neuraminyl-lactose-wheat germ agglutinin isolectin complexes, *J. Mol. Biol.* **215**, 635-651. (c) Wright, C. S. (1992) Crystal structure of a wheat germ agglutinin/glycophorin-sialo-glycopeptide receptor complex. Structural basis for cooperative lectin-cell binding, *J. Biol. Chem.* **267,** 14345-14352.
46. Martins, J., Maes, D., Loris, R., Pepermans, H. A. M., Wyns, L., Willen, R., Verheyden, P. (1996) 1H NMR study of the solution structure of Ac-AMP-2, a sugar binding antimicrobial protein isolated from Amaranthus caudatus, *J. Mol. Biol.*, **258,** 322-330.
47. Rodriguez-Romero, A., Ravichandran, K. G., and Soriano-Garcia, M. (1991), Crystal structure of hevein at 2.8 Å resolution, *Febs Lett.* **291**, 307-309.

APPLICATION OF NMR TO CONFORMATIONAL STUDIES OF AN HIV PEPTIDE BOUND TO A NEUTRALIZING ANTIBODY

ANAT ZVI AND JACOB ANGLISTER
Weizmann Institute of Science
Rehovot 76100, Israel

Abstract

NMR spectroscopy has become an important tool for structural studies of proteins and macromolecular complexes. In case of an antibody-antigen complex, different approaches are taken to alleviate the size limit, which is beyond NMR capabilities. The method of choice depends on the kinetic properties of the system under investigation and on the antibody fragment available. We have recently studied the conformation of RP135, a 24-residue HIV-1 peptide corresponding to the principal neutralizing determinant of the virus envelop glycoprotein gp120, in complex with 0.5β, a neutralizing antibody raised against gp120. The binding of the peptide to the antibody was too strong to observe TRNOE, therefore 2D-NOESY difference spectroscopy was applied using three strategies: (a) deuteration of specific residues of the peptide; (b) Arg→Lys replacement and; (c) truncation of the peptide antigen. The restraints on interproton distances within the bound peptide were used to calculate its conformation. The peptide forms a 10-residue loop, while the two segments flanking this loop interact extensively with each other and possibly form antiparallel β-strands.

1. Introduction

The application of NMR spectroscopy to studies of high molecular complexes is of outstanding interest as it can provide not only structural information on different aspects of a receptor/ligand complex, but also insights into the dynamics and kinetics of the system. In immunochemistry, the conformation of the bound antigen, antibody-antigen interactions and the dynamics of the system may provide a detailed understanding of the molecular basis of antibody-antigen recognition. In spite of the recent advances in the field of protein NMR, an antibody molecule is still beyond the NMR capabilities, due to its large size (150 kDa). A significant simplification would be achieved if a smaller fragment of the antibody is used. Indeed, the Fv fragment (25 kDa), which contains the entire antibody binding site is much more suitable for NMR studies and its complex with a peptide can therefore be investigated by modern multidimensional NMR techniques. Unfortunately, the Fv is not readily available and it can be obtained by proteolysis only for a very limited number of proteins. Moreover, for multidimensional NMR purposes, uniform ^{13}C and ^{15}N enrichment of the protein is required [1]. Genetic engineering techniques enable the expression of the Fv fragment in *E. Coli.* [2; 3].

Considering that an expression system for the Fv fragment of an antibody is not always available, as it is neither trivial nor always a successful task, several approaches have been followed to overcome the size limit of the whole molecule. The Fab fragment of the antibody (50 kDa), which is readily obtained by proteolytic cleavage of the entire molecule, is amenable to NMR analysis, however it is still too large for a complete structure determination. Most studies of Fab-ligand complexes have focused on the conformation of the bound ligand and have been carried out using transferred NOE (TRNOE), which requires a fast ligand off-rate [4-7]. Anglister and coworkers developed the 2D TRNOE difference spectroscopy to study antibody/peptide interactions and intramolecular interactions within the bound peptide [8]. TRNOE spectroscopy detects these interactions through the observation of NOE cross peaks between the sharp resonances of the free peptide and either other free peptide protons or protein proton resonances, rather than observing these interactions directly through the broad resonances of the bound peptide. The method is applicable in cases when the exchange between the bound and the free peptide is fast relative to the inverse of the longitudinal relaxation time and the mixing period used in NOESY experiments. The combining site structures of three monoclonal antibodies raised against a peptide of cholera toxin were studied by transferred NOE difference spectroscopy [9; 10]. 2D TRNOE was also used to study the conformation of carbohydrate epitopes bound to antibodies [11]. Isotope-edited experiments using ^{13}C and ^{15}N labeling of the ligand allow studies of tightly bound ligands. These experiments were applied to determine the conformation of ligands in complexes smaller than 35 kDa [12-15]. For larger proteins, the signal-to-noise ratio in these experiments is poor due to shorter ^{13}C and ^{15}N T_2 relaxation times. Despite this problem, Wright and coworkers successfully applied isotope-edited experiments to study some features of the secondary structure of ^{15}N- and ^{13}C-labeled peptide bound to a Fab fragment [16; 17]. Still, the data obtained was not sufficient to calculate the structure of the ligand. In studies of pepsin interactions with a tri-peptide inhibitor it was demonstrated that difference spectroscopy using deuterated peptides is an alternative method that provides the same type of structural information as TRNOE or isotope-editing techniques do [18]. The method involves the subtraction of two NOESY spectra measured for two complexes prepared with either a protonated or a specifically deuterated ligand. The difference spectrum reveals only NOEs involving ligand protons that have been replaced by deuterium.

The third variable region (V3) of the envelope glycoprotein (gp120) contains the principal neutralizing determinant (PND) of the human immunodeficiency virus type-1 (HIV-1) [19; 20]. Todate, V3 is the only gp120 epitope found to be the target of numerous HIV-neutralizing antibodies [21]. Segments of V3 or its immediate vicinity form the binding site for the gp120 co-receptors on T-cells and macrophages [22; 23], and its sequence determines the phenotype of the virus [24]. Antibody binding to V3 or

V3 deletion prevents the CD4-dependent binding of the virus to macrophages and T-cells [25] and the subsequent fusion of the virus with the target cell, thus abolishing the infectivity of the virus [22; 23]. Escape of HIV-1 from the immune system has been found to be associated with the lack of specific anti-V3 antibodies *in-vivo* [26]. Neutralization of a variety of HIV-1 strains, including field isolates, has been shown with serum obtained after immunization with antigens containing V3 peptides of different strains [27; 28].

The structure of the virus glycoprotein gp120 is not known. Wilson and coworkers determined the crystal structure of two anti-peptide HIV neutralizing antibodies in complex with a synthetic peptide derived from the V3 loop of gp120 (HIV-1$_{MN}$ strain) [29; 30]. Although the immunogen used was a cyclic 40-residue peptide, in each of the two complexes only a sequential 7-residue epitope was found to interact with the antibody.

The monoclonal antibody 0.5β was raised against gp120 of the HIV-1$_{IIIB}$ strain purified from infected cells [31]. The 0.5β is a very potent HIV neutralizing antibody. Neutralization of free virus is obtained at 100 ng/ml concentration of 0.5β and full neutralization of infected cells as measured by inhibition of syncytia formation is achieved at antibody concentrations as low as 50 μg/ml. A chimeric monoclonal antibody designated cβ1, which contains the intact variable region of 0.5β and human constant regions, was shown to protect chimpanzees from HIV-1 infection after passive immunization [32]. Measurements of the binding constant of 0.5β to gp120 and to RP135, the peptide representing the principal neutralizing of the virus in the V3 loop of gp120, revealed similar affinities of 2×10^8 and 7×10^7 M^{-1}, respectively [33]. The dissociation rate of the 0.5β antibody/RP135 peptide complex is very slow (10^{-3} sec^{-1} at 37 °C) [34]. It has been found that a slow dissociation rate of HIV-neutralizing antibody complexes with gp120 is correlated with the efficiency of virus neutralization [35]. Here we describe the solution conformation of RP135 when complexed with the antibody 0.5β.Fab fragment. NOESY difference spectroscopy, together with deuterium labeling of peptide residues, Arg→Lys replacements and truncation of the peptide revealed structural information on the conformation of the bound antigen. The NOEs were translated into distance restraints, which were then used to calculate the conformation of the bound peptide.

2. Conformation of RP135 bound to the 0.5β antibody

2.1. NOESY DIFFERENCE SPECTRA WITH DEUTERATED PEPTIDES

In NMR studies, the information on the three dimensional structure of a molecule is obtained from NOESY spectra, which detect through-space magnetization transfer between protons that are less than 5 Å apart. However, the NOESY spectrum of a 50 kDa Fab/peptide complex is totally unresolved due to the overlap between the large number of protons and due to the broadening of the lines (Fig. 1A). We applied difference spectroscopy to extract the information on the folding of the peptide and its interactions with the antibody from the broad background contributed by the macromolecule. The subtraction of the NOESY spectrum of the Fab in complex with an unlabeled peptide from the NOESY spectrum of the same complex but with a peptide deuterated at a specific residue reveals the NOE cross peaks of the labeled residue, arising from intra-residue, intra-peptide and Fab/peptide interactions; all other resonances of the 52 kDa complex are canceled out by the subtraction. To discern between NOESY cross peaks arising from intra-residue interactions and NOESY cross peaks arising from inter-residue interactions, the 1D difference spectrum of the complex with the unlabeled peptide and the complex with the deuterated peptide is calculated. This spectrum reveals the typical spin system of the labeled residue. In our study of the PND peptide bound to the HIV neutralizing antibody, the peptide used, designated P1053, comprises the full epitope recognized by 0.5β (RKSIRIQRGPGRAFVTIG). Fig. 1B shows the aliphatic region of a difference spectrum between the NOESY of the complex with an unlabeled peptide and the NOESY of the complex with a peptide deuterated at Thr-19. This difference spectrum reveals intra-residue interactions within Thr-19 as well as inter-residue interactions between Thr-19 and protons of the peptide or the Fab. The intensities of the cross peaks assigned to interactions between Ser-6 and Thr-19 backbone and side-chain protons indicate that these two residues are in close contact. In a similar manner we used specifically deuterated peptides to study the interactions of Ser-6, Ile-7, Ile-9, Gln-10, Gly-12, Pro-13, Gly-14, Ala-16, Phe-17, Val-18, and Ile-20.

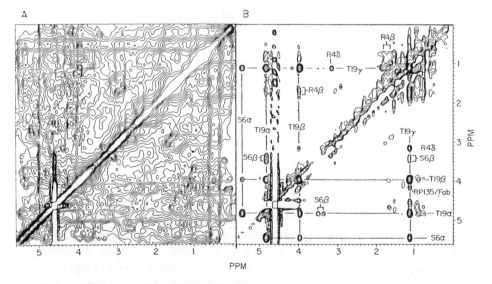

Figure 1. A comparison between the aliphatic region of (A) the NOESY spectrum of the 0.5β Fab complex with the peptide RP135a (RKSIRIQRGPGRAFVT) and (B) the NOESY difference spectrum of the 0.5β Fab complex with the peptide RP135a (RKSIRIQRGPGRAFVT) and the complex in which Thr-19 is deuterated. The difference spectrum shows the cross peaks of the deuterated Thr-19.

Fig. 2 shows the difference spectrum calculated from the NOESY spectrum of a complex with an unlabeled peptide and a NOESY spectrum of the complex with the same peptide deuterated at Ile-7, Ile-9, Gln-10, Gly-12, Pro-13, Gly-14, Ala-16, Phe-17, Val-18, and Ile-20. The assignment of the protons is based on the measurements with the specifically labeled residue. Interactions of Arg-4 and Arg-11, which were not deuterated, were observed in difference spectra using peptides in which the corresponding arginines were replaced by lysines (see below). In the aliphatic region of the difference spectrum (Fig. 2B), numerous hydrophobic long-range interactions between residues Arg-4, Lys-5, Ser-6, Ile-7 and Ile-9 at the N-terminus of the peptide and Phe-17, Val-18, Thr-19 and Ile-20 at the C-terminus of the peptide are observed with a remarkable signal-to-noise ratio. In addition to the intramolecular interactions of Phe-17 aromatic protons, the aromatic region of the difference spectrum (Fig. 2A) reveals several intermolecular NOEs between the aromatic protons of the Fab and Ile-7, Ile-9, Pro-13, Phe-17, and Ile-20 of the peptide.

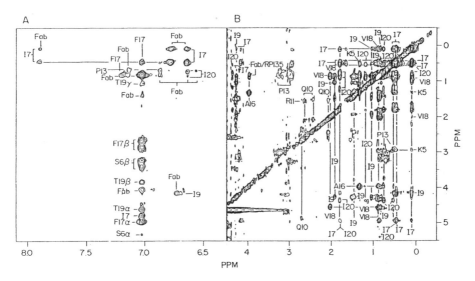

Figure 2. The difference between the NOESY spectrum of 0.5β Fab complex with the peptide RP135b (RKSIRIQRGPGRAFVTIG) and the spectrum of the Fab complex with a peptide molecule in which Ile-7, Ile-9, Gln-10, Gly-12, Pro-13, Gly-14, Ala-16, Phe-17, Val-18, and Ile-20 were deuterated. The assignments of the cross peaks are marked for the F_1 and the F_2 dimensions; (A) a section of the difference spectrum, showing the interactions of aromatic protons of 0.5β Fab and Phe-17 of the peptide with non aromatic protons of the peptide; (B) the aliphatic section of the difference spectrum, showing interactions of the peptide deuterated residues.

2.2. NOESY DIFFERENCE SPECTRA WITH ARG→LYS REPLACEMENTS

The peptide RP135 contains four arginine residues which were not deuterated for any of our experiments. To save the costs of using deuterated arginine, we introduced conservative replacements of these four residues. A similar approach was used in our studies of the anti-cholera-toxin-peptide antibody TE34, by transferred NOE difference spectroscopy [36]. To assess the influence of the replacements on the binding to the antibody, we measured the affinity of each of the five modified peptides (R4K, R8K, R8H, R11K and R15K) to the antibody by competitive ELISA binding measurements. While the binding of the peptides with R4K and R11K replacements is comparable to that of the unmodified peptide, the mutations R8K, R8H and R15K reduce the binding by more than two orders of magnitude (the mutation R8H was introduced to exclude the possibility that the reduction in binding of R8K was due to the β-breaking character of lysine in comparison with arginine. In this respect, histidine is a conservative mutation). In view of these results, only two of the five modified peptides were used in NOESY difference spectra measurements. The 1D difference spectra calculated for the R4K and for the R11K replacements contain almost solely the positive and negative resonances attributed to arginine and lysine, respectively (data not shown). However, the NOESY difference spectra contain not only cross peaks assigned to arginine (positive) and lysine (negative), but also some cross peaks which arise from local changes in the relative disposition of residues that occur both in the Fab and in the peptide as a result of the replacements. For example, the NOESY difference spectrum calculated for the R11K replacement shows also the resonances of Gln-10 (Fig. 3). A system of cross peaks between resonances at 2.37, 2.73 and 4.88 is typical of an AMX system and has not been assigned to any peptide protons. This system of cross peaks, that is accompanied by negative cross peaks of the Fab complex with the modified peptide, could be due to intra-residue cross peaks of the Fab. Examination of the part of the NOESY spectrum showing aromatic proton interactions reveals no cross peaks between the non-labile protons of Arg-4 and Arg-11 and the aromatic protons of the Fab.

2.3. NOESY DIFFERENCE SPECTRA WITH TRUNCATED PEPTIDES

We used truncated peptides to study the interactions of residues that were neither deuterated nor replaced (Lys-5, Ser-6, Arg-8 and Arg-15; Ser-6 was deuterated only at a latter stage). The difference between the spectrum of the Fab complex with the peptide **NNTR**KSIRIQRGPGRAFVTI and the spectrum of the Fab complex with the peptide KSIRIQRGPGRAFVTI is shown in Fig. 4. In addition to the cross-peaks of Thr-3 and Arg-4 that were truncated in the second peptide, this difference spectrum shows the cross peaks of Lys-5 which became the N-terminal residue, and the cross peaks of Ser-6, Thr-19 and Ile-20. Thr-3 was assigned on the basis of the HOHAHA spectrum of the 0.5β Fab/RP135 complex. (The $T_{1\rho}$ of residues that are outside the epitope, namely: Asn-1, Asn-2, Thr-3, Gly-21, Lys-22, Ile-23 and Gly-24, is considerably longer than that of the Fab protons and therefore they can be detected by the HOHAHA experiment with a good signal-to-noise ratio despite the molecular weight of the complex; [34]. Lys-5 was assigned on the basis of the typical cross-peak pattern of its spin system. The cross peaks of Ser-6 appear due to its proximity to the N-terminus of the truncated peptide. Its assignment was based on the typical spin system and was further verified by specific deuteration (data not shown). Thr-19 and Ile-20 were previously assigned (Fig. 1B and Fig. 2B, respectively). Strong interactions are again observed between: Arg-4 and Thr-19, Ser-6 and Thr-19, and between Lys-5 and Ile-20. The interactions of Thr-19 (excluding Thr-19/Arg-4 interaction) and the interactions of Ile-20 appear as both positive and negative cross peaks as both peptides contain these residues. The truncation of the four N-terminal residues of RP135 induces small but still observable changes in the chemical shifts of residues at the C-terminus of the determinant recognized by the antibody (Thr-19 and Ile-20) due to the long-range interactions between the N- and the C-termini of the epitope. Table 1 summarizes the intramolecular interactions observed in the bound peptide.

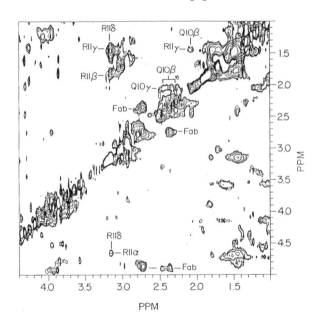

Figure 3. The difference between the NOESY spectrum of 0.5β Fab complex with the peptide RP135a (RKSIRIQRGPGRAFVT) and the spectrum of the Fab complex with a peptide molecule in which Arg-11 was replaced by Lys. Only positive cross peaks arising from the Fab complex with the unmodified peptide are shown.

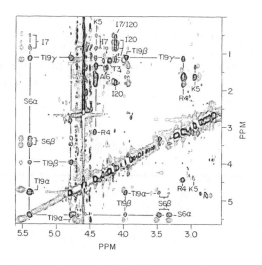

Figure 4. A section of the difference between the NOESY spectrum of 0.5β Fab complex with the peptide NNTRKSIRIQRGPGRAFVTI (the residues missing in the second peptide are in bold characters) and the complex with the peptide KSIRIQRGPGRAFVTI. Solid line: positive phase of the spectrum showing cross peaks arising from the first complex; dotted line: negative phase of the spectrum, contributed by the second complex.

2.4. STRUCTURE CALCULATION OF THE BOUND PEPTIDE

The observed NOE cross peak intensities were translated into distance constraints, and an overall of 122 constraints, of them 53 medium and long-range restraints were derived from the experiments. Distance geometry/simulated annealing calculations [37] using the X-PLOR program [38], revealed 12 refined lowest-energy structures with no NOE violations greater than 0.5 Å. The rms deviation between the individual structures and the average structure is 1.1 Å for backbone atoms of residues 5-9 and 17-20 and 1.6 Å for all 17 residues. The superposition (optimized for residues 4-9 and 17-20) of the 12 structures on the energy-minimized average structure is shown in Fig. 5. The peptide forms a 10-residue loop (residues 8-17), while the two segments flanking this loop (residues 5-7 and 18-20) interact extensively with each other (segments 5-7 and 18-20). Most of the residues in these strands are in an extended conformation and possibly form anti-parallel β-strands, thus defining, together with the loop, a β-hairpin. A very strong interaction between $C_\alpha H$ of Ser-6 and $C_\alpha H$ of Thr-19 ($d_{\alpha,\alpha}$ (i,j)) is indeed indicative of anti-parallel β-strands. Lys-5, Ile-7, Ile-9, Val-18 and Ile-20 side chains point to one side of the β-structure while the side chains of Arg-4, Ser-6, Phe-17 and Thr-19 are on the other side (Fig. 6). This is concluded from the strong interactions that appear between alternating residues: Lys-5 and Ile-7 side chains interact with Ile-20, and the side chains of both Ile-7 and Ile-9 interact with the side chain of Val-18, forming a mainly hydrophobic surface. Arg-4 interacts with Thr-19, and Ser-6 interacts with Phe-17 and Thr-19. The side chains of these four residues form a mostly polar surface.

Another evidence supporting the proposed antiparallel β-sheet structure is the deviations from random coil values of the $C_\alpha H$ chemical shifts. The $C_\alpha H$ chemical shift of residues in a β conformation is expected to experience downfield deviation in comparison to their random coil values [39]. Positive values are indeed observed for residues SI and FVTI. Due to lack of constraints, the conformation of the 7-residue segment, QRGPGRA, is poorly defined.

Table 1: Inter-residue interactions within the bound peptide

Arg-4	Thr-19
Lys-5	Ile-7, Ile-20
Ser-6	Ile-7, Phe-17, Thr-19
Ile-7	Lys-5, Ser-6, Ile-9, Phe-17, Val-18, Ile-20
Ile-9	Ile-7, Val-18
Gln-10	Arg-11
Arg-11	Gln-10
Phe-17	Ser-6, Ile-7, Thr-19
Val-18	Ile-7, Ile-9
Thr-19	Arg-4, Ser-6, Phe-17
Ile-20	Lys-5, Ile7

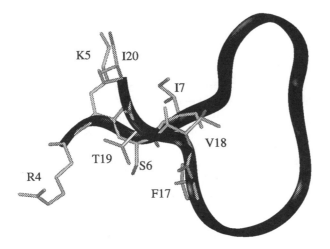

Figure 5. The mean structure of the 12 best-fit lowest energy calculated models for the peptide conformation when bound to 0.5β Fab. The mean structure was subjected to a short run of energy minimization; the peptide backbone is shown in a ribbon style; side-chains of interacting residues are shown in sticks. The figure was generated using the program Insight II (Biosym/MSI)

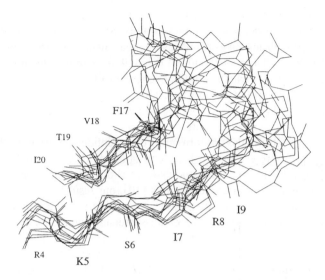

Figure 6. Superposition of the 12 lowest energy calculated models for the peptide conformation when bound to 0.5β Fab. Only backbone atoms are shown. The fit was optimized for the backbone of residues 4-9 and 17-20. The structures were superimposed on the mean structure, which was first subjected to a short run of energy minimization.

3. Discussion

NMR studies offer structural information that is not only site-specific, but also obtained at near physiological conditions and not subjected to crystal packing forces. Despite the large size of the antibody Fab/peptide complex, sufficient information regarding distances between interacting protons was obtained from the NOESY difference spectra to allow the calculation of the structure of the bound peptide. The method presented alleviate the need of labeling the ligand and/or the antibody by ^{13}C and ^{15}N. Also, the method uses the sensitivity of the ^{1}H NOESY spectrum and does not rely on coherence transfer as the isotope edited experiments using ^{13}C or ^{15}N labeled ligands do. The coherence transfer steps in the isotope-edited measurements reduce considerably the signal-to-noise ratio in spectra of large proteins (> 40 kDa) due to the decreasing transverse relaxation times. However, for a high resolution structure and better definition of the β-structure, information on the amide protons is needed to study backbone interactions.

The difference spectra measured using deuterated residues are the simplest and easiest to interpret. Difference spectra using truncated or modified peptides are usually more complicated. However, these last two types of spectra contain mainly the contribution of the truncated or modified residues if the modification is only minor and does not alter the binding considerably, as in the difference spectrum measured using the peptides NNTRKSIRIQRGPGRAFVTI and the peptide KSIRIQRGPGRAFVTI.

Arg-4, which is at the N-terminus of the epitope, is truncated. This modification does not cause any significant change in the binding constant of 0.5β to the peptide antigen. However, the difference spectrum calculated using RP135 and the peptide RKSIRIQRGPGRAFVT (data not shown) resulted in a complicated difference spectrum as the truncation of Ile-20 which is at the C-terminus of the epitope caused a two orders of magnitude decrease in the binding constant. Since Ile-20 interacts with Lys-5 and Ile-7 at the N-terminal part of the peptide, upon its truncation these residues experienced changes in their proton chemical shifts.

In our present study, the conformation of a peptide representing the PND of the virus envelope glycoprotein gp120 has been studied when bound to its neutralizing antibody 0.5β. Although previous studies have shown that the free peptide is in fast equilibrium between different conformations [40], our results indicate a highly ordered secondary structure of the bound peptide. Considering the extensive network of hydrogen bonds and long-range sidechain-sidechain interactions between the two strands, it is very unlikely that such a conformation would be induced by the antibody upon gp120 binding or that this part of the V3 loop would be flexible. The β-strand conformation suggested for the two regions flanking the GPGR loop was also predicted by LaRosa et al. [41] who have studied structural elements in the PND of different HIV strains by means of a neural network. In view of the fact that LaRosa's prediction is based on homology between proteins with known structure, it most likely reflects the conformation of V3 within gp120 better than studies of the free peptides representing V3. Of the 245 sequences examined by LaRosa et al., [41], 93% were predicted to have a β strand around position 6-9 and 49% show a β strand near position 18. LaRosa prediction and our results may indicate that the PND conformation that we found is common to most HIV-1 strains

The loop conformation of the PND peptide was observed in neither of the two complexes of HIV-neutralizing anti-peptide antibodies studied by Wilson and his coworkers [29; 30]. The PND peptide of the HIV-1$_{MN}$ strain was found to adopt a double-turn conformation in the GPGRAF sequence when it is bound to an HIV-neutralizing anti-peptide antibody [29]. The 8-residue antigenic determinant, KRIHIGPG (residues 312-321 in gp120), recognized by the 50.1 and the 9-residue antigenic determinant, HIGPGRAFYT (residues 315-326 in gp120), recognized by 59.1, studied by Wilson and his co-workers [29; 30] were too short to allow the detection of long range interactions in the PND. In the case of 0.5β, the epitope consists of 17 residues, RKSIRIQRGPGRAFVTI (residues 311-327 in gp120), thus enabling observations of long-range interactions in the antigenic determinant. It is expected that the conformation of a flexible PND peptide bound to an antibody raised against gp120 is similar to the conformation of the corresponding part of gp120 when bound to the same antibody.

Short linear peptides representing antigenic fragments of proteins of interest may represent potential candidates for effective synthetic vaccines. In the case of HIV, the structure of V3 peptide is useful in determining the optimal length of peptide immunogens to be used in potential peptide-based vaccine and in engineering the peptide conformation for better cross reactivity with the antibody. Our studies have shown that the β-structure of the peptide is composed of a hydrophobic face (residues Ile-7, Ile-9, Val-18 and Ile-20) and a mostly polar face (made of residues Arg-4, Ser-6, Phe-17, Thr-19, and possibly Arg-8). Ile-9 and Val-18 side-chains have extensive interactions within the peptide but do not interact with the antibody, implying that these residues are probably buried in gp120. On the other side, residues that interact with the Fab are probably exposed to the solvent. Of the exposed residues, those which are conserved among different strains may have an important role in gp120 activity (may interact with the coreceptors found on the target cells or be involved in virus fusion). These residues should be conserved in a designed vaccine based on a "cocktail" of V3 peptides from different strains. Our results show that Ile-7, Ile-9, Pro-13, Phe-17 and Ile-20 interact with protons of the Fab (Fig. 2A). Ile-7, Ile-9, and Pro-13 are conserved in 94%, 82% and 95% of gp120 sequences, respectively [41]. Considering the complexity of HIV neutralization, it is likely that an efficient HIV vaccine will consist of a combination of several subunits and that V3 peptides will be vital components in such a future multicomponent HIV-1 vaccine.

4. Acknowledgments

We are greatly indebted to Dr. Shuzo Matsushita for the 54'CB1 hybridoma cell line producing the 0.5β monoclonal antibody, to Mr. Vitali Tugarinov for his help with XPLOR, to Prof. Michael Levitt for most fruitful discussions, to Prof. Mati Fridkin and Mrs. Aviva Kapitkovsky for peptide synthesis, to Mr. Yehezkiel Hayek for peptide purification and to Mrs. Rina Levy for excellent technical assistance. This work was supported by a grant from the Minerva foundation.

5. References

1. McIntosh, L. P., and Dahlquist, F. W. (1990) Biosynthetic incorporation of ^{15}N and ^{13}C for assignment and interpretation of nuclear magnetic resonance spectra of proteins, *Q Rev Biophys* **23**, 1-38.

2. Skerra, A., and Plückthun, A. (1988) Assembly of a functional immunoglobulin Fv fragment in Escherichia coli, *Science* **240**, 1038-41.

3. Ward, E. S., Gussow, D., Griffiths, A. D., Jones, P. T., and Winter, G. (1989) Binding activities of a repertoire of single immunoglobulin variable domains secreted from Escherichia coli [see comments], *Nature* **341**, 544-6.

4. Behling, R. W., Yamane, T., Navon, G., and Jelinski, L. W. (1988) Conformation of acetylcholine bound to the nicotinic acetylcholine receptor, *Proc Natl Acad Sci U S A* **85**, 6721-5.

5. Clore, G. M., Gronenborn, A. M., Carlson, G., and Meyer, E. F. (1986) Stereochemistry of binding of the tetrapeptide acetyl-Pro-Ala-Pro-Tyr-NH2 to porcine pancreatic elastase. Combined use of two-dimensional

transferred nuclear Overhauser enhancement measurements, restrained molecular dynamics, X-ray crystallography and molecular modelling, *J Mol Biol* **190**, 259-67.

6. Ni, F., Ripoll, D. R., Martin, P. D., and Edwards, B. F. (1992) Solution structure of a platelet receptor peptide bound to bovine alpha-thrombin, *Biochemistry* **31**, 11551-7.

7. Sykes, B. D. (1993) Determination of the conformations of bound peptides using NMR-transferred nOe techniques, *Curr Opin Biotechnol* **4**, 392-6.

8. Levy, R., Assulin, O., Scherf, T., Levitt, M., and Anglister, J. (1989) Probing antibody diversity by 2D NMR: comparison of amino acid sequences, predicted structures, and observed antibody-antigen interactions in complexes of two antipeptide antibodies, *Biochemistry* **28**, 7168-75.

9. Scherf, T., Hiller, R., Naider, F., Levitt, M., and Anglister, J. (1992) Induced peptide conformations in different antibody complexes: molecular modeling of the three-dimensional structure of peptide-antibody complexes using NMR-derived distance restraints, *Biochemistry* **31**, 6884-97.

10. Zilber, B., Scherf, T., Levitt, M., and Anglister, J. (1990) NMR-derived model for a peptide-antibody complex, *Biochemistry* **29**, 10032-41.

11. Glaudemans, C. P., Lerner, L., Daves, G. D., Jr., Kovac, P., Venable, R., and Bax, A. (1990) Significant conformational changes in an antigenic carbohydrate epitope upon binding to a monoclonal antibody, *Biochemistry* **29**, 10906-11.

12. Bax, A., Grzesiek, S., Gronenborn, A. M., and Clore, M. G. (1994) Isotope-filtered 2D HOHAHA spectroscopy of a peptide-protein complex using heteronuclear Hartmann-Hahn dephasing, *J Magn Res* **106**, 269-73.

13. Fesik, S. W., Luly, J. R., Erickson, J. W., and Abad Zapatero, C. (1988) Isotope-edited proton NMR study on the structure of a pepsin/inhibitor complex, *Biochemistry* **27**, 8297-301.

14. Ikura, M., and Bax, A. (1992) Isotope-filtered 2D NMR of a protein-peptide complex: study of a skeletal muscle myosin light chain kinase fragment bound to calmodulin, *J Am Chem Soc* **114**, 2433-40.

15. Weber, C., Wider, G., von Freyberg, B., Traber, R., Braun, W., Widmer, H., and Wüthrich, K. (1991) The NMR structure of cyclosporin A bound to cyclophilin in aqueous solution, *Biochemistry* **30**, 6563-74.

16. Derrick, J. P., Lian, L. Y., Roberts, G. C., and Shaw, W. V. (1992) Analysis of the binding of 1,3-diacetylchloramphenicol to chloramphenicol acetyltransferase by isotope-edited ^{1}H NMR and site-directed mutagenesis, *Biochemistry* **31**, 8191-5.

17. Tsang, P., Rance, M., Fieser, T. M., Ostresh, J. M., Houghten, R. A., Lerner, R. A., and Wright, P. E. (1992) Conformation and dynamics of an Fab'-bound peptide by isotope-edited NMR spectroscopy, *Biochemistry* **31**, 3862-71.

18. Fesik, S. W., and Zuiderweg, E. R. P. (1989) An approach for studying the active site of enzyme/inhibitor complexes using deuterated ligands ánd 2D NOE difference spectroscopy, *J Am Chem Soc* **111**, 5013-15.

19. Rusche, J. R., Javaherian, K., McDanal, C., Petro, J., Lynn, D. L., Grimaila, R., Langlois, A., Gallo, R. C., Arthur, L. O., Fischinger, P. J., and et al. (1988) Antibodies that inhibit fusion of human immunodeficiency virus-infected cells bind a 24-amino acid sequence of the viral envelope, gp120 [published errata appear in Proc Natl Acad Sci U S A 1988 Nov;85(22):8697 and Proc Natl Acad Sci U S A 1989 Mar;86(5):1667], *Proc Natl Acad Sci U S A* **85**, 3198-202.

20. Palker, T. J., Clark, M. E., Langlois, A. J., Matthews, T. J., Weinhold, K. J., Randall, R. R., Bolognesi, D. P., and Haynes, B. F. (1988) Type-specific neutralization of the human immunodeficiency virus with antibodies to env-encoded synthetic peptides, *Proc Natl Acad Sci U S A* **85**, 1932-6.

21. Moore, J., and Trkola, A. (1997) HIV type 1 coreceptors, neutralization serotypes, and vaccine development, *AIDS Res Hum Retroviruses* **13**, 733-6.

22. Trkola, A., Dragic, T., Arthos, J., Binley, J. M., Olson, W. C., Allaway, G. P., Cheng Mayer, C., Robinson, J., Maddon, P. J., and Moore, J. P. (1996) CD4-dependent, antibody-sensitive interactions between HIV-1 and its co-receptor CCR-5 [see comments], *Nature* **384**, 184-7.

23. Wu, L., Gerard, N. P., Wyatt, R., Choe, H., Parolin, C., Ruffing, N., Borsetti, A., Cardoso, A. A., Desjardin, E., Newman, W., Gerard, C., and Sodroski, J. (1996) CD4-induced interaction of primary HIV-1 gp120 glycoproteins with the chemokine receptor CCR-5 [see comments], *Nature* **384**, 179-83.

24. Hwang, S. S., Boyle, T. J., Lyerly, H. K., and Cullen, B. R. (1992) Identification of envelope V3 loop as the major determinant of CD4 neutralization sensitivity of HIV-1, *Science* **257**, 535-7.

25. Valenzuela, A., Blanco, J., Krust, B., Franco, R., and Hovanessian, A. G. (1997) Neutralizing antibodies against the V3 loop of human immunodeficiency, *J Virol* **71**, 8289-98.

26. Schreiber, M., Muller, H., Wachsmuth, C., Laue, T., Hufert, F. T., Van Laer, M. D., and Schmitz, H. (1997) Escape of HIV-1 is associated with lack of V3 domain-specific, *Clin Exp Immunol* **107**, 15-20

27. Honda, M., Matsuo, K., Nakasone, T., Okamoto, Y., Yoshizaki, H., Kitamura, K., Sugiura, W., Watanabe, K., Fukushima, Y., Haga, S., and et al. (1995) Protective immune responses induced by secretion of a chimeric soluble, *Proc Natl Acad Sci U S A* **92**, 10693-7.

28. Hamajima, K., Bukawa, H., Fukushima, J., Kawamoto, S., Kaneko, T., Sekigawa, K., Tanaka, S., Tsukuda, M., and Okuda, K. (1995) A macromolecular multicomponent peptide vaccine prepared using the glutaraldehyde conjugation method with strong immunogenicity for HIV-1, *Clin Immunol Immunopathol* **77**, 374-9.

29. Ghiara, J. B., Stura, E. A., Stanfield, R. L., Profy, A. T., and Wilson, I. A. (1994) Crystal structure of the principal neutralization site of HIV-1, *Science* **264**, 82-5.

30. Rini, J. M., Stanfield, R. L., Stura, E. A., Salinas, P. A., Profy, A. T., and Wilson, I. A. (1993) Crystal structure of a human immunodeficiency virus type 1 neutralizing antibody, 50.1, in complex with its V3 loop peptide antigen, *Proc Natl Acad Sci U S A* **90**, 6325-9.

31. Matsushita, S., Robert Guroff, M., Rusche, J., Koito, A., Hattori, T., Hoshino, H., Javaherian, K., Takatsuki, K., and Putney, S. (1988) Characterization of a human immunodeficiency virus neutralizing monoclonal antibody and mapping of the neutralizing epitope, *J Virol* **62**, 2107-14.

32. Emini, E. A., Schleif, W. A., Nunberg, J. H., Conley, A. J., Eda, Y., Tokiyoshi, S., Putney, S. D., Matsushita, S., Cobb, K. E., Jett, C. M., and et al. (1992) Prevention of HIV-1 infection in chimpanzees by gp120 V3 domain-specific monoclonal antibody, *Nature* **355**, 728-30.

33. Skinner, M. A., Ting, R., Langlois, A. J., Weinhold, K. J., Lyerly, H. K., Javaherian, K., and Matthews, T. J. (1988) Characteristics of a neutralizing monoclonal antibody to the HIV envelope glycoprotein, *AIDS Res Hum Retroviruses* **4**, 187-97.

34. Zvi, A., Kustanovich, I., Feigelson, D., Levy, R., Eisenstein, M., Matsushita, S., Richalet Secordel, P., Regenmortel, M. H., and Anglister, J. (1995) NMR mapping of the antigenic determinant recognized by an anti-gp120, human immunodeficiency virus neutralizing antibody, *Eur J Biochem* **229**, 178-87.

35. VanCott, T. C., Bethke, F. R., Polonis, V. R., Gorny, M. K., Zolla Pazner, S., Redfield, R. R., and Birx, D. L. (1994) Dissociation rate of antibody-gp120 binding interactions is predictive of V3-mediated neutralization of HIV-1, *J Immunol* **153**, 449-59.

36. Anglister, J., and Zilber, B. (1990) Antibodies against a peptide of cholera toxin differing in cross-reactivity with the toxin differ in their specific interactions with the peptide as observed by ^{1}H NMR spectroscopy, *Biochemistry* **29**, 921-8.

37. Nilges, M., Clore, G. M., and Gronenborn, A. M. (1988) Determination of three-dimensional structures of proteins from, *FEBS Lett* **239**, 129-36.

38. Brunger, A. T. (1992). XPLOR 3.1 Manual, Y. U. Press, ed. (New Haven, CT.

39. Williamson, M. P. (1990) Secondary-structure dependent chemical shifts in proteins, *Biopolymers* **29**, 1423-31.

40. Zvi, A., Hiller, R., and Anglister, J. (1992) Solution conformation of a peptide corresponding to the principal neutralizing determinant of HIV-1IIIB: a two-dimensional NMR study, *Biochemistry* **31**, 6972-9.

41. LaRosa, G. J., Davide, J. P., Weinhold, K., Waterbury, J. A., Profy, A. T., Lewis, J. A., Langlois, A. J., Dreesman, G. R., Boswell, R. N., Shadduck, P., and et al. (1990) Conserved sequence and structural elements

in the HIV-1 principal neutralizing determinant [published erratum appears in Science 1991 Feb 15;251(4995):811], *Science* **249**, 932-5.

PARAMAGNETIC NMR EFFECTS OF LANTHANIDE IONS AS STRUCTURAL REPORTERS OF SUPRAMOLECULAR COMPLEXES

CARLOS F.G.C. GERALDES
*Department of Biochemistry, Faculty of Science and Technology, and Center of Neurosciences, University of Coimbra,
P.O. Box 3126, 3000 Coimbra, Portugal*

Abstract

The presence of paramagnetic metal centers causes enhanced shift and relaxation effects on the resonances of NMR-active nuclei of ligand molecules. In this work we focus on the uses of such paramagnetic NMR effects of the lanthanide(III) ions as structural reporters of supramolecular complexes, with a particular emphasis on the aspects relevant to the development of powerful NMR probes for biomedical applications, including design of shift agents for *in vivo* MRS of alkali ions (eg. $^{23}Na^+$, $^7Li^+$) and of Gd^{III} based MRI contrast agents.

Certain paramagnetic Ln^{III} complexes have useful shift reagent capabilities. The origins of these lanthanide induced shift (LIS) effects are briefly reviewed. The contact and dipolar shift mechanisms are described as well as the methods for the separation of those contributions to the paramagnetic shift. The use of LIS values and lanthanide induced relaxation (LIR) enhancements to determine the solution structure of such lanthanide chelates, and their binding to other species will be discussed. Examples will include inner and outer sphere binding to small inorganic and organic cations and anions and interaction with cyclodextrins. The relation of the efficiency of the shifting ability of the chelate with the binding site(s) localization and strenght will be pointed out. The structural factors affecting the formation and geometry of the supramolecular arquitectures emerging from these model studies, are further exploited in 2D-NMR investigations of the interaction of relaxation agents with protein surfaces in aqueous solution. This is a convenient method to evaluate their binding specificities, with potential lessons to help design tissue targeting approaches leading to improved performance of later generation contrast agents.

Optimization of relaxivity is a major goal in MRI contrast agents research. Several examples of the use of supramolecular structures to manipulate inner-sphere and outer-sphere relaxation parameters towars this goal will be described, including Gd^{III} chelates interacting with small molecules, micellar systems or proteins such as HSA, present in blood plasma.

1. Introduction.

Paramagnetic species have been extensively used in NMR spectroscopy [1-4]. Interest in the paramagnetic properties of lanthanide cations (Ln^{3+}) began through the use of Eu^{III} chelates as paramagnetic shift reagents (SR's) for adducts formed in organic solvents. This led to extensive studies on the use of aqueous lanthanide ions and their chelates as MR shift and relaxation probes in biological systems, including structural studies of nucleotides, aminoacids and proteins in solution [5-7]. The introduction of high magnetic fields and multidimensional techniques in NMR spectroscopy allowed the development and extensive use of quantitative structural methods for large biomolecules, based mainly in interproton distance constraints derived from NOE effects [8]. Thus, structural applications of Ln^{III} NMR effects have become much more restricted, the most popular use being as chiral shift reagents for the NMR separation of enantiomers [9]. Other important recent applications of lanthanide complexes are in biomedical Magnetic Resonance Spectroscopy (MRS), as shift reagents for the separation of NMR signals from intra- and extracellular alkali metal ions [10]. The induced shift of extracellular metal ions results from binding of a given cation in the second coordination sphere of a negatively charged Ln^{III}-complex, which cannot permeate cell membranes. More effective and more selective SR's for these applications are being developed.

With the rapid development of Magnetic Resonance Imaging (MRI), there has been an enormous growth in studies of Ln^{III} complexes for application as contrast agents (CA's) [11-14]. These agents can improve image contrast by enhancing the relaxation rates ($1/T_1$ and $1/T_2$) of protons of tissue water, which contribute to the MRI signal intensity. This allows mapping regions in tissue with different accessibility to these reagents, resulting from varying barrier permeabilities, flow or perfusion, allowing an easier distinction of normal from deseased tissue. Specific targetting of such agents to certain tissues in normal or pathological situations is also of great importance.

2. Lanthanide Ions as NMR Structural Probes of Supramolecular Complexes.

The trivalent lanthanide ions, which constitute a family with useful electronic and magnetic properties and fairly similar coordination chemistry, make up a set of potentially powerful extrinsic probes [6,7]. However, the various lanthanide complexes with the same ligand are usually not completely isostructural, as the chemical behaviour of the lanthanides is determined by four physical properties: ion size, charge, coordination number and kinetics of ligand and water exchange. Their radii exhibit the well known lanthanide ion contraction, from about 1.06 Å for La^{3+} to 0.85 Å for Lu^{3+}, with consequent variation along the series of the hydration numbers of their aqueous ions and of the coordination numbers of their complexes (most commonly between 9 and 8), as well as of their thermodynamic and kinetic properties. The first protein NMR structural study based on proton LIS and LIR methods was on lysozyme [15], where the experimental data were compared with those calculated on the basis of the protein crystal structure [16], assuming axial symmetry for the LIS. The agreement for many resonances was found to be excellent, indicating general accord between the crystal structure of lysozyme and the solution structure as determined by the LIS method. However, the assumption of axial symmetry of the LIS values was later found not to be generally correct [17], and therefore the possibility of obtaining a more detailed structure of the protein from the paramagnetic data independently of the crystal structure was not warranted.

The replacement of a natural occuring cation by the probe in a biological macromolecule (eg. a Ca^{2+}-binding protein) should be not only isomorphous but also functional. The fact that the ionic radii of the lanthanides are very similar to the ionic radius of Ca^{2+} (0.99 Å), with similar coordination numbers and preferences for ionic interactions, has led to many studies in which the former have been used as isomorphic replacements for Ca^{2+}. However only in few cases, like thermolysin, perfect isomorphous replacement has actually been shown to occur. Lanthanide ions have been demonstrated to be good kinetic substitutes for Ca^{2+} in α-amylase, but act as competitive inhibitors for the Ca^{2+} enzymes staphylococal nuclease and concanavalin A [7]. In this respect, besides structural similarities, kinetic and equilibrium complexation similarities are also very important. The extra charge could cause them to distort the metal ion coordination polyhedron enough to destroy the catalytic ability of an enzyme. Lanthanide ions also have lifetimes of water molecules in the coordination sphere an order of magnitude longer than Ca^{2+}. A few Ca^{2+}-binding proteins have been been studied using the LIS method, but the most extensive study so far was of parvalbumin [18,19]. This protein, of molecular mass around 11 KDa and known X-ray crystal structure, binds two equivalents of Ca^{2+} in two distinct binding domains called the "CD and EF hands". Proton NMR studies of the apoprotein in the presence of increasing amounts of Yb^{3+} showed sequential loading of the EF and CD sites by Yb^{3+}, as shown by large shifts of proton resonances in a slow-exchange regime with the diamagnetic positions. The proton paramagnetic shifts and broadenings induced by Yb^{3+} binding on several of the shifted signals were assigned on the basis of the known X-ray structure of the protein, after locating the direction and the magnitude of the Yb^{3+} susceptibility tensors using the X-ray coordinates for five nuclei which had been assigned with certainty. With these parameters, LIS values for all other nuclei in the protein could be predicted. It was noted that the calculated LIS values for protons close to the EF binding site (5-10 Å) were generally larger than the observed shifts, suggesting that the solution structure in this region was less compact than that predicted from the crystal structure. However, a full structural analysis of the protein on the basis of the lanthanide effects was not possible.

However, structural studies of small rigid paramagnetic Ln^{III} chelates in solution has had much more success. Upon binding of any paramagnetic lanthanide cation except Gd^{3+}, each of the ligand nuclei experiences a hyperfine shift, which may be expressed as:

$$LIS_t = LIS_c + LIS_{pc} \qquad (1)$$

Here LIS_{pc}, the dipolar or pseudo-contact shift, which is caused by a local dipolar magnetic field transmitted through space from the cation anisotropic magnetic moment, is given by the dipolar equation, which is quite simple in the case of axial symmetry:

$$LIS_{pc} = (A^2_0 C_j) G = (A^2_0 C_j)(3\cos^2\theta - 1/r^3) \qquad (2)$$

where C_j is Bleaney's constant, characteristic of the Ln(III) ion [20], A^2_0 a crystal field parameter characteristic of the Ln(III) chelate, θ is the angle between the main axis of symmetry of the susceptibility tensor and the metal-nucleus vector and r is the length of that vector. LIS_c, the contact shift due to a local magnetic field at the nucleus transmitted through the chemical bonds from the cation magnetic moment, is given by:

$$LIS_c = F <S_z> \qquad (3)$$

where F is the hyperfine coupling constant and $<S_z>$ is also characteristic of the lanthanide [21]. Therefore, assuming axial symmetry one has:

$$LIS_t = (A^2_0 C_j) G + F <S_z> \qquad (4)$$

where G, the geometrical term in eq. 2, contains structurally useful geometrical information. Therefore, for nuclei whose hyperfine shifts contain a contact contribution, this must be separated out in order to be able to use the dipolar term in structural analysis. The most general separation method [22], under the assumptions of axial symmetry of the dipolar shifts and constancy of the A^2_0 parameter along the Ln(III) series, relies upon the use of both theoretically calculated C_j and $<S_z>$ values to separate those components. This equation may be rearranged into two linear forms:

$$LIS_t/<S_z> = G\,(A^2_0 C_j / <S_z>) + F \qquad (5a)$$
$$LIS_t/C_j = G\,A^2_0 + F\,(<S_z>/C_j) \qquad (5b)$$

The first equation should be used in linear regression analysis when LIS_t is dominated by dipolar shifts (calculated G/F>>1) and the second when LIS_t is dominated by contact effects (calculated G/F<<1).

Gd^{3+} differs from the other paramagnetic lanthanides in that it has 7 unpaired electrons symmetrically distributed in its 4f orbitals and hence cannot induce a pseudocontact shift upon binding. Instead, it produces a selective line-broadening or relaxation, LIR, of each resonance that is inversely proportional to the sixth power of the metal-nucleus distance [23]:

$$LIR = 1/T_{2p} = C/r^6 \qquad (6)$$

The early work on the use of LIS and LIR data from aqueous Ln(III) ions or complexes as structural probes of small flexible biological molecules, such as nucleotides and peptides, demonstrated that the combination of lanthanide induced pseudocontact shift (LIS) data, yielding angular and distance information, plus Gd^{3+} relaxation (LIR) data which provides distance information, allows the determination of the solution structure of any molecule that has a well defined lanthanide ion binding site, as long as the averaging effect of the conformational equilibrium is taken into account by simultaneous use of independent conformational information from other NMR parameters, such as vicinal coupling constants and NOEs [5,23]. More recently, this lanthanide-based solution structural method has been applied extensively to study the solution structures of fairly rigid linear and macrocyclic chelates useful as MRS shift reagents or MRI contrast agents (Figs.1,7 and 10). [23]. Separation of contact contributions to the observed shifts is necessary [5, 22] and a more general procedure to calculate geometric factors from the LIS data has been proposed [24,25]. We will discuss briefly some of these studies, referring to the literature a more complete overview [23].

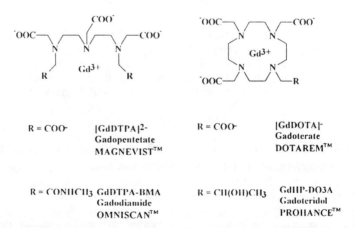

Fig. 1. Chemical formulae and trade names of the four extracellular MRI contrast agents currently available for use in humans.

The structure and internal conformational equilibria of the Ln^{III} complexes with DTPA [26] and DTPA-PA$_2$ (DTPA-PA$_2$ = DTPA-bis(propylamide)) [27] have been defined in detail. The dipolar Nd^{III} induced relaxation rate enhancements of ^{13}C ligand nuclei, gave the metal-C distances in the complexes, showing that the metal is in both cases coordinated by the ligand in an octadentate fashion, with one water also present in the first coordination sphere, as shown by ^{17}O NMR LIS measurements [26,27]. Evaluation of G values with the methods outlined above, gave results consistent with those calculated from X-ray crystal structures [28,29]. Since the three N atoms of the diethylenetriamine backbone are bound to the Ln^{III} ion, their inversion is inhibited. The DTPA complexes have two isomers, in accordance with their NMR spectra, which shows, at low temperature, two sets of resonances in slow exchange [25]. In the bis(amide) derivatives, the three N atoms are chiral which results in 8 possible forms of the complex (4 pairs of enantiomers), all of which have been observed by ^{13}C NMR [27]. The dynamics of the interchange of the various isomers in these complexes have been studied by line-shape analysis [26] and by 2D exchange spectroscopy [30]. The racemization process for the central backbone N atom is faster than that for the terminal ones, and does not require decoordination of the concerning N and its neighbouring acetate groups, but rather a non-dissociative interconversion ("wagging") of the backbone ethylene groups between two gauche conformations [26].

The structures of macrocyclic Ln^{III} chelates have also been extensively studied by NMR. Here we discuss a couple of representative systems. The Ln(DOTA) complexes (DOTA=1,4,7,10-tetraazacyclododecane-N',N'',N''',N''''-tetraacetate) have been investigated in solution using analysis of LIS and LIR NMR data [31-38]. The reported crystal structures of the complexes show that the ligand provides eight donor sites (four oxygens and four nitrogens), with one inner-sphere water molecule, but, while the smaller Ln^{III} ions (Ln = Eu, Gd, Lu) give a monocapped square antiprismatic geometry (M), the larger La^{III} ion prefers a monocapped inverted square antiprismatic geometry with smaller distorsion from the square prismatic structure (m)[39,40] (Fig.2).

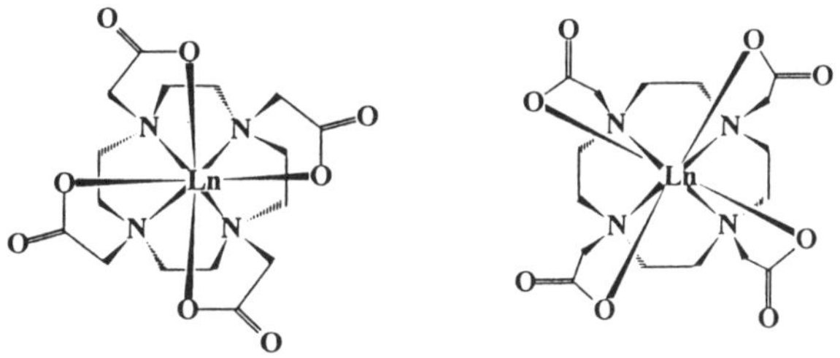

Fig. 2. Schematic antiprismatic arrangements of the Ln(DOTA)⁻ chelates observed in X-ray crystal structures [40].

The Ln(DOTA) chelates have also been extensively studied by NMR. The ^{17}O LIS values for water have shown that the hydration number is one along the lanthanide series [31]. Two sets of resonances have been observed in the ^1H and ^{13}C NMR spectra of the Ln(DOTA)⁻ complexes in solution, showing the presence of two structural isomers in slow exchange on the NMR time scale [32-38]. Fig. 3 illustrates the proton spectra of some paramagnetic complexes, one isomer having larger induced shifts than the other. The relative populations of the two species changes markedly along the Ln^{III} series and is also quite dependent on the concentration of added inorganic salts.

Analysis of the proton LIS values for the Yb^{3+} complex, which are pseudocontact [32,33], of the LIR values for the heavier lanthanides [34] and the full proton and ^{13}C LIS data analysis [37] led to a definition of the ligand structures of these two isomers. The structure of the 12-membered macrocyclic ring is the same in both isomers, adopting a very rigid square conformation, with all ethylene groups in the same gauche conformation, either δ or λ. The cation is bound to the four N atoms of the ring and to the four carboxylate groups, and is located slightly above the macrocyclic ring cavity.

The major isomer for Eu(DOTA)⁻ in solution is similar to the monocapped square antiprismatic structure M found in the solid by X-ray diffraction [39], and this has been found to be the case for all Ln ions from Sm^{III} to Lu^{III} [37,38]. The other isomer, which predominates in solution from La^{III} to Nd^{III}, has the reversed square antiprismatic structure m found in the x-ray crystal structure of La(DOTA)[40].

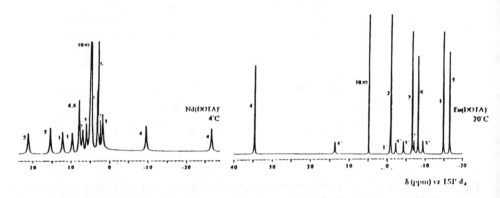

Fig. 3. Proton NMR spectra of 10 mM Ln(DOTA)⁻ complexes in D_2O, pH 7.0: Nd(DOTA)⁻ at 4°C; Eu(DOTA)- at 20°C. The unprimed and primed numbers refer to the isomer with larger (M) and smaller (m) magnetic anisotropy, respectively [37].

The dynamic interconversion processes of the two isomers have been studied by variable temperature 1D lineshape analysis and 2D exchange methods [32,33,35,36]. The structure and dynamics of this system are summarized in Fig. 4. It involves δ/λ interconversion of the conformations of the ethylene bridges of the macrocyclic ring, as well as conformational interconversion of the acetate arms through rotation around the C-N bonds of the pendant groups, without bond dissociation. This model involves interconversion of two enantiomeric pairs of diastereoisomers that differ in the ligand conformations.

A variable temperature, pressure and ionic strength ^1H NMR study of the DOTA complexes of different trivalent cations (Sc, Y, and La =>Lu, except Gd) yielded data that were in contradiction a two-isomer equilibrium M⇔ m, as it could not explain the reversal of the isomer ratio at the end of the series (Er => Lu) [38]. Superimposed on this conformational equilibrium is a coordination equilibrium, as both conformers are able to lose the inner sphere water molecule, forming eight-coordinate species m' and M'(Fig. 5).

The presence of this coordination equilibrium was proven by large positive reaction volumes for the isomerization of $[Yb(DOTA)(H_2O)_x]^-$ and $[Lu(DOTA)(H_2O)_x]^-$ (x=1,0). The isomerization of $[Nd(DOTA)(H_2O)_x]^-$ and $[Eu(DOTA(H_2O)_x]^-$ is purely conformational, as proven by near zero reaction volumes. A concordant behaviour of the reaction enthalpies and entropies was observed: near zero enthalpies and entropies for the isomerization of the lighter lanthanide DOTA complexes up to Dy, moderately positive enthalpies and strongly positive entropies for Tm and the later ones. The thermodynamic data allowed a detailed analysis of the thermodynamics and kinetics of the equilibria present (Fig. 6). Only in the case of the very small cation Sc^{III} could the m' ⇔ M' equilibrium be detected.

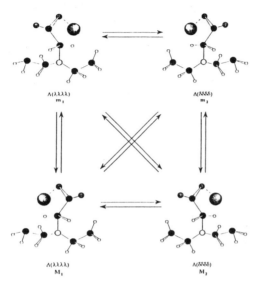

Fig. 4. Schematic representation of the structure and dynamics of the diastereoisomers of the Ln(DOTA)⁻ complexes. Only the geometry around one of the bound N-CH$_2$-COO⁻ groups is given. The symbols Λ and Δ refer to the helicity of the acetate arms, whereas λλλλ and δδδδ to the cycle. Here the conformational process is shown for the M and m forms, but is equally applicable to M' and m'. M$_1$ and M$_2$, as well as m$_1$ and m$_2$ [35,36], are NMR indistinguishable enantiomeric pairs respectively for the M and m forms detected by NMR.

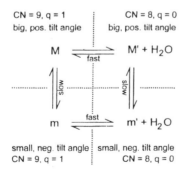

Fig.5. Conformational and coordination equilibria for the [Ln(DOTA)(H$_2$O)$_x$]⁻ complexes in aqueous solution. Enantiomeric pairs are not represented [38].

The shift of the isomerization equilibria by a variety of non-coordinative salts favours one of the ligand conformations relative to the other, irrespectively of the presence or absence of the inner sphere water molecule. This results from weak ion binding and water solvent stabilization of that conformation rather than from the decrease of the activity of the bulk water in the solution, favorable to the formation of complexes without inner sphere water. Fluoride ions are a special case, being able to enter the inner coordination sphere substituting the water molecule, with relative preference m >M (not for m') [38]. Inner sphere coordination of carboxylate ligands to Ln(DOTA) forming mixed complexes had been observed before [31].

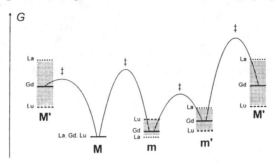

Fig.6. Gibbs free energies $-RT \ln K_i$ for equilibria m <=> M ($1/K_1$), M <=> m' (K_2)., m <=> m', and M <=> M' and activation barriers for isomerisations and dissociation of inner sphere water (semiquantitative) [38].

We also studied in detail the solution structure of the Ln^{III} complexes of the tetraazamacrocyclic phosphonate ligand $DOTP^{8-}$, $Ln(DOTP)^{5-}$ (Fig.7), as obtained by analysis of the LIS and LIR values and MMX molecular mechanics calculations [41]. The solution structure of all the Ln^{III} chelates of this series is of the m' type (Fig.2), an 8-coordinate reversed square antiprism (with no inner-shere water, as shown by ^{17}O Dy^{III} induced shifts), which was confirmed by a yet unpublished X-ray crystal structure of $(NH_4)_5Tm(DOTP)$ [42]. Thus, the DOTP ligand is capable of forming more compact structures around the Ln^{III} ions than DOTA. The thermodynamics of $DOTP^{8-}$

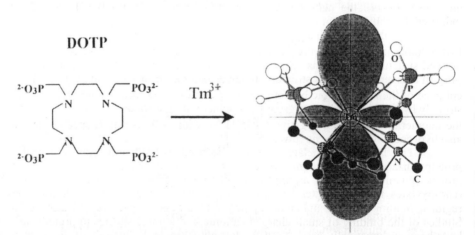

Fig.7. Chemical structure of $DOTP^{8-}$ and 3D structure of its complex $Tm(DOTP)^{5-}$ obtained from analysis of LIS and LIR data [41]. The gray contour is the geometrical pseudocontact shift function, $(3\cos^2\theta-1/r^3)$ [10].

complexation by the Ln^{III} ions and of protonation of the complexes has been also studied in detail by potentiometry and ^{31}P NMR, showing that the chelates can monoprotonate the four phosphonate groups, with pK_a values of about 7.9, 6.6, 5.6 and 4.4, forming protonated species of lower negative charge up to $Ln(DOTPH_4)^-$ [43].

The paramagnetic ^{13}C NMR LIR and ^1H LIS of a series of positively charged organic molecules containing amino groups, such as adamantanamine (ADA), propylamine (PA), cyclen (CY) and N-methyl-glucamine (meglumine, MEG), induced by a series of paramagnetic LnIII chelates, such as $Ln(DTPA)^{2-}$, $Ln(DTPA$-bisamides), $Ln(DOTA)^-$ and $Ln(DOTP)^{5-}$, were measured in aqueous solution and used to model the supramolecular complexes resulting from their non-covalent interactions [44]. These complexes showed a varying degree of binding specificity and strength. Weak, non-specific binding was observed for the neutral DTPA-bis(amide) complexes, possibly due to hydrophobic interactions, whereas $Ln(DTPA)^{2-}$ and $Ln(DOTA)^-$ complexes showed weak interactions with positive ammonium functions (eg., the association constant of $Tm(DTPA)^{2-}$ with ADA at pH 7.0 is K = 14 M^{-1}), due to their low negative charge. The LIS values are also very small, because the weak negative binding sites of these chelates for the ammonium functions, which are negative oxygen atoms of four (for $Ln(DOTA)^-$[39,40]) or five (for $Ln(DTPA)^{2-}$[45]) coordinated carboxylates, aevarge out the dipolar geometric term G (eq. 2). The strongest and most specific interactions occurred between the highly negatively charged $Ln(DOTP)^{5-}$ chelates and the protonated linear and macrocyclic amines. In the case of $Tm(DOTP)^{5-}$ADA (association constant K = 310 M^{-1}), the experimental ^1H LIS and ^{13}C LIR values were fitted to eqs. 2 and 6, defining a geometry for the 1:1 adduct where the ammonium group of ADA interacts with the Ln-unbound negatively charged oxygen(s) of one phosphonate group of $Tm(DOTP)^{5-}$ (Fig. 8). The adduct molecule is within the dipolar cone define by Tm^{III}, but is not centrally located above the four phosphonate groups. Two $Ln(DOTP)^{5-}$ are able to sandwich the diprotonated tetraazamacrocyclic amine CY, forming a 2:1 adduct. In the polyhydroxyammonium compound MEG, the strong electrostatic interaction is assisted by hydrogen bonding of hydroxyl groups to the Ln-unbound phosphonate oxygens of DOTP, forming an adduct $[Tm(DOTPH)(MEG)_4]$. In all cases, the strong pH dependences of the observed paramagnetic NMR effects chelates on the ^1H and ^{13}C nuclei of the ligands, with a maximum at pH ~10 (where the ammine ligand is protonated and the chelate has a 5- charge), and a large decrease at higher (where thc ligand is deprotonated) and lower (where the negative charge of the chelate is reduced) pH values, clearly indicated the dominance of the electrostatic interactions [44].

We also found that $Tm(DOTP)^{5-}$ is capable of weak (K=4 M^{-1}) but specific hydrophobic interactions with γ-cyclodextrin, forming a 1:1 inclusion complex of well defined structure, which was again obtained from analysis of LIS and LIR data (Fig.9) [46]. In this supramolecular structure, the hydrophobic surface of $Tm(DOTP)^{5-}$ fits at the entrance of the hydrophobic basket on the top rim of γ-CD, with the negatively charged, strongly hydrated phosphonate groups in contact with the solvent. Recent data show that the interaction of the less negative $Ln(DOTA)^-$ with γ-CD is not as specific, involving also hydrogen bonding to the lower rim [47].

The lanthanide chelates can also be used as shift and relaxation probes of protein surfaces, by interacting non-covalently with their surface residues. Cationic groups (guanidinium from arginines or ammonium from lysines) or anionic groups (carboxylates from aspartates and glutamates) may be found concentrated in certain regions, yielding cationic and anionic patches. Hydrophobic patches may also be found. Studies of the binding of small charged cationic and anionic species to protein surfaces have been systematically carried out for ferricytochrome c. Analysis of the location of their binding sites was possible through observation of NMR effects on specific proton resonances, which function as reporter groups [48]. A different LIS and LIR approach has been used more recently to study the bacteriophage ss-DNA binding gene 5 protein

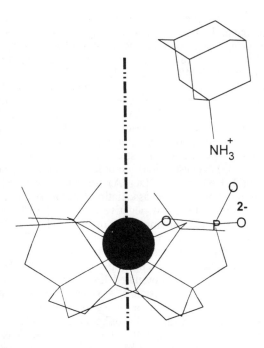

Fig. 8. Structure of the supramolecular complex formed by Tm(DOTP)$^{5-}$ and ADAH^{+}, as defined by NMR LIS and LIR values. The Ln-N(ADA) distance is 5.2 Å, the Ln-N vector makes an angle of 46° with the C_4 axis of the chelate, and the Ln-N-C_1 angle is 138°[44].

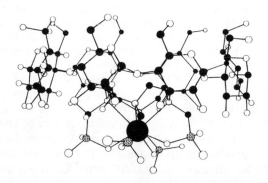

Fig. 9. Structure of the Tm(DOTP)$^{5-}$ -γ-CD inclusion complex [46].

(G5P). This protein, which in solution is a dimer of about 20 KDa molecular mass, may be reductively methylated to introduce ^{13}C-enriched methyl groups into all six lysyl residues without significantly disrupting its ability to bind ss-DNA [49]. The negatively charged lanthanide chelates, Ln(DOTP)$^{5-}$ were shown to bind rather selectively at positively charged lysine residues at its surface, by measuring the induced LIS (Ln=Tb) and LIR (Ln=Gd) of the ^{13}C resonances of the modified G5P, and also of the proton resonances of the intact protein [49]. These lysine residues are those involved in the binding to ss-DNA, as shown by competition experiments. A later study of the interaction of related G5P proteins, M13 and IKe, with Gd(DOTP)$^{5-}$ [50] used 2D NOESY and TOCSY difference spectra to show that at pH > 7 the chelate binds with high affinity at the ss-DNA binding domain consisting of conserved, phosphate binding, positively charged clusters at the surface of the proteins.

Recently we also studied the interaction of Gd(DOTP)$^{5-}$ and of other negative, positive and neutral/hydrophobic Gd^{III} chelates with hen egg white lysozyme (HEWL), using the specific bleaching of TOCSY cross-peaks shown in 2D TOCSY difference spectra to map the sites of interaction of those relaxation probes with the positive, negative and hydrophobic surface patches of the protein. Initial slopes of plots of intensities of TOCSY cross-peaks as a function of the probe to protein concentration ratio, allowed the definition of histograms of the degree of interaction of the probes with the aminoacid side chains along the protein sequence. The fact that the observed interactions do not correlate with the degree of residue surface exposure, but rather with the charge or hydrophobicity of the aminoacid side chains present in the 3D protein surface, shows that some degree of specificity is obtained in those interactions [51].

3. Shift Reagents for MRS of Alkali Metals in Biological Systems.

Water soluble SR's for discriminating intra- and extracellular cations were introduced in the earlier 1980's [52,53]. This is accomplished by designing a paramagnetic anionic shift reagent chelate which remains extracellular and binds to the ion of interest, thereby inducing an isotropic shift in those nuclei. The nuclei of interest in intracellular compartment(s) remains unshifted (except for bulk susceptibility effects), as the exchange of cations between compartments is not fast enough to average their chemical shifts. This results in two resolved NMR signals which may be monitored to examine ionic concentrations in each compartment. Most studies of this type have so far been concerned with alkali metal ion NMR (^{23}Na, ^{39}K, as well as ^{7}Li and ^{133}Cs) measurements, involving reports of transmembrane ion transport and compartmentation [5,10,23,52-54]. The shift reagents used are anionic lanthanide chelates which bind the alkali ions by forming ion-pairs with coordinated or pendant negatively charged groups in the chelate at physiological pH. A large through space dipolar shift is produced if those ion-pairing sites are located in favorable geometric positions relative to the axis of the chelate susceptibility tensor. Dy(DTPA)$^{2-}$ is a very poor SR for ^{7}Li NMR because it binds two Li^{+} counterions exchanging between five locations at the carboxylate oxygens, averaging out the dipolar shift function G (see Fig. 10A [45]). A similar averaging is found with Dy(DOTA)$^{-}$ [31]. The most efficient reagents, on the basis of the ability to shift the metal ion resonance per unit concentration, are Dy^{III}-bis(tripolyphosphate), Dy(PPP)$_2^{7-}$ [52] and Tm(DOTP)$^{5-}$ [55], whose structures in solution (Figs. 7 and 10) have been determined using multinuclear NMR shift and relaxation data [41,56]. The large shifts result from preferential metal ion binding near the main symmetry axis of the chelate (eg. 6 Na+ ions bind to Dy(PPP)$_2^{7-}$ and 4 to Tm(DOTP)$^{5-}$). This later complex is axially symmetric and the magnitude and direction of the ^{23}Na^{+} shifts for the various lanthanides indicate that the Na^{+} binding region is near the 4-fold axis of symmetry of the complex (see Fig.10). Recently, multinuclear NMR studies of the binding of Na^{+} and other alkali metal ions, Mg^{2+}, Ca^{2+} and Co(en)$_3^{3+}$ to Tm(DOTP)$^{5-}$ have been undertaken [43,57] It was found that the SR has one strong (log K_1 = 2.58) preferential Na^{+} binding site A, near the C_4 symmetry axis, where the dipolar term G is maximum (average θ = 24°, r = 3.7 Å), leading to a very large fully bound ^{23}Na^{+} shift of 423 ppm. This is the site which also binds Ca^{2+} and Mg^{2+} with high affinity, leading to a marked decrease of the ^{23}Na$^+$ shift in tissues, as it is displaced from site A. However, it has another 3-4 weaker (log K_i ~ 1.5) sites B, with less favorable geometric parameters ($\theta = 34°$, r = 4.5 Å) and smaller fully bound shifts (162 ppm), which are only blocked by divalent ions when they fully occupy site A, and this is the reason why the Tm(DOTP)$^{5-}$ ^{23}Na induced shift is much less sensitive to the presence of those ions than Dy(PPP)$_2^{7-}$ [57].

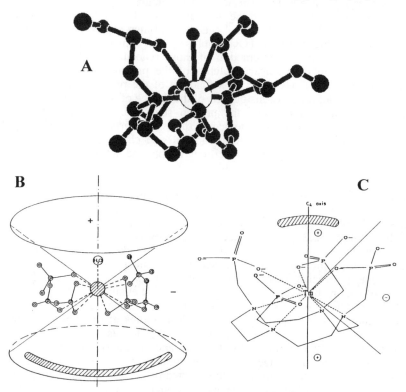

Fig. 10. Structure of shift reagents for *in vivo* MRS of alkali metal ions and location of the binding sites for the counterions, obtained from LIS and LIR data: (A) Dy(DTPA)$^{2-}$: 2 Li$^+$ ions exchanging between five sites [45]; (B) Dy(PPP)$_2^{7-}$ and location of the six Na$^+$ sites [56]; (C) Tm(DOTP)$^{5-}$ and location of the four Na$^+$ sites [41,57].

Fig.11 illustrates the excellent resolution of intra- and extracellular ^{23}Na$^+$ resonances of human erythrocytes achieved with 3.6 mM Dy(PPP)$_2^{7-}$ or Dy(DOTP)$^{5-}$ [55].

The use of shift reagents in tissues, organs or live animals involves much more stringent conditions than in isolated cell systems. For example, Dy(PPP)$_2^{7-}$ is known to dissociate in the presence of tissue Ca^{2+} and degrade to Dy^{3+} and inorganic phosphate in tissues that contain active pyrophosphatases [57]. The use of Dy(TTHA)$^{3-}$ has been advocated by Springer [53,54] because of its high stability constant. It is much less sensitive to the presence of Ca^{2+} than the others and the ^{23}Na$^+$ shifts are nearly independent of the pH between 5.5 and 12. Unfortunately, its ability as a shift agent per unit concentration is about a factor of five less than that of Dy(PPP)$_2^{7-}$. This means that much higher concentration (typically 10 mM) of Dy(TTHA)$^{3-}$ must be used to resolve the ^{23}Na$^+$ resonances in different compartments, which cause much larger bulk susceptibility shift interferences. The observed shift is much smaller (and of opposite sign) in this complex relative to Dy(PPP)$_2^{7-}$ mainly due to its lower negative charge and

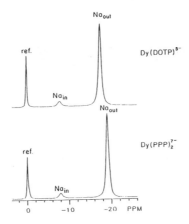

Fig. 11. ^{23}Na NMR spectra of human erythrocytes in solution containing 3.6 mM Dy(PPP)$_2^{7-}$ or Dy(DOTP)$^{5-}$ and 150 mM NaCl. The reference signal comes from a capillary containing 100 mM NaCl and 50 mM DyCl$_3$ [55].

the unfavorable spacial locations of the Na$^+$ binding sites (the free carboxylate group is outside the dipolar cone) relative to the main symmetry axis of the complex.

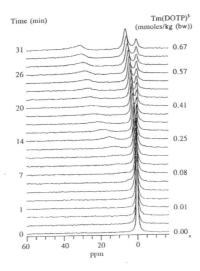

Fig. 12. ^{23}Na NMR spectrum of an *in vivo* rat kidney during infusion of Tm(DOTP)$^{5-}$, illustrating the high chemical shift dispersion of the intra and extracellular ^{23}Na resonances. The broad signal at 35 ppm in the top spectrum is from filtrate Na$^+$ [10, 61].

Tm(DOTP)$^{5-}$ is thermodynamically and kinetically very stable, is not degraded by phosphatases, produces isotropic $^{23}Na^+$ shifts comparable to those of $Dy(PPP)_2^{7-}$ and much lower bulk susceptibility shifts [62]. The $^{23}Na^+$ shifts are also less sensitive to Ca^{2+} ions than for $Dy(PPP)_2^{7-}$. These properties make it an ideal shift reagent [10] for ^{23}Na NMR studies in perfused hearts [59], or in *in vivo* spectroscopic studies in rat brain, liver or kidney (Fig.12) [60-62].

Finally, cationic Ln^{III} chelates can be devised which act as SR's for in vivo NMR of anionic species. A cationic macrocyclic tetramide complex $Eu(DTMA)^{3+}$, when added to human blood, was found to cause a separation of the intra- and extracellular 31P NMR resonances of inorganic phosphate [14]. Similar agents can be devised for carbonate, chloride, lactate, etc.

4. MRI Contrast Agents

The presence of a paramagnetic ion of spin S in solution generates large local magnetic fields associated with the unpaired electron magnetic moment which fluctuate in time, due to the dynamics of the system. The frequency components of those fields at the nuclear and electronic Larmor frequencies, ω_I and ω_S, induce spin transitions at the nucleus I, thus causing very efficient relaxation processes in nuclei of neighbouring molecules (either a ligand or the solvent). Some types of efficient paramagnetic relaxation compounds have potentially useful clinical applications to improve the contrast of magnetic resonance images by enhancing the relaxation rates ($1/T_1$ and $1/T_2$) of protons of tissue water, which contribute to the signal intensity of those images [10-14,63]. Small complexes of Gd^{III} have been the most popular, since this metal ion, with a S electronic ground state, combines a large molecular moment with a long electron spin relaxation time ($\sim 10^{-9}$ s at magnetic field strenghts of interest for MRI, $B_o \sim 1.5$ T), which lead to an optimum efficiency of nuclear paramagnetic spin relaxation. Such contrast agents (CA's), when injected into humans should have high efficiency (defined by maximum *in vivo* tissue relaxivity) for minimal effective dose, minimal toxicity (low osmolality and high thermodynamic and kinetic stability are the important properties for this purpuse), and high excretion rates (displaying no retention). Biodistribution factors, such as high tissue specificity and specific excretion pathway, are also important for the design of more specific CAs [11-13]. Some of these properties have opposite requirements: complexation of Gd^{3+} reduces the toxicity associated with the free cation, but also reduces relaxivity due to the replacement of inner-sphere water molecules by ligand donor atoms. So far, only four contrast agents are in clinical use (Fig.1), namely the negatively charged Gd^{3+} chelates $Gd(DTPA)^{2-}$ (MAGNEVIST) and $Gd(DOTA)^-$ (DOTAREM), and the neutral ones Gd(DTPA-BMA) (OMNISCAN) and Gd(HP-DO3A) (PROHANCE) [10,14]. These small, very hydrophilic chelates, are extracellular fluid (ECF) agents. After i.v. injection, they are rapidly 100% renal excreted ($t_{1/2} \sim 90$ min.) after an early intravascular distribution and diffusion into the extracellular (intersticial) space ($t_{1/2} \sim 10$ min). They do not bind to plasma proteins or cross cell membranes or the blood-brain barrier (BBB). They are quite useful to study brain perfusion abnomalies, such as disruption of the BBB by tumors.

However, other potential MRI contrast media are under intense research and development. These include other small ionic chelates of Gd^{3+} or Mn^{2+} bearing an hydrophobic group which are hepatobiliary agents, such as the hydrophobic Gd(DTPA) derivatives $Gd(EOB-DTPA)^{2-}$ and $Gd(BOPTA)^{2-}$, as well as $Mn(DPDP)^-$. Because they contain negative charge(s) and hydrophobic moieties, they are taken up by the hepatocyte anion transport system and have hepatobiliary excretion, and are thus used to image the liver [64]. Paramagnetic chelates entrapped or bound to lipossomes are also useful as liver agents [13]. Another type of compounds are the blood pool (BP) agents, which include natural macromolecules or synthetic polymers (eg. albumin, polylysine, dextran, Starburst[a] dendrimers) of molecular weight higher than 50,000 Daltons covalently labelled with paramagnetic chelates. Their size restricts their diffusion into the interstitium, causing prolongued stay in the vasculature. This allows mapping of the intravascular space using Magnetic Resonance Angiography (MRA) [65]. Their excretion half-life is of several hours, as it is dependent on polymer biodegradation. Paramagnetic particulate iron oxide particles coated with polymers (eg. dextran, starch, polyethyleneglycol) have biodistributions (blood pool, liver, spleen, lymph nodes, bone marrow) which depend on their size and type of coating, and they cause MRI contrast through shortening of tissue T_2 values [13]. Other compounds include CA's for the gastrointestinal tract, such as Gd^{3+} ions incorporated into zeolites, used as oral contrast agents [10], complexes which concentrate or bind specifically to tumors, such as paramagnetic metalloporphyrins (however, the use of labeled monoclonal antibodies specific to a particular tumor line, although promising in radiotherapy, seems not to be feasible in MRI, due to lack of sensitivity to their contrast effect.[66]), organic free radicals or even the deoxy form of hemoglobin which is the basis for functional MRI.

We now briefly review the structural and dynamic determinants to optimize the relaxivity of Gd^{III} complexes as potential MRI contrast agents, in particular those aspects relevant to supramolecular chemistry.

In an aqueous solution of a paramagnetic agent, the observed longitudinal and transverse relaxation rates R_i of the solvent protons are the sum of the paramagnetic (p) and diamagnetic (d) (the water relaxation rate in the absence of the paramagnetic compound) contributions [67]:

$$R_i^{obs} = R_{ip} + R_{id} \qquad (i=1,2) \qquad (7)$$

The overall paramagnetic relaxation enhancement R_{ip} referred to a 1 mM concentration of a Gd^{III} chelate is called its relaxivity. The fluctuating local magnetic fields originating from the paramagnetic center decrease sharply with the distance r. Therefore, the corresponding relaxation effects reach all the solvent molecules both through translational diffusion and specific chemical interactions. The inner sphere (is) relaxation effect results from binding of the ligand or solvent molecule in the first coordination sphere of the metal ion, whereas the outer sphere (os) effect results from mutual translational diffusion of the species in the absence of chemical interaction. The total relaxivity is:

$$R_{ip} = R_{ip}^{is} + R_{ip}^{os} \qquad (i=1,2) \qquad (8)$$

A schematic representation of these two relaxation mechanisms of solvent water protons in solution by Gd^{III} complexes is shown in Fig.13. The outer sphere contribution, which is the only one when the solvent coordination number is zero, decreases when the distance of closest approach (d) between the two species decreases, and also depends on the electron spin relaxation time (τ_S) of the cation and the mutual translational diffusion time (τ_D), which modulate the electron-nucleus dipolar interaction [67]. The inner sphere contribution to R_{1p} results from the process of chemical exchange of the solvent molecules between the first coordination sphere of the metal ion and the bulk solution:

$$R_{ip}^{is} = p_M \, q \, R_{1M} = (c \, q / 55.6) / (T_{1M} + \tau_M) \qquad (9)$$

where c is the molar concentration of the paramagnetic complex, q is the inner sphere water coordination number of the chelate, τ_M is their mean residence lifetime and T_{1M} is their longitudinal relaxation time. For the case of S-ground state ions, the bound rate R_{1M} is given by the contribution from two mechanisms of electron-nuclear magnetic interaction, the scalar or contact and the dipolar contributions, which are described by the generalized Solomon-Bloembergen-Morgan (SBM) equations [11,23,67]. These processes are modulated, respectively, by the correlation times τ_c and τ_e, which are given by:

$$1/\tau_{ci} = 1/\tau_R + 1/\tau_{iS} + 1/\tau_M \qquad (10a)$$
$$1/\tau_e = 1/\tau_{2S} + 1/\tau_M \qquad (10b)$$

where τ_R is the rotational correlation time of the complex, τ_{1S} and τ_{2S} are the longitudinal and transverse electron spin relaxation times of the cation. However, the dipolar contribution to the nuclear relaxation rates is dominant for Gd^{III}, and is given by:

$$R_{1M} = [(2/15)(\mu_0/4\pi)^2 (h^2\gamma_S^2\gamma_H^2) S(S+1)/r^6][3\tau_{c1}/(1+\omega_H^2\tau_{c1}^2) + 7\tau_{c2}/(1+\omega_S^2\tau_{c2}^2)] \qquad (11)$$

where γ_S and γ_H are the electron and proton nuclear magnetogyric ratios, respectively. The electron spin relaxation times τ_{1S} and τ_{2S} are frequency dependent,

$$(1/\tau_{1S})^{ZFS} = [(12/5)\Delta^2\tau_v][1/(1+\omega_S^2\tau_v^2) + 7\tau_{c2}/(1+4\omega_S^2\tau_v^2)] \qquad (12a)$$
$$(1/\tau_{2S})^{ZFS} = [(12/10)\Delta^2\tau_v][3 + 5/(1+\omega_S^2\tau_v^2) + 2/(1+4\omega_S^2\tau_v^2)] \qquad (12b)$$

and characterized by the correlation time τ_v of the modulation of the transient zerofield splitting, with squared trace Δ^2.

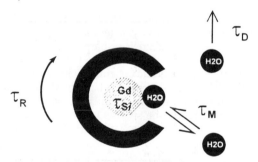

Fig. 13. Schematic view of the paramagnetic relaxation mechanisms and the main relaxation parameters for an aqueous solution of a Gd^{III} chelate.

In extreme narrowing conditions ($\omega_S\tau_{c2} \ll 1$ and $\omega_H\tau_{c1} \ll 1$):

$$R_{1M} = R_{2M} = C\, r^{-6}\, \tau_c \qquad (13)$$

where C is a constant characteristic of the Ln^{III}. This equation can be used, under appropriate conditions, to obtain absolute values of r by specifying the values of τ_c. For the $Ln^{3+}_{(aq)}$ ions, typical values of the correlation times are $\tau_R \sim 10^{-10}\text{-}10^{-11}$s and $\tau_M \sim 10^{-9}$s. $Gd^{3+}_{(aq)}$ (and Mn^{2+}) has long electronic relaxation times, $\tau_S \sim 10^{-8}\text{-}10^{-9}$s, τ_c is dominated by τ_R and the relaxation effect is large (line broadening of 20-200 KHz), so it is an efficient relaxation probe and gives the most efficient contrast agents. For ions with short τ_S, such as the other paramagnetic Ln^{3+} ions ($\tau_S \sim 10^{-8}$-10^{-9}s), τ_c is dominated by τ_S. These ions do not cause significant NMR broadening effects (1-100 Hz), but may cause shifts if they form magnetically anisotropic chelates- they are known as shift probes [3].

Relaxivity studies of paramagnetic chelates, including contrast agents, are optimally performed using the experimental technique known as nuclear magnetic relaxation dispersion (NMRD), which consists of studying the magnetic field dependence of the nuclear relaxivities of the solvent protons in the chelate solutions over a range corresponding to proton Larmor frequencies of 0.01-50 MHz [67]. This technique yields a set of experimental data points which afford, through comparison with a theoretical treatment based on the SBM equations, the parameters describing the system under study [67]. As R_1 depends on a variety of structural parameters, in the multiparameter fitting procedure, it is advisable to fix the values of the parameters which can be determined through independent experiments [14,23] Their effects can be evaluated separately [67]: a) in conditions of fast exchange, R_1 increases linearly with the inner sphere water coordination number q of the cation in the chelate under study (eq. 9). This has been verified experimentally for series of linear and macrocyclic chelates of Gd^{3+}, where the other molecular factors did not change appreciably [68,69]; b) due to the r^{-6} dependence of the dipolar interaction (eq. 11), R_1 increases when the distance r of the water protons to the electron spins decreases [67].

We now discuss in more detail the most important parameters affecting the relaxivity: a) An increase of the electron spin relaxation time at zero field (τ_{S0}) causes an increase of R_1 at low fields. Fig.14 illustrates this for a series of Gd^{3+} chelates [70]. Longer τ_{S0} values are associated with more symmetric and more rigid chelates, such as Gd(DOTA)$^-$.

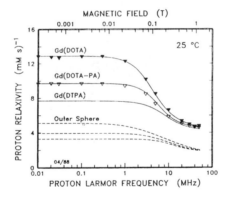

Fig. 14. Dependence of low field R_1 values in NMRD profiles on τ_{S0}, for various Gd^{3+} chelates (τ_{S0} is 660 ps for Gd(DOTA)$^-$, 132 ps for Gd(DOTA-PA) and 66 ps for Gd(DTPA)$^{2-}$ [70].

b) The residence time of the water molecules in the cation coordination sphere (τ_M) may affect R_1 by limiting the efficiency of the water exchange process. It has been shown by pressure, magnetic field and temperature dependent ^{17}O water R_{2p} determinations [70,71] that the water exchange rates for various Gd^{III} chelates are considerably longer (19 μs- 68 ns range) than for the aquo complex (1.2 ns), due to a dissociatively activated exchange mechanism. For systems with q =1, exchange rates increase with positive charge of the chelate and decrease with the presence of bulky substituent groups in the ligand [14]. Thus, long τ_M in the μs range could be a limiting factor for the proton relaxivity when τ_{S0} and τ_R are large.

c) for ions with long τ_{S0} values, an increase of τ_R causes an increase of R_1, with a maximum at ω_H ~40 MHz. However, Fig. 15 shows, for q=1 and r = 3 Å, that the result of the immobilization of the complex with a τ_R = 30 ns on R_1 depends on the values of τ_M and τ_{1S}. For τ_M = 20 ns and τ_{1S}= 10 ns, an optimal value of R_1 ~140 mM-1 can be obtained, a relaxivity much above the presently available values. Increased relaxivity has been observed through covalent or non-covalent immobilization of Gd^{III} chelates to macromolecules, such as human serum albumin (HSA), or polymeric compounds (eg. polysacharides, dendrimers) [11,14,73]. However, the R_1 increase is lower than expected from the τ_R value of the macromolecule, as local flexibility and non-ideal τ_M values limit that increase. Non-covalent binding of Gd^{III} chelates to HSA seems to be much more efficient [11,14,74,75] to increase R_1. Fig. 16 shows the large increase of relaxivity of $Gd(DOTA-BOM_3)^-$ in the presence of BSA, as Gd^{III} chelate of the DOTA-like ligand bearing three hydrophobic β-benzyloxy-α-propionic substituents binds strongly (K_a=1.7x10^3 M^{-1}) to the two protein sites in subdomains IIA and IIIA, as shown by Scatchard plots in the absence and presence of the site specific drugs warfarin and ibuprofen [75].

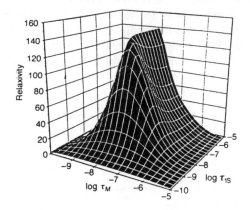

Fig. 15. 3D plot of the dependence of R_1 on τ_M and τ_{1S} for an immobilized (τ_R = 30 ns) Gd^{III} chelate (q=1, r = 3 Å) [14].

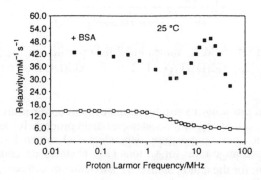

Fig. 16. R_1 NMRD profiles for $Gd(DOTA-BOM_3)^-$ with (■) and without (□) BSA (25°C, pH 6.9) [75].

d) Outer-sphere relaxation is about half of the total relaxivity contribution of a contrast agent in the frequent case of q = 1. The large relaxivity value of R_{1p}^B = 53. mM^{-1}.s^{-1} obtained for the above chelate Gd(DOTA-BOM$_3$)$^-$ with q =1 cannot be explained solely by its immobilization, and it has been proposed that a large contribution to the relaxivity originates in the structure and dynamics of mobile protons at the protein-chelate interface which are dipolarly relaxed by the GdL center [14]. This second sphere relaxation effect is even more evident in the case of the large (ε_b=4.3) relaxation enhancement effect of binding of Gd(BzDOTP)$^-$, with no inner-sphere water (q = 0), to BSA (K_a = 3.6x10^3 M^{-1}) [76]. The presence of exchangeable protons on the proximity of the coordination sphere of the Gd^{III} ion could provide efficient relaxation pathways to the solvent, such as in the interaction of Gd(DOTP)$^{5-}$ (a chelate with q =0 and a strong second sphere hydration of the phosphonate oxygens) with meglumine [44,77] or of contrast agents with the side-chains of HSA [75]. In the protein surface-chelate interface, ordered hydration water molecules with long residence times near GdL have higher outer-sphere relaxation effects. Fig. 17 shows that NOE effects from protein non-exchangeable surface protons to these hydration water molecules (a) or to protein exchangeable surface protons (b), followed by chemical exchange, could provide efficient magnetization transfer pathways [78] to the vicinity of the GdL chelate and increase the non-first sphere relaxivity. The importance of the exchange process is illustarted by the marked increase of relaxivity of GdL chelates bearing N-H protons with q =0 or 1 (long τ_M) at high pH, as the base catalysed prototropic N-H exchange provides a new efficient relaxation pathway to the water protons [14].

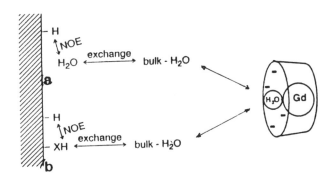

Fig. 17. Magnetization transfer pathways between protein surface protons and bulk water, which can mediate non-first sphere paramagnetic relaxation of GdL chelates in the proximity of the protein surface [78].

In conclusion, the optimization of relaxivity is of fundamental importance to increase the sensitivity of targeted MRI contrast agents to levels nearer to that available to Nuclear Medicine through the use of radiolabelled compounds. This will allow the design of new generation CA's consisting of a Gd^{III} chelate attached to a recognition synthon which binds specifically to natural targets such as neuropeptide receptors or the transport system of hepatocytes, or available in increased concentrations in pathological conditions, such as linearly glycated sidechains, or enzymes such as trypsin [14].

Acknowledgements. This work was supported by FCT, Portugal (project Praxis 2/2.2/SAU/1194/95).

5. REFERENCES

1. La Mar, G.N., Horrocks, Jr., W. de W. and Holm, R. H. (1973) *NMR of Paramagnetic Molecules*, Academic Press, New York.
2. Dwek, R.A., (1973) *NMR in Biochemistry*, Clarendon Press, Oxford.
3. Banci, L., Bertini, I. and Luchinat, C. (1991) *Nuclear and Electron Relaxation*, VCH, Weinheim.
4. Bertini, I. and Luchinat, C. (1996) *NMR of Paramagnetic Substances, Coord. Chem. Rev.*, **150**, 1-296.
5. Sherry, A. D. and Geraldes, C.F.G.C. (1989) in *Lanthanide Probes in Life, Chemical and Earth Sciences*, Bünzli, J.C.G. and Choppin, G.R. (eds.), Elsevier, Amsterdam, pp. 93-126.
6. Evans, C.H. (1990) *Biochemistry of the Lanthanides*, Vol.8, New York, Plenum Press.
7. Geraldes, C.F.G.C. (1993) *Methods Enzymol.*, **227**, 43.
8. Wuthrich, K. (1989) *Science*, **243**, 45; (1989) *Acc. Chem. Res.*,**22**, 36.
9. Wenzel, T.J. (1987) *NMR Shift Reagents*, CRC Press, Boca Raton.
10. Sherry, A. D. (1996). in *Magnetic Resonance and the Kidney: Experimental and Clinical Applications*, Endre, Z. H.(ed.), Marcel Dekker, New York.
11. Lauffer, R.B. (1987) *Chem.Rev.*, **87**, 901.
12. Kumar, K. and Tweedle, M. F. (1993) *Pure Appl. Chem.*, **65**, 515.
13. Rocklage, S. M., Watson, A.D. and Carvlin, M.J. (1992) in *Magnetic Resonance Imaging*, 2nd edn, Stark, D.D. and Bradley, W.G. (eds.), Mosby Year Book, St. Louis, Chapter 14.
14. Aime, S., Botta, M., Fasano, M. and Terreno, E. (1998) *Chem. Soc. Rev.*, **27**, 19.
15. Campbell, I.D., Dobson, C. M. and Williams, R.J.P. (1975) *Proc. Roy. Soc. Lond. Ser. A*, **345**, 41.
16. Imoto, T., Johnson, L. N., North, A.C.T., Phillips, D.C. and Rupley, J.A. (1972) in *The Enzymes*, Vol. III, 3rd edn., Boyer, P.D. (ed.), Academic Press, New York, pp-666.
17. Agresti, D.G., Lenkinski, R.E. and Glickson, J.D. (1977) *Biochem. Biophys. Res. Commun.*, **76**, 711.
18. Lee, L. and Sykes, B.D. (1980) *Biochemistry*, **19**, 3208; (1981) **20**, 1156;. (1983) **22**, 4366
19. Williams, T.C., Corson, D.C. and Sykes, B.D. (1984) *J.Am.Chem.Soc.*,**106**, 5698.
20. Bleaney, B. (1972) *J. Magn. Reson.*, **8**, 91.
21. Golding, R.M. and Halton, M.P. (1972) *Aust. J. Chem.*, **25**, 2577.
22. Reilley, C.N., Good, B. W. and Allendoerfer, R.D. (1976) *Anal. Chem.*, **48**, 1446.
23. Peters, J.A., Huskens, J. and Raber, D.J. (1996) *Progr. NMR Spectr.*, **28**, 283.
24. Kemple, M.D., Ray, B.D., Lipkowitz, K.B., Prendergast, F.G and Rao, B.D.N. (1988) *J. Am. Chem. Soc.*, **110**, 8275.
25. Forsberg, J. H., Delaney, R.M., Zhao, Q., Harakas, G. and Chandran, R. (1995) *Inorg. Chem.*, **34**, 3705.
26. Peters, J.A. (1988) *Inorg. Chem.*, **27**, 4686.
27. Geraldes, C.F.G.C., Urbano, A.M., Hoefnagel, M.A. and Peters, J.A. (1993) *Inorg.*

Chem., **32**, 2426.
28. Stekowski, J.J. and Hoard, J. L. (1984) *Isr. J. Chem.*, **24**, 323.
29. Konings, S.M., Dow, W.C., Love, D.B., Raymond, K.N., Quay, S.C. and Rocklage, S.M. (1990) *Inorg. Chem.*, **29**, 1488.
30. Jenkins, B.G. and Lauffer, R.B. (1988) *Inorg. Chem.*, **27**, 4730.
31. Bryden, C.C., Reilley, C.N. and Desreux, J.F. (1981) *Anal. Chem.*, **53**, 1418.
32. Desreux, J. F. (1980) *Inorg. Chem.*, **19**, 1319.
33. Aime, S., Botta, M. and Ermondi, G. (1992) *Inorg. Chem.*, **31**, 4291.
34. Aime, S., Barbero, L., Botta, M. and Ermondi, G. (1992) *J. Chem. Soc. Dalton. Trans.*, 225.
35. Hoeft, S. and Roth, K. (1993) *Chem. Ber.*, **126**, 869.
36. Jacques, V. and Desreux, J. F. (1994) *Inorg. Chem.*, **33**, 4048.
37. Marques, M.P.M., Geraldes, C.F.G.C., Sherry, A.D., Merbach, A.E., Powell, H., Pubanz, D., Aime, S. and Botta, M. (1995) *J. Alloys Comp.*, **225**, 303.
38. Aime, S., Botta, M., Fasano, M., Marques, M.P.M., Geraldes, C.F.G.C., Pubanz, D. and Merbach, A.E. (1997) *Inorg. Chem.*, **36**, 2059.
39. Spirlet, M.R., Rebizant, J., Desreux, J.F. and Loncin, M.F. (1984) *Inorg. Chem.*, **23**, 359.
40. Aime, S., Barge, A., Benetollo, F., Bombieri, G., Botta, M. and Uggeri, F. (1997) *Inorg. Chem.*, **36**, 4287.
41. Geraldes, C.F.G.C., Sherry, A.D. and Kiefer, G.E. (1992) *J. Magn. Reson.*, **97**, 290.
42. Sherry, A.D., personal communication.
43. Sherry, A.D., Ren, J., Huskens, J., Brücher, E., Tóth, É., Geraldes, C.F.G.C., Castro, M.M.C.A. and Cacheris, W.P. (1996) *Inorg. Chem.*, **35**, 4604.
44. Carvalho, R.A., Peters, J.A. and Geraldes, C.F.G.C. (1997) *Inorg. Chim.Acta*, **262**, 167.
45. Peters, J. A., (1988) *Inorg. Chem.*, **27**, 4686
46. Sherry, A.D., Zarzycki, R. and Geraldes, C.F.G.C. (1994) *Magn. Reson. Chem.*, **32**, 361.
47. Zitha-Bovens, E., van Bekkum, H., Peters, J. A. and Geraldes, C.F.G.C., manuscript in preparation.
48. Arean, C.O., Moore, G.R., Williams, G. and Williams, R.J.P. (1988) *Eur. J. Biochem.*, **173**, 607.
49. Dick, C.R., Geraldes, C.F.G.C., Sherry, A.D., Gray, C.W. and Gray, D.M. (1989) *Biochemistry*, **28**, 7896.
50. van Duynhoven, J.P.M., Nooren, I.M.A., Swinkels, D.W., Folkers, P.J., Harmsen, B.J.M., Konings, R.N.H., Tesser, G.I. and Hilbers, C.W. (1993) *Eur. J. Biochem.*, **216**, 507
51. Geraldes, C.F.G.C., Carvalho, R.A., Rodrigues, R.C. and Brito, R.M., manuscript in preparation.
52. Gupta, R.K. and Gupta, P. (1982) *J.Magn.Reson.*, **47**, 344.
53. Pike, M.M. and Springer, C.S. (1982) *J.Magn.Reson.*, **46**, 348.
54. Springer, C.S. (1987) *Annals N.Y. Acad. Sci.*, **508**, 130.
55. Sherry, A.D., Malloy, C.R., Jeffrey, F.M.H., Cacheris, W.P. and Geraldes, C.F.G.C. (1988) *J. Magn.Reson.*, **76**, 528.
56. Ramasamy, R., Freitas, D.M., Geraldes, C.F.G.C. and Peters, J.A. (1991) *Inorg. Chem.*, **30**, 3188.
57. Ren, J., Springer, Jr., C.S. and Sherry, A.D. (1997) *Inorg.Chem.*, **36**, 3493.

58. Matwiyoff, N.A., Gasparovic, C., Wenk, R., Wicks, J.D. and Rath, A. (1986) *Magn. Reson. Med.*, **3**, 164.
59. Buster, D.C., Castro, M.M.C.A., Geraldes, C.F.G.C., Malloy, C.R., Sherry, A.D. and Siemers, T. (1990) *Magn.Reson.Med.*, **15**, 25.
60. Bansal, N., Germann, M.J., Lazar, I., Malloy, C.R. and Sherry, A.D. (1992) *J. Magn. Reson. Imaging*, **2**, 385
61. Bansal, N., Germann, M.J., Seshan, V., Shires III, G.T., Malloy, C.R. and Sherry, A.D. (1993) *Biochemistry*, **32**, 5638.
62. Seshan, V. , Germann, M.J., Preisig, P., Malloy, C.R., Sherry, A.D. and Bansal, N. (1995) *Magn. Reson. Med.*, **34**, 25.
63. *New Developments in Contrast Agent Research* (1993) Rinck , P.A. and Muller, R.N. (eds.), European Magnetic Resonance Forum Foundation, Locarno.
64. Geraldes, C.F.G.C., Sherry, A.D., Lazar, I., Miseta, A., Bogner, P., Berenyi, E., Sumegi, B., Kiefer, G.E., McMillan, K., Maton, F. and Muller, R.N. (1993) *Magn. Reson. Med.*, **30**, 696.
65. Vogl, T.J., Juergens, M. and Balzer, J.O. (1995) *Advances in MRI Contrast*, **3**, 84.
66. Macri, M.A., de Luca, F., Maraviglia, B., Polizio, F., Garreffa, G., Cavallo, S. and Natali, P.G. (1989) *Magn. Reson. Med.*, **11**, 283.
67. Koenig, S. H. and Brown III, R.D. (1990) *Progr. Nucl. Magn. Reson. Spectr.*, **22**, 487.
68. Geraldes, C.F.G.C., Brown III, R. D., Brucher, E., Koenig, S. H., Sherry, A. D. and Spiller, M. (1992) *Magn. Reson. Med.*, **27**, 284.
69. Zhang, T., Chang, C.A., Brittain, H.G., Garrison, J. M., Telser, J.and Tweedle, M.F. (1992). *Inorg. Chem.*, **31**, 5597
70. Sherry, A.D., Brown III, R.D., Geraldes, C.F.G.C., Koenig, S.H., Kuan, K.-T. and Spiller, M. (1989) *Inorg. Chem.*, **28**, 620.
71. Micskei, K., Helm, L., Brucher, E. and Merbach, A. E. (1993) *Inorg. Chem.*,**32**, 3844.
72. Pubanz, D., González, G., Powell, D.H. and Merbach, A.E. (1995) *Inorg. Chem.*, **34**, 4447.
73. Lauffer, R. B., Brady, T. J., Brown III, R. D., Baglin, C. and Koenig, S. H. (1986) *Magn. Reson. Med.*, **3**, 541.
74. Geraldes, C.F.G.C., Urbano, A.M., Alpoim, M.C., Sherry, A.D., Kuan, K.-T., Rajagopalan, R., Maton, F. and Muller, R.N. (1995) *Magn. Reson. Imaging*, **13**, 401.
75. Aime, S., Botta, M., Fasano, M., Crich, S.G. and Terreno, E. (1996) *J. Biol. Inorg. Chem.*, **1**, 312.
76. Aime, S., Batnasov, A.S., Botta, M., Howard, J.A.K., Parker, D., Senanayake, K. and Williams, G. (1994) *Inorg. Chem.*, **33**, 4696.
77. Aime, S., Botta, M., Terreno, E., Anelli, P.L. and Uggeri, F. (1993) *Magn. Reson. Med.*, **30**, 583.
78. Liepinsh, E. and Otting, G. (1996) *Magn. Reson. Med.*, **35**, 30.

CHARACTERIZATION OF REACTION COMPLEX STRUCTURES OF ATP-UTILIZING ENZYMES BY HIGH RESOLUTION NMR

B. D. Nageswara Rao
Department of Physics
Indiana University-Purdue University at Indianapolis (IUPUI)
402 N. Blackford St., Indianapolis, IN 46202 USA

1. Introduction

A paradigm of biomolecular science is that the elucidation of the molecular basis of enzyme catalysis requires knowledge of the active-site structures of enzymes i.e., the precise conformational arrangement of the enzyme-bound substrates and their amino-acid environment. In this article, a review of the recent and ongoing efforts to characterize the active-site structures of a group of ATP-utilizing enzymes, by the use of high-resolution NMR methods, is presented. ATP-utilizing enzymes occur in a variety of important biochemical pathways and may be divided into three categories on the basis of the transferable moiety from ATP (see Fig. 1): 1. phosphoryl transfer (kinases), 2. nucleotidyl transfer (e.g. amino-acyl tRNA synthetases), and 3. pyrophosphoryl transfer (e.g. phosphoribosyl pyrophosphate synthetase), listed in decreasing order of their abundance in biochemical pathways. All the three categories of enzymes require a divalent cation, Mg(II) *in vivo*, as an obligatory component. The diversity in their catalytic roles, the obligatory cation requirement, and their ubiquity in biochemical pathways make ATP-utilizing enzymes an attractive group for structural investigations aimed at gathering insight into the molecular basis of enzyme catalysis.

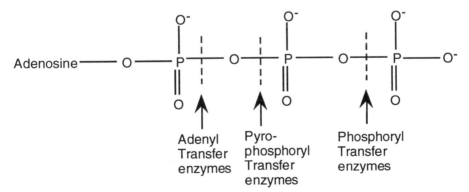

Figure 1. The phosphoryl chain cleavage points of the various ATP-utilizing enzymes

In the matter of obtaining structural data on the amino-acid residues at the active site, x-ray diffraction is the method of choice (especially for proteins of high molecular weight). A significant body of impressive crystallographic evidence, accumulated in the past decade, shows that nucleotide binding sites in a number of ATP-utilizing enzymes possess conserved residues and consensus sequences such as GX_4GKT/S or GX_2GXGKT/S in the ATP binding region that suggest mechanistic roles for specific residues at the active site [1-5]. Similar consensus sequences such as KMSKS have been found at the ATP-binding sites for the amino-acyl tRNA synthetases which are adenyl transfer enzymes [5]. Crystallographic information on the conformation of the substrates at the active site is sparse in comparison, because of difficulties associated with co-crystallizing enzyme-substrate complexes and with soaking methods. Furthermore, the presence of anions such as sulfate interfere with the binding of cations to the nucleotides, and there appears to be no simple means to ascertain that the conformation assumed by the substrate in the crystalline state is the active or productive conformation.

NMR and EPR methods offer an advantage in that the measurements can be made in the liquid state, and the interpretation of the results is not complicated by issues such as anionic interference and crystal packing effects. However, these methods possess their own limitations, e.g., EPR techniques [6-8] are limited to the immediate environment of the cation, and the distance information is not explicitly obtained; with NMR methods the information is related to a few (usually spin 1/2) nuclei, which sometimes necessitates extensive isotopic labeling (especially ^{13}C and ^{15}N), and the quantity of protein required is large. Nevertheless, NMR methods are capable of providing explicit and reliable distance information on the reaction complexes that will be an excellent and necessary complement to the protein structural data obtained by crystallographic methods. The obligatory presence of the cation in the reaction complexes of ATP-utilizing enzymes not only raises the specific question of the role of the cation in catalysis, but also provides a means to obtain structural information on these complexes by NMR and EPR methods. Most of these enzymes can be activated by substituent paramagnetic cations such as Mn(II) and Co(II) in place of Mg(II). Accordingly, the distances of the substrate nuclei from the cation may be determined on the basis of distance-dependent paramagnetic contribution of the cation to the spin-lattice relaxation rates of the substrate nuclei [9, 10]. Such structural information can be obtained, irrespective of direct chelation with the cation, as long as the relaxation rates of the substrate nuclei are measurably enhanced.

A strategy for comprehensive structural characterization of the reaction complex was devised using the following reasoning. An isolated ATP molecule has three distinct internal mobilities viz., those associated with the flexible phosphate chain, the rotation of the adenine base about the glycosidic bond, and the sugar pucker. The flexibility may also facilitate the choice of different points of cleavage on the phosphate chain in the three categories of reactions described in Fig. 1. Assuming that the cation-nucleotide complex has a unique conformation when bound to the enzyme, a determination of this conformation requires structural data that will reveal how the internal motions are arrested in the bound species. Our step-wise NMR strategy for structural characterization of the enzyme-bound reaction complex is to:

1. determine the location of the cation with reference to the phosphate chain of the nucleotides on the basis of distance-dependent paramagnetic contribution to the T_1 of the ^{31}P nuclei,
2. define the glycosidic torsion angle of the adenine base with respect to the ribose, and the sugar pucker, on the basis of transferred nuclear Overhauser effect spectroscopy (TRNOESY) measurements,
3. determine the orientation of the phosphate chain with respect to adenosine making T_1 measurements on ^{13}C and ^{15}N labeled nucleotides in the presence of paramagnetic cations, and finally
4. determine the conformation of the second substrate from similar T_1 measurements made on labeled second substrates.

That the enzyme-bound substrates on which NMR measurements are made are in active conformations is vividly demonstrated by the fact that spectra of enzyme-bound equilibrium mixtures can be readily observed as the interconversion of substrates and products, E•MS$_1$•S$_2$↔E•MP$_1$•P$_2$, is in progress. Furthermore, relaxation measurements on selected nuclei in the presence of paramagnetic cations may be performed on enzyme-bound equilibrium mixtures to determine structural changes in the reaction complexes accompanying enzyme turnover.

The basis for the two methods used *viz.*, relaxation effects due to paramagnetic cations and TRNOESY measurements, is straightforward. However, the measurements require deliberate and incisive methodological scrutiny both from NMR and biochemical points of view. This observation may seem redundant at first. However, insouciant application of the methodologies in the past have led to anomalous results and incorrect published structures. In this article, we present a discussion of the pitfalls in these experimental methods and strategies to overcome them, followed by a summary of the results obtained thus far, and the prognosis for the future.

2. Experimental Methods

2.1 NUCLEAR SPIN RELAXATION DUE TO PARAMAGNETIC CATIONS

The theory behind this method has been extensively reviewed in the literature [9, 10]. In order to appreciate the essential features of the experimental method, we present the key equations below. The spin relaxation rate of a nucleus due to dipolar interaction with a cation located at a distance r from it is given by

$$T_{1M}^{-1} = \left(\frac{C}{r}\right)^6 f(\tau_C) , \qquad (1)$$

in which

$$C = \left[(2/15)S(S+1)g^2\gamma_I^2\beta^2\right]^{1/6} , \qquad (2)$$

$$f(\tau_C) = \frac{3\tau_{C1}}{1+\omega_I^2\tau_{C1}^2} + \frac{7\tau_{C2}}{1+\omega_S^2\tau_{C2}^2} , \qquad (3)$$

$$\tau_{Ci}^{-1} = \tau_R^{-1} + \tau_S^{-1} \qquad (4)$$

In Eqs. (1)-(4) S, g, ω_S, τ_{S1}, and τ_{S2} are the spin, the g-factor, the Larmor precession frequency, the longitudinal, and the transverse relaxation times, respectively, of the paramagnetic cation, γ_I, and ω_I are the gyromagnetic ratio and Larmor frequency, respectively, of the relaxing nucleus, β is the Bohr magneton, and τ_R is the isotropic rotational correlation time of the enzyme complex. Furthermore, the following assumptions are implicit: (i) g is isotropic, (ii) contributions of scalar interaction between the nucleus and the cation are negligible, (iii) the zero-field splitting of the cation is much smaller than its Zeeman interaction, and (iv) the electron relaxation is uniquely described by τ_{S1}, and τ_{S2}. These assumptions are justifiable for Mn(II) complexes and are questionable for Co(II) complexes. Nevertheless, the form of Eq. (1) is valid for enzyme-bound Co(II)-nucleotide complexes, and the contribution of the scalar interaction was recently shown to be negligible [11].

Paramagnetic effects on nuclear spin relaxation are measured on samples in which the cation concentration is a small fraction of the ligand and the observed relaxation is altered by exchange between the diamagnetic and paramagnetic complexes. The paramagnetic contribution, T_{1P}^{-1}, to the observed relaxation rate, $T_{1,obsd}^{-1}$, is given by

$$T_{1P}^{-1} = T_{1,obsd}^{-1} - T_D^{-1} = \frac{p}{T_{1M} + \tau_M} \qquad (5)$$

with

$$p = [\text{cation}]/[\text{ligand}] \qquad (6)$$

where T_{1D}^{-1} is the relaxation rate in the diamagnetic complex and τ_M is the lifetime of the paramagnetic complex. Equation (5) is valid for two exchanging complexes such that $T_{1D}^{-1} \ll T_{1M}^{-1}$ and $p \ll 1$. Clearly, if $\tau_M \gg T_{1M}$, T_{1P}^{-1} is bereft of structural information. Conversely, T_{1M} can be obtained from measurements of T_{1P}^{-1} if $\tau_M \ll T_{1M}$, or if $\tau_M \approx T_{1M}$ and can be quantitatively taken into account.

TABLE 1. Exchange effects on ^{31}P relaxation rates in creatine kinase•ATP complexes [From ref. 12]

		^{31}P NMR Frequency	
		121.5 MHz	190.2 MHz
Complex	ΔE (Kcal/mole)	$(pT_{1p})^{-1}$ s^{-1}	$(pT_{1p})^{-1}$ s^{-1}
E•MnATP		± 200 s^{-1}	
α-P	6.5	1840	2010
β-P	8.1	2170	2210
γ-P	5.1	2280	2550
E•CoATP		± 20 s^{-1}	
α-P	2.4	85	170
β-P	1.0	120	260
γ-P	3.1	170	265

The experimental strategy [12] for obtaining the quantities T_{1M}^{-1} and $f(\tau_C)$ (see Eq. (1)) required for evaluating cation-nucleus distances in enzyme-nucleotide complexes consists of making these measurements exclusively on enzyme-bound complexes as a function of frequency and temperature, and with each of the cations Mn(II) and Co(II). Measuring T_{1P} exclusively on enzyme-bound complexes ensures conformity with Eq. (5), and maximizes the contribution of the E•M•S (enzyme•metal•substrate) complex. Since T_{1M} depends on frequency and τ_M does not, and furthermore, the activation energies of T_{1M} and τ_M are expected to be in the range of 1-3 Kcal/mole and >5 Kcal/mole, respectively, measuring T_{1P} as a function of frequency and temperature allows a separation of the contributions of τ_M and T_{1M} to T_{1P}. An example of the frequency and temperature dependent T_{1P} measurements of ^{31}P nuclei of nucleotide complexes bound to rabbit muscle creatine kinase in the presence of Mn(II) and Co(II) [12] is shown in Table 1. Results of this kind, obtained with several different enzymes, showed that $\tau_M > T_{1M}$ in the presence of Mn(II) for cation-^{31}P distances up to about 6.5 Å. To determine shorter distances, Co(II), which induces weaker nuclear relaxation than Mn(II) (at the same distance), should be used [12].

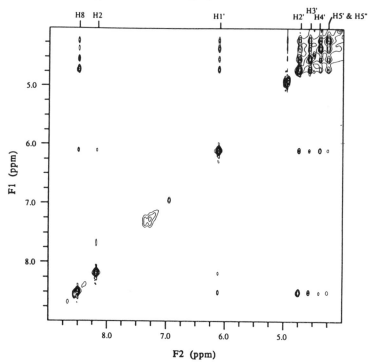

Figure 2. Proton TRNOESY (500 MHz, mixing time 120 ms) of 1.03 mM creatine kinase with 10.18 mM MgADP [From ref. 16]

2.2 TRNOESY

In a TRNOESY experiment, the conformation of a ligand (small molecule) bound to a macromolecule is investigated by making NOE measurements when the ligand is in exchange between the bound and free states. Since the rotational correlation time of the bound complex is typically longer (by a factor of about a thousand) than that of the free ligand, the cross relaxation rates in the bound state are correspondingly larger. If the ligand is in fast exchange between bound and free states i.e., the exchange rates are much faster than the cross relaxation rates, it can be shown that the composite relaxation matrix reduces to a weighted average of that in the two states [13]. Thus, if the experiment is performed with a ten-fold excess of ligand concentration over that of the macromolecule, the cross relaxation rates of the exchanging ligand will still be dominated by the bound complex by a factor of about 100. Thus, the observed signal is almost entirely due to free ligand (with a ten-fold excess), whereas the NOE is predominantly due to that in the bound complex [14, 15].

In view of the above, it was considered a major advantage of the TRNOESY method that the experiments could be performed on samples containing a significant excess of substrate over the enzyme [14, 15]. Thus, in virtually all of the early experiments aimed at determining the conformation of the adenosine moiety in the nucleotide

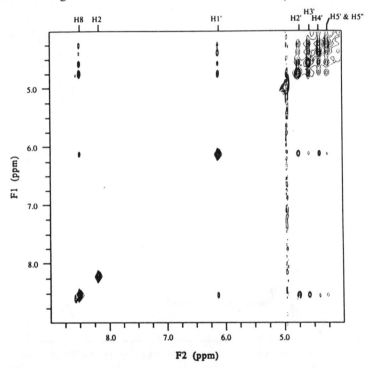

Figure 3. Proton TRNOESY (500 MHz, mixing time 120 ms) of 0.45 mM γ-globulin with 10.18 mM MgADP [From ref. 16]

complexes of ATP-utilizing enzymes, typical sample protocols contained 1 mM enzyme sites, and 10 mM nucleotide. A TRNOESY spectrum with this kind of sample for the MgADP (10.18 mM) complex of creatine kinase (1.03 mM), for a mixing time of 120 ms, is shown in Fig. 2. In order to investigate the possible contribution of adventitious binding of the nucleotide, a similar experiment was performed for a sample containing 10.18 mM MgADP and 0.45 mM γ-globulin, which is not known to have a specific binding site for MgADP [16]. The resulting spectrum, shown in Fig. 3, bears a striking resemblance to Fig. 2. Addition of 100 mM KCl to the sample resulted in a 25% reduction in the NOE's indicating that binding of MgADP to γ-globulin does occur, and is diminished in the presence of high salt concentration. A third experiment was performed with bovine serum albumin (BSA) which also does not have a specific MgADP binding site, instead of γ-globulin, and a very similar spectrum was obtained. These experiments demonstrate that nonspecific binding contributes significant NOE's [16].

Returning to the sample protocol of 10 mM nucleotide and 1 mM enzyme sites, assuming a dissociation constant of ~50 μM, it is readily seen that the specific nucleotide binding sites are saturated under these conditions, and there is still ~9 mM nucleotide in the sample in the presence of 1 mM enzyme. Since the nucleotides have a negatively charged phosphate chain, adventitious association with the protein is expected to occur in the vicinity of positively charged regions on the surface of the protein. The extent of such binding, the multiplicity of such sites, and the relative dissociation constants depend on the particular enzyme-substrate complexes. However, with a nine fold excess of nucleotide over the enzyme, the concentration of adventitiously bound nucleotide may even exceed that bound at the active site in some cases.

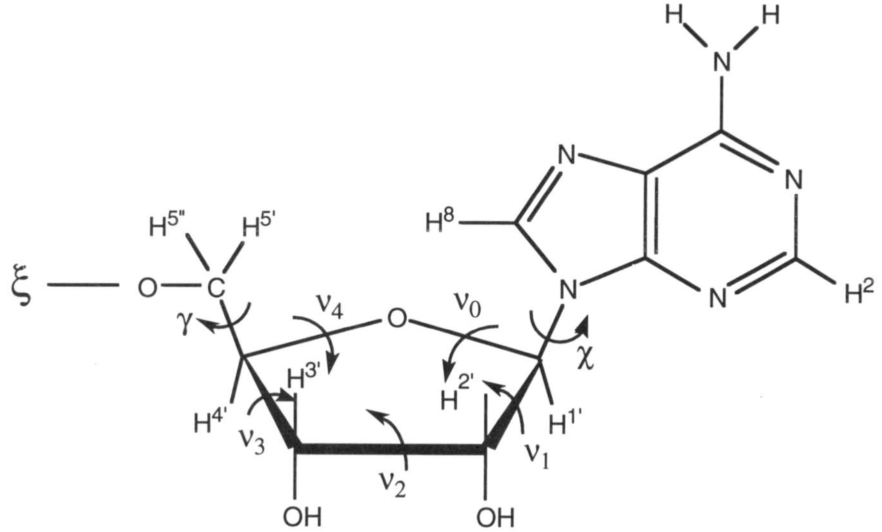

Figure 4. Adenosine proton numbering and torsion angles

In order to assess the extent of weak nonspecific binding in the creatine kinase-MgADP sample, and to determine a sample protocol that will minimize the adventitious binding, TRNOESY measurements were made, at a constant mixing time, for a series of samples in which the ligand concentration was varied from 1.5 mM to 10 mM while keeping the ratio of ligand and enzyme concentrations fixed at 10:1. A plot of the normalized NOE for the H1'-H2' proton pair (see Fig. 4 for proton numbering in adenosine) as a function of ligand concentration is shown in Fig. 5. Since the H1'-H2' distance is considered to be invariant in any conformation of adenosine [17], the increase in

NOE at high ligand concentrations clearly and unequivocally demonstrates the presence of adventitious binding. Further details are given in reference 16. This weak nonspecific binding is not serious for ligand concentrations up to about 3 mM (see Fig. 5). Thus, adventitious binding occurs with dissociation constants larger than about 2 mM. A sample protocol was chosen with 1.5 mM MgADP and 0.36 mM enzyme for the TRNOESY measurements, so that the NOE's arise exclusively from the nucleotide bound at the active-site, and these were used to determine the nucleotide conformation. The TRNOESY spectrum of this sample is shown in Fig. 6. The NMR parameters for this spectrum are the same as those for the spectra in Fig. 2 and 3. There are some qualitatively discernible differences between Fig. 6 and Fig. 2. Especially interesting are the NOE's between nucleotide and enzyme protons. The TRNOE buildup curves for all

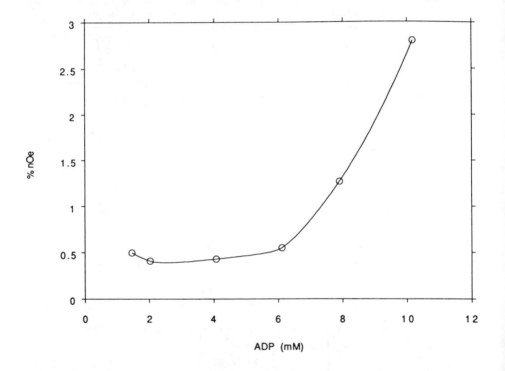

Figure 5. Dependence of TRNOE of H1'-H2' on [ADP] for constant creatine kinase:ADP ratio (1:10) [From ref. 16]

ligand proton pairs were analyzed using a complete relaxation matrix approach, formulated earlier in a form applicable to TRNOESY measurements [13]. The interproton distances were then used as constraints in modeling the ligand structure using the program CHARMm. The data for MgATP and MgADP could be fit with the same conformation. The concentration dependence of TRNOE described above suggests the following points regarding the method:
  1. Weak nonspecific binding effects, hitherto ignored in all investigations, contribute significantly at sample protocols typically used in practically all nucleotide con-

formation determinations by the TRNOE method prior to 1993. Thus, most of these previously published conformations [17-26] are likely to be incorrect.

2. The practice of using a substrate concentration far in excess of the enzyme, which was perceived as an advantage in TRNOESY studies [14, 15] has to be sacrificed if nonspecific binding effects are to be minimized in these measurements.

3. By measuring the ligand concentration dependence of the observed TRNOE, while keeping the concentration ratio of ligand to enzyme constant and sufficiently large to ensure complete saturation of the active site throughout the range of concentrations studied, it is possible to devise a sample protocol in which these adventitious binding effects are minimized if not eliminated. The optimum sample conditions are likely to be different for different enzyme-substrate complexes.

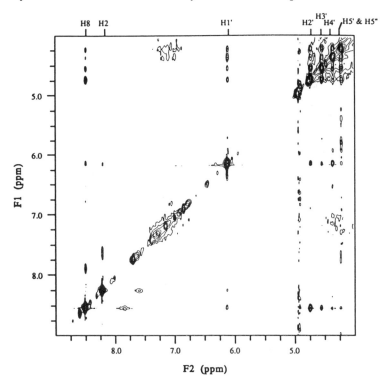

Figure 6. Proton TRNOESY (500 MHz, mixing time 120 ms) of 0.36 mM creatine kinase with 1.5 mM MgADP [From ref. 16]

3. Structural Information Obtained Thus Far

3.1 LOCATION OF THE CATION

These data were obtained for various kinases on the basis of ^{31}P relaxation measurements of enzyme-bound nucleotides in the presence of Co(II). As shown in Table 1, the relaxation rates with Mn(II) are exchange limited. The calculation of Co(II)-^{31}P distances is beset with theoretical problems associated with $f(\tau_C)$. As stated earlier, the assumptions implicit in Eq. (3) may not be valid for Co(II) complexes. However, on the basis of qualitative features of the EPR spectra of Co(II) complexes, and of relaxation measurements and their frequency dependence, it is plausible to choose the range of 10^{-12} s $< f(\tau_C) < 5 \times 10^{-12}$ s [9, 10, 12]. The distances calculated on this basis, and with g=4.33 and s=3/2 (in Eq. (2)), are listed in Table 2. For specific details regarding the experimental protocols, references 12, 27-29 may be consulted. These Co(II)-^{31}P distances

TABLE 2. Distances in Å between Co(II) and the ^{31}P nuclei of ATP and ADP in various enzyme complexes.

Enzyme	α-P	β-P	γ-P	reference
Creatine Kinase (ADP)	3.0-4.0	2.9-3.8		12
Creatine Kinase (ATP)	3.2-4.2	3.0-4.0	2.9-3.8	12
Arginine Kinase (ADP)	3.2-4.2	3.4-4.5		29
Arginine Kinase (ATP)	3.5-4.6	3.2-4.2	3.2-4.2	29
3-P-glycerate Kinase (ADP)	2.8-3.6	2.7-3.5		28
3-P-glycerate Kinase (ATP)	3.1-4.1	2.9-3.8	2.8-3.7	28
Adenylate Kinase (GDP)	3.4-4.5	2.7-3.6		27
Adenylate Kinase (GTP)	3.6-4.7	3.0-3.9	3.0-4.0	27
Adenylate Kinase (ATP)	3.9-5.1	2.9-3.8	2.7-3.6	27

suggest direct coordination of the cation with all the phosphate groups of the nucleotides except in the case of α-P(ATP) bound to adenylate kinase. In this case, the Co(II)- α-P distance is too large for direct coordination. For creatine kinase and 3-P-glycerate kinase, the data agree with the conclusions reached on the basis of ^{17}O superhyperfine structure of selectively labeled ligands on Mn(II) EPR spectra [6, 7].

3.2 ADENOSINE CONFORMATION

TRNOESY measurements were made to determine interproton distances in the adenosine moieties of nucleotides bound to four phosphoryl transfer enzymes, creatine kinase [16], arginine kinase [30], pyruvate kinase [31], and adenylate kinase [32], one adenyl transfer enzyme, methionyl tRNA synthetase [33], and one pyrophosphoryl transfer enzyme, phosphoribosyl pyrophosphate synthetase [34]. In all these cases, the sample conditions were optimized to minimize the nonspecific binding effects discussed earlier. Glycosidic orientations and ribose puckers corresponding to the energy minimized structures compatible with the distances obtained from the NOE data are given in Table 3. It should be noted that the sugar pucker is not particularly sensitive to most of the interproton distances of the ribose moiety other than those between H3'-H4' and H1'-H4' [35], and therefore the deduced value may not be particularly accurate. On the other hand, the glycosidic torsion angles in Table 3 fall in a rather narrow range of 52±8°. Such an adenosine conformation is depicted in Fig. 7. This similarity suggests a possible recognition and binding motif for the adenosine moiety at the active sites of ATP-utilizing enzymes irrespective of whether they catalyze phosphoryl transfer, adenyl transfer, or pyrophosphoryl transfer reactions. It should be pointed out that x-ray crystallographic data on ATP-utilizing enzymes co-crystallized with nucleotides yield a broad range of glycosidic orientations and do not reveal such similarity as the NMR data does [33]. The NMR measurements are made on enzyme-substrate complexes under conditions in which the enzymatic reactions readily occur and enzyme-bound equilibrium mixtures are observable. In the crystalline state, this may or may not be the case, since the substrate may sometimes be trapped in an unproductive conformation. Therefore, the NMR determined conformations, within the limitations of their accuracy, are more likely to represent the active form.

TABLE 3. Adenosine conformations of bound nucleotides in various enzyme complexes

Complex	χ(deg)	Sugar Pucker	Reference
ArgK·MgADP	51	^0E	30
ArgK·MgADP·NO_3^-·Arg	52	$_1E - {}_1^2T$	30
ArgK·MgATP	50	$_1E - {}_1^2T$	30
CK·MgADP	51	^0_4T	16
CK·MgATP	51	^0_4T	16
AdK·M(II)ATP	48	^2E	32
AdK·M(II)GDP·AMP	40	$^0_4T - {}^0E$	32
PyK·MgATP (active site)	44	$^3_4T - {}_4E$	31
PyK·MgATP (ancillary)	46	$_1E - {}_1^2T$	31
PRPPS·MgATP	50	$^0_1T - {}_1E$	34
MTRS·MgATP	53	$^3_4T - {}_4E$	33
MTRS·MgATP·L-Methioninol	59	$^3_4T - {}_4E$	33
MTRS·AMP·L-Methioninol	55	$^3_4T - {}_4E$	33
Free ATP	5	$^3E - {}^3_4T$	

Enzyme abbreviations: ArgK, arginine kinase; CK, creatine kinase; AdK, adenylate kinase; PyK, pyruvate kinase; PRPPS, 5'-phosphoribosyl pyrophosphate synthetase; and MTRS, methionyl tRNA synthetase.

In structure determinations from TRNOESY data, potential problems may arise from finite on-off rates, ligand motions at the active site, ligand-protein cross relaxation, and protein mediated spin diffusion [36-39]. The last two of these effects can be handled in some fashion if the amino-acid environment is independently known. It is found that these factors reduce the effective rotational correlation time of the bound complex as reflected in TRNOESY build-up. Qualitatively, this reduction increases with the size of the protein. Nevertheless, internally consistent conformations could be deduced for the ligands, presumably because the ratios of cross-relaxation rates are unaffected by a change in the correlation times. For a further discussion of this question, references 16, 30-34 may be consulted.

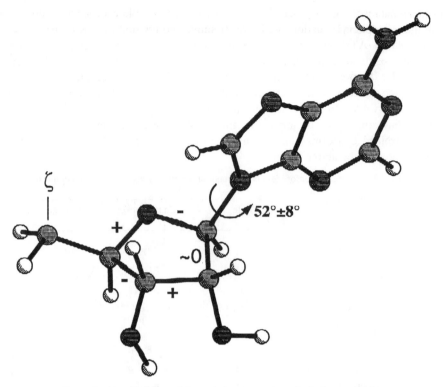

Figure 7. Observed glycosidic angle in enzyme-bound adenine nucleotides

3.3 ORIENTATION OF THE PHOSPHATE CHAIN

As stated earlier, once the location of the cation with respect to the phosphate chain, and the conformation of the adenosine moiety are known, the orientation of the phosphate chain with reference to adenosine can be measured through T_{1P} measurements on ^{13}C or ^{15}N labeled nucleotides bound to the enzyme. Since most of the distances are expected to be larger than 6.5 Å, Mn(II) may be used as the cation, which allows an evaluation of these distances with some precision based on frequency-dependent measurements. The first such measurements were made on creatine kinase complexes of [2-^{13}C]ATP and [2-^{13}C]ADP synthesized in our laboratory, which yielded distances of 10.0±0.5 Å and 8.6±0.5 Å, respectively [40]. While these distances alone are not adequate to precisely determine the orientation of the phosphate chain, they do significantly reduce the allowed ranges for it. Furthermore, it was possible to show that in a coordinate frame fixed on the adenosine moiety, Mn(II) moves through 1.7 Å as the enzyme turns over and ATP changes to ADP on the enzyme surface [40].

Nucleotides uniformly labeled with ^{13}C and ^{15}N have now become commercially available, and measurements are currently in progress with several different enzymes. A T_1-stack plot of such a measurement with [ul-^{13}C]ATP bound to 3-P-glycerate kinase is shown in Fig. 8 [41]. It is clear that all the 10 carbon signals are resolved, and Mn(II)-

C distance data may be obtained for all of them. A complete characterization of the bound nucleotide conformations will soon be published for some of these enzymes.

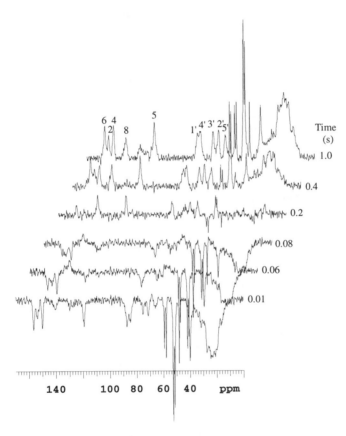

Figure 8. ^{13}C T_1-stack-plot of [ul-^{13}C]ATP bound to 3-P-glycerate kinase with 3% Mn(II). Resonances from the adenosine carbons are numbered. All other resonances are from the protein or the buffer. [From ref. 41]

3.4 CATION-NUCLEUS DISTANCES FOR THE SECOND SUBSTRATE

Only two such distances have been published thus far, and these involve second substrates containing a ^{31}P nucleus. These are the Mn(II)-P(3-P-glycerate) distance in the E•MnADP•3-P-glycerate complex of 3-P-glycerate kinase [28] and Mn(II)-P(AMP) distance in E•MnGDP•AMP of adenylate kinase [27]. More information of this kind is expected in the future, once the nucleotide conformations are clearly established.

3.5 STRUCTURE MEASUREMENTS IN ENZYME-BOUND EQUILIBRIUM MIXTURES

We have recently begun making paramagnetic relaxation measurements on enzyme-bound equilibrium mixtures in order to probe the structural alterations in the reaction complex accompanying the enzyme turnover. The data were analyzed on the basis of a theory which treats the complications arising from the fact that there is an additional exchange process due to the interconversion of the reactants and products on the enzyme surface [42], thus allowing one to determine, for example, how far the itinerant phosphoryl group moves on the surface of a phosphoryl-transfer enzyme (see Fig. 9). It is found that the Co(II)-γ-P(ATP) distance of 3.0 Å in E•CoATP•creatine complexes changes to a Co(II)-P-creatine distance of about 4.4 Å in the E•CoADP•P-creatine complex. The itinerant phosphoryl group thus moves at least 1.4 Å as the enzyme turns over. We do not know of any other means to determine dynamic structural information of this kind. Details of this work will be published soon.

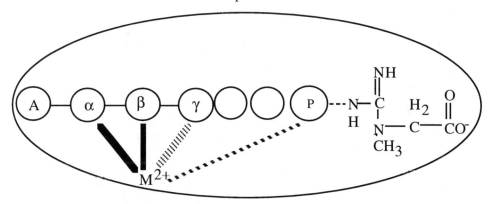

Figure 9. Movement of the itinerant phosphoryl group on creatine kinase from ATP to phosphocreatine

4. Concluding Remarks

The foregoing is a summary of the recent and ongoing efforts to characterize the structures of the reaction complexes of ATP-utilizing enzymes. Experience has shown that the strategies for obtaining reliable structures need to be carefully devised. The ultimate goal of these structures is to provide some of the critical information relevant to understanding the mechanisms of these enzymes. Such an understanding requires knowledge of the structures of the reaction complexes and their amino-acid environment. While x-ray crystallography provides the most reliable information of the protein structure, NMR methods appear to be the best means to obtain reaction complex structures, especially for ATP-utilizing enzymes. The two methodologies thus eminently complement each other.

As our structure determination of the reaction complexes attains a level of completeness, we have recently begun making attempts to combine these structures with x-ray structures of the protein active sites. Although it is premature to forecast the insights that such an approach will yield, there is reason to be hopeful because it has the merit of pooling together the strengths of the two most popular methodologies for the study of macromolecular structure.

5. Acknowledgments

I am grateful to Dr. Bruce D. Ray, Dr. Gotam K. Jarori, Dr. Nagarajan Murali, Dr. Mei Hing Chau, Dr. Steven B. Landy, Dr. Vidya Raghunathan, and Ms. Yan Lin, all of whom have significantly contributed to this research.

This research was supported by the National Science Foundation in the initial phase, by the NIH (GM 43966) since 1989, and by IUPUI. I am thankful to Dr. B. D. Ray for considerable help and advice in preparing this article.

6. References

1. Higgings, C. F., Hiles, I. D., Salmond, G. P. C., Gill, D. R., Downie, J. A., Evans, I. J., Holland, I. B., Gray, L., Buckel, S. D., Bell, A. W. and Hermodson, M. A. (1986) A Family of Related ATP-Binding Subunits coupled to Many Distinct Biological Processes in Bacteria, *Nature* **323**, 448-450.
2. Saraste, M., Sibbald, P. R. and Wittinghofer, A. (1990) The p-loop - A Common Motif in ATP- and GTP-Binding Proteins, *Trends in Biochem. Sci.* **15**, 430-434.
3. Schulz, G. E. (1992) Binding of Nucleotides by Proteins, *Curr. Opin. Struct. Biol.* **2**, 61-67.
4. Traut, T. W. (1994) The Functions and Consensus Motifs of Nine Types of Peptide Segments that Form Different Types of Nucleotide Binding Sites, *Eu. J. Biochem.* **222**, 9-19.
5. Hountondji, C. Dessen, P. and Blanquet, S. (1993) The SKS and KMSKS Signature of Class I Amino Acyl tRNA Synthetases Correspond the GKT/S Sequence Characteristic of the ATP-Binding Site of Many Proteins, *Biochemie* **75**, 1137-1142.
6. Leyh, T. S., Goodhart, P. J., Nguyen, A. C., Kenyon, G. L. and Reed, G. H. (1985) Structures of Mn(II) Complexes with ATP, ADP and Phosphocreatine in the Reactive Central Complexes with Creatine Kinase, *Biochemistry* **24**, 308-316.
7. Moore, J. M. and Reed, G. H. (1985) Coordination Scheme and Stereochemical Configuration of Mn(II) Adenosine 5'-Diphosphate at the Active Site of 3-P-glyceratekinase, *Biochemistry* **24**, 5328-5333.
8. Buchbinder, J. L. and Reed, G. H. (1990) Electron Paramagnetic Resonance Studies of the Coordination Scheme and Site Selectivities for Divalent Metal Ions in Complexes with Pyruvate Kinase, *Biochemistry* **29**, 1799-1806.
9. Villafranca, J. J. (1984) Paramagnetic Probes of Enzyme Complexes with Phosphorous Containing Compounds, in D. G. Gorenstein (ed.), *Phosphorous-31 NMR: Principles and Applications*, Academic Press, New York, pp. 155-174.
10. Mildvan, A. S. and Gupta, R. K. (1978) A Nuclear Relaxation Measurement of the Geometry of Enzyme-Bound Substrates and Analogs, *Methods Enzymol.* **49G**, 322-359.
11. Ray, B. D., Jarori, G. K. and Nageswara Rao, B. D. (1999) Paramagnetic Effects on Nuclear Relaxation in Enzyme-Bound Co(II)-Adenine Nucleotide Complexes: Relative Contributions of Dipolar and Scalar Interactions, *J. Magn. Reson.* **136** (in press).
12. Jarori, G. K., Ray, B. D. and Nageswara Rao, B. D. (1985) Structure of Metal-Nucleotide Complexes Bound to Creatine Kinase: ^{31}P NMR Measurements Using Mn(II) and Co(II), *Biochemistry* **24**, 3487-3494.
13. Landy, S. B. and Nageswara Rao, B. D. (1989) Dynamical NOE in Multiple-Spin Systems Undergoing Chemical Exchange, *J. Magn. Reson.* **81**, 371-377.
14. Campbell, A. P. and Sykes, B. D. (1991) Theoretical Evaluation of 2D TRNOE, *J. Magn. Reson.* **93**, 77-92.
15. Campbell, A. P. and Sykes, B. D. (1993) The 2D TRNOE: Theory and Practice, *Ann. Rev. Biophys. Biomol. Struct.* **22**, 99-122.
16. Murali, N., Jarori, G. K., Landy, S. B. and Nageswara Rao. B. D. (1993) TRNOESY Studies of Nucleotide Conformation in Creatine Kinase Complexes: Effects Due to Weak Nonspecific Binding, *Biochemistry* **32**, 12941-12948.
17. Rosevear, P. R., Bramson, H. N., O'Brian, C., Kaiser, E. T. and Mildvan, A. S. (1983) NOE Studies of the Conformations of Tetraaminocobalt(III)ATP Free and Bound to Bovine Heart Protein Kinase, *Biochemistry* **22**, 3439-3447.

18. Clore, G. M. and Gronenborn, A. M. (1982) Theory and Applications of the TRNOE to the Study of Small Ligands Bound to Proteins, *J. Magn. Reson.* **48**, 402-417.
19. Gronenborn, A. M., Clore, G. M., Brunori, M., Giardina, B., Falcioni, G. and Perutz, M. F. (1984) Stereochemistry of ATP and GTP Bound to Fish Haemoglobins, *J. Mol. Biol.* **178**, 731-742.
20. Ferrin. L. J. and Mildvan, A. S. (1985) NOE Studies of the Conformations and Binding Site Environments of Deoxynucleoside Triphosphate Substrates Bound to DNA Polymerase I, and its Large Fragment, *Biochemistry* **24**, 6904-6913.
21. Fry, D. C., Kuby, S. A. and Mildvan, A. S. (1987) NMR Studies of the AMP-Binding Site and Mechanism of Adenylate Kinase, *Biochemistry* **26**, 1645-1655.
22. Rosevear, P. R., Fox, T. L. and Mildvan, A. S. (1987) NOE Studies of the Conformations of MgATP Bound to the Active Site and Secondary Sites of Muscle Pyruvate Kinase, *Biochemistry* **26**, 3487-3493.
23. Rosevear, P. R., Powers, V. M., Dowhan. D., Mildvan, A. S. and Kenyon, G. L. (1987) NOE Studies on the Conformation of MgATP Bound to Rabbit Muscle Creatine Kinase, *Biochemistry* **26**, 5338-5344.
24. Williams, J. S. and Rosevear, P. R. (1991) NOE Studies on the Conformation of Mg(α,β-methylene)ATP Bound to *Escherichia coli* Methionyl tRNA Synthetase, *J. Biol. Chem.* **266**, 2089-2098.
25. Williams, J. S. and Rosevear, P. R. (1991) NOE Studies on the Conformation of Mg(α,β-methylene)ATP Bound to *E. coli* Isoleucyl tRNA Synthetase, *Biochem. Biophys. Res. Comm.* **176**, 682-689.
26. Landy, S. B., Plateau, P., Ray, B. D., Lipkowitz, K. B. and Nageswara Rao, B. D. (1992) Conformation of MgATP Bound to Nucleotidyl and Phosphoryl Transfer Enzyme: 1H TRNOE Measurements on Complexes of Methionyl tRNA Synthetase and Pyruvate Kinase, *Eur. J. Biochem.* **205**, 59-69.
27. Ray, B. D., Rösch, P. and Nageswara Rao, B. D. (1988) ^{31}P NMR Studies of Cation-Nucleotide Complexes Bound to Porcine Muscle Adenylate Kinase, *Biochemistry* **27**, 8669-8676.
28. Ray. B. D. and Nageswara Rao, B. D. (1988) ^{31}P NMR Studies of Enzyme-Bound Substrate Complexes of Yeast 3-Phosphoglycerate Kinase. 2. Structure Measurements Using Paramagnetic Relaxation Effects of Mn(II) and Co(II), *Biochemistry* **27**, 5574-5578.
29. Jarori, G. K., Ray, B. D. and Nageswara Rao, B. D. (1989) ^{31}P and ^{1}H NMR Studies of the Structure of Enzyme-Bound Substrate Complexes of Lobster Muscle Arginine Kinase, *Biochemistry* **28**, 9343-9350.
30. Murali, N., Jarori, G. K. and Nageswara Rao, B. D. (1994) TRNOESY Studies of Nucleotide Conformations in Arginine Kinase Complexes, *Biochemistry* **33**, 14227-14236.
31. Jarori, G. K., Murali, N. and Nageswara Rao, B. D. (1994) TRNOESY Study of the Conformation of MgATP Bound at the Active and Ancillary Sites of Rabbit Muscle Pyruvate Kinase, *Biochemistry* **33**, 6784-6791.
32. Lin, Y. and Nageswara Rao, B. D. (to be published) and Lin, Y. (1999) Ph.D. Thesis, Indiana University-Purdue University at Indianapolis (IUPUI).
33. Murali, N., Lin, Y., Mechulam, Y. Plateau, P. and Nageswara Rao, B. D. (1997) Adenosine Conformations of Nucleotides Bound to Methionyl tRNA Synthetase by Transferred Nuclear Overhauser Effect Spectroscopy, *Biophys. J.* **70**, 2275-2284.
34. Jarori, G. K., Murali, N., Switzer, R. L. and Nageswara Rao, B. D. (1995) Conformation of MgATP Bound to 5-Phospho-α-D-ribose 1-diphosphate Synthetase by Two-Dimensional Transferred Nuclear Overhauser Spectroscopy, *Eur. J. Biochem.* **230**, 517-524.
35. Housur, R. V., Govil, G. and Todd Miles, H. (1988) Application of Two Dimensional NMR Spectroscopy to the Determination of Conformation of Nucleic Acids, *Magn. Reson. in Chem.* **26**, 927-944.
36. Ni, F. and Zhu, Y. (1994) Accounting for Ligand-Protein Interaction in the Relaxation-Matrix Analysis of Transferred Nuclear Overhauser Effects, *J. Magn. Reson.* **B102**, 180-184.
37. Moseley, H. N. B., Curto, E. V. and Krishna, N. R. (1995) Complete Relaxation and Conformation Exchange Matrix (CORCEMA) Analysis of NOESY Spectra of Interaction Systems: Two-Dimensional Transferred NOESY, *J. Magn. Reson.* **B108**, 243-261.
38. Jackson, P. L., Moseley, H. N. B. and Krishna, N. R. (1995) Relative Effects of Protein-Mediated and Ligand-Mediated Spin-Diffusion Pathways on Transferred NOESY, and Implications on the Accuracy of the Bound Ligand Conformation, , *J. Magn. Reson.* **B107**, 289-292.
39. Zheng, J. and Post, C. (1993) Protein Indirect Relaxation Effects in Exchange-Transferred NOESY by a Rate Matrix Analysis, , *J. Magn. Reson.* **B101**, 262-270.
40. Ray, B. D., Chau, M. H., Fife, W. K., Jarori, G. K. and Nageswara Rao, B. D. (1996) Conformation of Mn(II) Nucleotide Complexes Bound to Rabbit Muscle Creatine Kinase: ^{13}C NMR Measurements using [2-^{13}C]ATP and [2-^{13}C]ADP, *Biochemistry* **35**, 7239-7246.
41. Raghunathan, V., Chau, M. H. and Nageswara Rao, B. D. (unpublished results).
42. Nageswara Rao, B. D. (1995) Measurements of Structural Changes Accompanying Enzyme Turnover by Nuclear Relaxation Rates in Enzyme-Bound Equilibrium Mixtures Containing Paramagnetic Cations, *J. Magn. Reson.* **108B**, 289-293.

SPIN RELAXATION METHODS FOR CHARACTERIZING PICOSECOND-NANOSECOND AND MICROSECOND-MILLISECOND MOTIONS IN PROTEINS

ARTHUR G. PALMER III AND CLAY BRACKEN
Department of Biochemistry and Molecular Biophysics
Columbia University
630 West 168th Street, New York, NY 10032

Abstract

Time-dependent conformational fluctuations of proteins and other macromolecules are related by statistical mechanical principles to diverse biophysical phenomena, including thermodynamic stability, folding, molecular recognition, and catalysis. Heteronuclear (^{2}H, ^{13}C, and ^{15}N) NMR spin relaxation spectroscopy constitutes a powerful experimental approach for globally characterizing conformational dynamics of macromolecules in solution, and consequently for probing biological function. Molecular dynamics on picosecond-nanosecond time scales can be characterized using the laboratory frame spin-lattice relaxation rate constant, the spin-spin relaxation rate constant and heteronuclear Overhauser enhancement. The model-free formalism parameterizes the relaxation data in terms of an overall rotational diffusion tensor for the molecule and a generalized order parameter and effective internal correlation time for each nuclear spin. The order parameter and correlation time characterize the amplitude and time scale for reorientational motions of the principal axes of the dipolar, chemical shift anisotropy and quadrupolar interactions responsible for relaxation of the nuclear spin. The use of the generalized order parameter to investigate entropic contributions to ligand binding and energy landscapes for peptide backbone motions in bovine calbindin D_{9k}, *Escherichia coli* ribonuclease H, and human fibronectin type III domains are described. Conformational exchange processes occurring on microsecond-millisecond time scales in biological macromolecules can be studied by Carr-Purcell-Meiboom-Gill and $R_{1\rho}$ experiments. The measurement of exchange rates using these experiments is hindered by the limited range of spin-echo delays that are feasible in the former experiment and the limited range of radiofrequency field strengths accessible in the latter. Nonetheless, the temperature dependence of the transverse relaxation rate constant determined by the Carr-Purcell-Meiboom-Gill technique can be used to define the activation barrier for the exchange process without precise knowledge of the exchange rate constant. Off-resonance $R_{1\rho}$ experiments provide effective fields in the rotating frame that are much larger than the applied radiofrequency field strengt and can be used to measure exchange time constants as short as 10 μs. The methods are applied to *Escherichia coli* ribonuclease H and the third fibronectin type III domain from human tenascin.

1. Introduction

All macromolecules, including proteins, are subject to structural fluctuations that depend on the Boltzmann distribution of conformational substates at thermal equilibrium. As a result, information about the time- or ensemble-average structure of a macromolecule generally is insufficient for a complete understanding of stability and reactivity. Since the advent of NMR spectroscopy, nuclear magnetic spin relaxation has been recognized as a rich source of information on molecular dynamic processes [1]. Recent advances in NMR techniques and biosynthetic labeling strategies have made heteronuclear (^{2}H, ^{13}C, and ^{15}N) NMR spin relaxation spectroscopy into a powerful approach for measuring conformational dynamics of proteins in solution [1, 2]. ^{15}N spin relaxation is particularly convenient for characterizing protein backbone dynamics because biosynthetic enrichment with ^{15}N is easily performed, the ^{15}N spins in protein backbones are isolated from each other, and ^{1}H-^{15}N heteronuclear correlation spectra are sensitive and well-resolved.

This article discusses ongoing efforts to comprehensively elucidate the intramolecular backbone dynamics on multiple time scales in bovine calbindin D$_{9k}$ [3-5], *Escherichia coli* ribonuclease H (RNase H) [6, 7] and two homologous fibronectin type III domains, derived from the third domain of human tenascin (TNfn3) [8, 9] and from the tenth domain of human fibronectin (FNfn10) [8].

2. Picosecond-nanosecond motions

2.1 EXPERIMENTAL METHODS

Investigations of intramolecular protein dynamics by solution NMR spectroscopy usually measure the longitudinal relaxation rate constant, R_1; the transverse relaxation rate constant, R_2, or the on-resonance rotating-frame relaxation rate constant, $R_{1\rho}$; and the steady state {I}-S nuclear Overhauser effect, *NOE*, or cross-relaxation rate constant, σ_{IS} [1]. For radiofrequency field amplitudes available in high resolution NMR spectrometers, the on-resonance $R_{1\rho} = R_2$ [10]. The relaxation rate constants for each ^{15}N spin in the molecule are determined from a series of two dimensional ^{1}H-^{15}N heteronuclear correlation spectra as has been described in detail [11, 12].

The relaxation rate constants for ^{15}N nuclei are given by [13],

$$R_1 = (d^2/4)[J(\omega_H - \omega_N) + 3J(\omega_N) + 6J(\omega_H + \omega_N)] + c^2 J(\omega_N) \tag{1}$$

$$R_2 = (d^2/8)[4J(0) + J(\omega_H - \omega_N) + 3J(\omega_N) + 6J(\omega_H) + 6J(\omega_H + \omega_N)] \\ + (c^2/6)[J(0) + 3J(\omega_N)] + R_{ex} \tag{2}$$

$$\sigma_{IS} = (NOE - 1)R_1 \gamma_N / \gamma_H = (d^2/4)[6J(\omega_H + \omega_N) - J(\omega_H - \omega_N)] \tag{3}$$

in which

$$d = \mu_0 h \gamma_N \gamma_H r_{NH}^{-3}/(8\pi^2) \quad (4)$$

$$c = \omega_N \Delta\sigma/\sqrt{3} \quad (5)$$

μ_0 is the permeability of free space; h is Planck's constant; γ_H and γ_N are the gyromagnetic ratios of ^1H and ^{15}N, respectively; $r_{NH} = 1.02$ Å is the N-H bond length; $\Delta\sigma$ is the chemical shift anisotropy; and $J(\omega)$ is the power spectral density function. In most cases, a value of $\Delta\sigma = -160$ ppm has been used, based on measurements of the ^{15}N tensor by solid-state NMR spectroscopy [14]; recent solution NMR studies suggest that $\Delta\sigma = -170$ ppm may be more appropriate [15]. A contribution to R_2 from conformational exchange on microsecond-millisecond time scales, R_{ex}, has been included in Eq. 2; measurement and interpretation of this quantity is described in Section 3.

The amplitudes and time scales of the intramolecular motions of the N–H bond vectors are determined from the relaxation rate constants by using the model-free formalism [16-18]. Alternatively, the values of $J(\omega)$ appearing in Eqs. 1-3 can be determined using spectral density mapping [19-21] and subsequently interpreted [22].

The model-free spectral density function affords a natural separation between overall rotational diffusion and internal motions that is particularly useful for globular proteins. In the original formulation of Lipari and Szabo [17, 18],

$$J(\omega) = \frac{2}{5}\left\{\frac{S^2\tau_m}{1+\omega^2\tau_m^2} + \frac{(1-S^2)\tau}{1+\omega^2\tau^2}\right\} \quad (6)$$

in which $\tau_m = 1/(6D)$ is the isotropic rotational correlation time of the molecule, D is the isotropic rotational diffusion constant of the molecule, $\tau = \tau_e\tau_m/(\tau_e+\tau_m)$, S^2 is the square of the generalized order parameter, commonly termed the order parameter, and τ_e is the effective internal correlation time. The latter two quantities characterize the amplitude and time scales of motions of the N-H bond vector for dipolar relaxation, or principal axis system for ^{15}N CSA relaxation, on picosecond-to-nanosecond time scales. For ^{15}N spins, the angle between the N-H bond vector and the principal axis of the CSA tensor is <20°; consequently, the same spectral density function normally is used for both interactions. As indicated by Eq. 6, the relaxation rates are insensitive to internal motions for which $\tau_e \gg \tau_m$. The relationships between overall rotation of the molecule, internal motions characterized by S^2 and τ_e, and internal motions characterized by R_{ex} are depicted schematically in Figure 1.

Deviations of the hydrated molecular shape of a protein from spherical symmetry results in different rotational diffusion rates around the three principal axes of the rotational diffusion tensor. Consequently, overall rotational motion even for globular proteins generally cannot be described by a single rotational diffusion constant D, but must be modeled using an axially-symmetric (characterized by diffusion constants $D_{\|}$ and $D_{\perp}$) or anisotropic diffusion tensor (characterized by diffusion constants D_{xx}, D_{yy}, and D_{zz}). Expressions for $J(\omega)$ have been derived that incorporate rotational diffusion anisotropy [23]; however, the data presented herein has been analyzed assuming a single rotational correlation time. Although determination of S^2 generally is robust, neglect of rotational diffusion anisotropy can introduce spurious τ_e and R_{ex} dynamical parameters [24]; consequently, care must be taken when interpreting these parameters. In particular, R_{ex} values < 2 s^{-1} can arise from modest diffusion anisotropy [7].

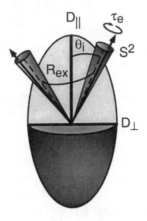

Figure 1. Dynamical parameters in proteins. Overall rotational diffusion of the molecule is represented using an axially symmetric diffusion tensor for an ellipsoid of revolution. The diffusion constants are $D_\parallel$ for diffusion around the symmetry axis of the tensor and $D_\perp$ for diffusion around the two orthogonal axes. For isotropic rotational diffusion, $D_\parallel = D_\perp$. The equilibrium position of the ith N-H bond vector is located at an angle θ_i with respect to the symmetry axis of the diffusion tensor. Picosecond-nanosecond dynamics of the bond vector are depicted as stochastic motions within a cone with amplitude characterized by S^2 and time scale characterized by τ_e. Slower conformational exchange motions are depicted as larger amplitude reorientations of the bond vector and are characterized by R_{ex}.

The original model-free formalism has been extended to include a second time scale for internal motions [25]. In this analysis $J(\omega)$ is modeled as,

$$J(\omega) = \frac{2}{5}\left[\frac{S^2 \tau_m}{1+\omega^2 \tau_m^2} + \frac{(S_f^2 - S^2)\tau}{1+\omega^2 \tau^2}\right] \tag{7}$$

in which τ_f and τ_s are the effective correlation times for internal motions on fast and slow time scales ($\tau_f < \tau_s < \tau_m$), respectively, S_f^2 and S_s^2 are the squares of the order parameters for the internal motions on the fast and slow time scales, respectively; $\tau = \tau_s \tau_m / (\tau_s + \tau_m)$; and $S^2 = S_f^2 S_s^2$. The value of $\tau_f < 10$ ps is assumed to be sufficiently fast that τ_f does not appear explicitly in the expression for $J(\omega)$. The parameter τ_e in Eq. 4 and τ_s in Eq. 5 are formally equivalent; Eq. 7 reduces to Eq. 6 as $S_f^2 \to 1$.

The square of the generalized order parameter characterizes the distribution of thermally accessible orientations of the bond vector or tensor principal axis system, defined by the polar and azimuthal angles $\{\theta, \phi\}$ in a molecular reference frame. For a small-amplitude axially-symmetric distribution of orientational fluctuations,

$$S^2 \approx 1 - 3 <\theta^2> \tag{8}$$

in which $<\theta^2>$ is the mean square fluctuation in the polar angle. Other expressions have been given for orientational fluctuations in a parabolic potential on the surface of a cone [26], lattice jump models [17], and combinations of diffusion and jump models [27].

Spectral density mapping is powerful particularly in applications to disordered systems in which the assumptions of the model-free formalism are not satisfied. Reduced spectral density mapping for ^{15}N spins provides values of $J(0)$, $J(\omega_N)$, and $J(0.87\ \omega_H)$ from R_1, R_2, and σ_{IS} data by assuming that the variation in $J(\omega)$ is smooth between $J(\omega_H+\omega_N)$ and $J(\omega_H-\omega_N)$ [19-21]. The resulting dipolar and CSA relaxation rate constants are given by,

$$R_1 = (d^2/4)[3J(\omega_N) + 7J(0.921\omega_H)] + c^2 J(\omega_N) \tag{9}$$

$$R_2 = (d^2/8)[4J(0) + 3J(\omega_N) + 13J(0.955\omega_H)]$$
$$+ (c^2/6)[J(0) + 3J(\omega_N)] + R_{ex} \tag{10}$$

$$\sigma_{IS} = (5d^2/4)[J(0.870\omega_H)] \tag{11}$$

in which $J(\varepsilon\omega_H) = (0.870/\varepsilon)^2\ J(0.870\omega_H)$. The spectral density values are obtained by inverting Eqs. 9-11. Reduced spectral density mapping avoids the use of relaxation rate constants measured for two-spin operators, such as $2I_zS_z$ and $2I_zS_x$, necessary for full spectral density mapping, and appears to give less systematically biased, higher precision, estimates of the spectral densities [21]. The spectral density values contain contributions from both overall rotational motion and intramolecular dynamics; a novel analysis of the spectral density values based on the correlation between $J(0)$, $J(\omega_N)$, and $J(0.87\ \omega_H)$ provides information on internal picosecond-nanosecond and overall motional correlation times [22]. In addition, the apparent $J(0)$ obtained by solving Eqs. 9-11 contains contributions from chemical exchange effects on R_2 (*vide infra*).

2.2 ENTROPIC EFFECTS ON CALCIUM BINDING BY CALBINDIN D$_{9K}$

Calbindin D$_{9k}$ is a small protein (75 amino acid residues) that belongs to the S-100 subgroup of the calmodulin superfamily of calcium binding proteins [28]. The superfamily is characterized by a common helix-loop-helix motif called the EF-hand that constitutes the calcium binding site. Calbindin D$_{9k}$ contains a single pair of EF-hands (the minimal functional unit) that exhibit cooperative binding of two calcium ions. As in all S-100 proteins, the N-terminal EF-hand contains a variant loop with 14 residues, rather than the 12 residues of a consensus calmodulin-like motif found in the C-terminal loop. The structure of calbindin D$_{9k}$ is illustrated in Figure 2.

The changes in backbone dynamics of calbindin D$_{9k}$ consequent to calcium binding were assessed by measuring ^{15}N spin relaxation for the apo and calcium loaded states of the protein [4, 5]. The order parameters determined from the relaxation rate constants are shown as a function of amino acid sequence in Figure 3. The order parameters have an average value of ~0.85 in the helical elements of secondary structure, which corresponds to a root-mean square angular fluctuation of ~13° using Eq. 8. This value of S^2 is typical of ordered elements of α-helix and β-sheet secondary structures in proteins [2]. The major difference between the two states of the molecule occurs in calcium binding loop II. In the apo state, this loop is highly flexible, with S^2 <0.7. In the calcium loaded state, this loop is essentially as rigid as the secondary structure elements. The C-terminal seven residues of the protein become less rigid upon calcium ligation.

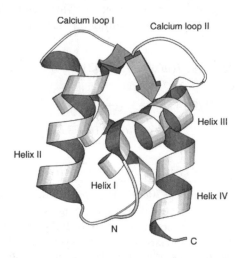

Figure 2. Ribbon diagram of the three-dimensional structure of calbindin D_{9k}. The figure was using the structural coordinates taken from the PDB file 2bca.

The change in backbone configurational entropy due to calcium binding can be estimated from changes in order parameters, as originally outlined by Akke et al. [3] for classical motional models. The differences between classical and quantum mechanical models for motions of a bond vector have been explored [29, 30]. The change in free energy due to entropic changes that are reflected in the order parameters is given by [3],

$$\Delta G_o = -kT \sum_i \ln\left[\frac{1-S_{i2}^2}{1-S_{i1}^2}\right] \qquad (12)$$

in which S_{ij} denotes the order parameter for spin i in state j ($j = 1$ is the apo state and $j = 2$ is the calcium loaded state), and the summation extends over all spins for which the order parameter changes significantly between the two states. Application of Eq. 12 to calbindin D_{9k} yields a free energy change of ~13 kJ/mol associated with the loss of backbone configurational freedom upon calcium binding. This quantity compares favorably in magnitude to the overall free energy of cooperativity of -7.7 kJ/mol [31]; consequently, the changes in configurational free energy are large enough to contribute significantly to the thermodynamics of cooperativity. This entropic effect, which opposes binding, must be compensated by other thermodynamic contributions to calcium ion binding.

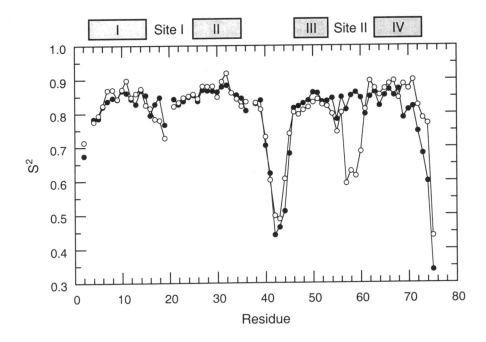

Figure 3. Comparison of the generalized order parameters for (○) apo and (●) calcium loaded calbindin D_{9k}. The order parameters are shown as a function of sequence. The error bars are on the order of the size of the plotted points. The location of the four helices in the protein are indicated by gray rectangles. The calcium binding sites are labeled.

2.3 COMPARISIONS OF HOMOLOGOUS FIBRONECTIN TYPE III DOMAINS

Fibronectin [32] and tenascin [33] are large extracellular matrix (ECM) proteins that bind to several different cell-surface receptors known as integrins, among other molecules. This interaction is facilitated by a short tripeptide Arg-Gly-Asp (RGD) motif that is located in a loop between strands F and G of a autonomously folded domain called the fibronectin type III domain [34-37], which is illustrated in Figure 4. This interaction serves a structural role by anchoring cells to the ECM and is implicated functionally in cell migration during development, wound healing, and metastasis. The tenth type III domain of fibronectin (FNfn10) and the third type III domain of tenascin (TNfn3) both contain RGD motifs; however, the greatest structural difference between the FNfn10 and TNfn3 domains occurs in this loop [34, 36, 37]. In TNfn3, the FG loop forms a tight type II' β-turn. In FNfn10, two additional amino acids flank each side of the RGD motif and cause the FG loop to protrude from the body of the protein.

As suggested by the investigation of calcium binding by calbindin D_{9k}, the dynamical nature of interacting molecules can be an important factor in molecular recognition. The degree to which a ligand and its receptor are constrained to productive conformations can modulate binding affinity and specificity by reducing the entropic cost of binding and increasing the ethalpic penalty for binding to an alternative receptor

Figure 4. Ribbon diagram of the fibronectin type III domain drawn using the structural coordinates from the PDB file 1fna. The RGD motif is drawn in black in the upper right.

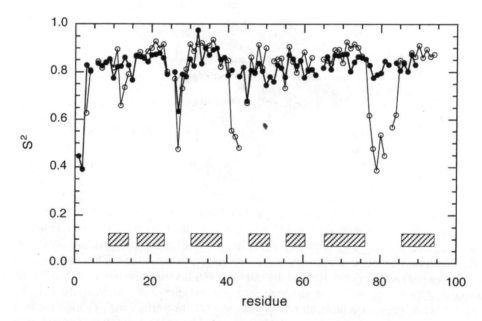

Figure 5. Order parameters for (○)TNfn3 and (●) FNfn10 type III domains. The location of the seven β-strands are shown as hatched rectangles.

or ligand. To begin to address these considerations in the interactions between RGD-containing proteins and integrins, the dynamic properties of the FNfn10 and TNfn3 domains were characterized by ^{15}N NMR spin relaxation [8]. The order parameters for TNfn3 and FNfn10 are presented in Figure 5 as a function of sequence position. Most notably, the order parameters are significantly lower in the C-C' and F-G loops of FNfn10 than TNfn3. Although the increased flexibility of the F-G loop in FNfn10 might have been anticipated from the greater length of the loop, the C-C' loops are the same length in the two proteins. The increased flexibility of the F-G loop, which contains the RGD motif, in FNfn10 may contribute to the relaxed specificity of fibronectin for a number of different integrin receptors.

2.4 TEMPERATURE DEPENDENCE OF DYNAMICS IN RIBONUCLEASE H

Ribonuclease H is an ubiquitous endonuclease that hydrolyzes the RNA strand in RNA-DNA hybrid oligonucleotides [38]. *Escherichia coli* RNase H is a single-chain polypeptide of 155 amino acid residues (M_r = 17,600). It inhibits replication from sites other than oriC and removes Okazaki fragments during lagging strand synthesis. As shown in Figure 6, RNase H is an α/β protein with five α-helices and five β-strands. The region between α_C and α_D (residues 90-99) has been termed the handle region.

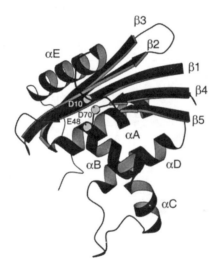

Figure 6. Ribbon diagram of *E. coli* RNase H drawn using the structural coordinates from PDB file 1rnh. The α-helices (α_A to α_E) and β-strands (β_1 to β_5) are labeled. The three absolutely necessary catalytic resides, D10, E48, and D70, are drawn as spheres.

The ^{15}N spin relaxation rate constants were measured for *E. coli* RNase H at temperatures of 285 K, 300 K, and 310 K [6, 7]. The order parameters derived from these measurements for the two extreme temperatures are mapped onto the backbone ribbon diagram of the protein in Figure 7. The weighted mean order parameters for secondary structure elements decrease as a function of increasing temperature from 0.879 ± 0.002 at 285 K, to 0.853 ± 0.001 at 300 K, to 0.851 ± 0.001 at 310 K. The loop regions surrounding the active site in RNase H contain residues with reduced order parameters, compared with secondary structure elements, at all temperatures. As discussed for calbindin D$_{9k}$ and the fibronectin type III domains, changes in order parameters in the binding loops of RNase H may contribute entropically to the thermodynamics of ligand binding. The order parameters for residues in helix α_E decrease as a function of temperature more strongly than for other secondary structure elements. The increased amplitude of motion in this helix may be a harbinger of thermal denaturation of the molecule.

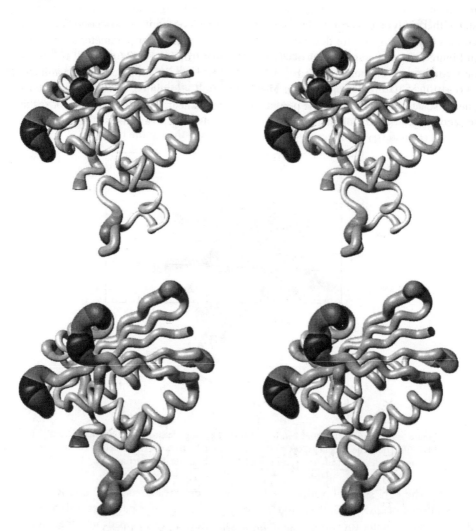

Figure 7. Order parameters for *E. coli* RNase H at (top) 285 K and (bottom) 310 K. The values of $1-S^2$ are mapped onto a stereographic backbone tube representation of the protein. The width of the tube ranges from 0.4 Å to 3.4 Å and the color scheme ranges from white to black for $1-S^2$ ranging from 0.05 to 0.6. The widths and colors are interpolated for residues for which relaxation data were not available.

The temperature dependence of order parameters defines the characteristic energy scale for librational motions [7]. If the N-H bond vectors are restricted to small excursions within an axially symmetric parabolic potential well, then, using Eq. 8,

$$\frac{d(1-S)}{dT} = \frac{3}{2T^*} \quad (13)$$

in which T^* defines a characteristic temperature that describes the density of energy states thermally accessible to the bond vector on picosecond-nanosecond time scales. Graphs of the temperature dependence of the generalized order parameter are presented in Figure 8. The average values of $d(1-S)/dT$ range from $(4.1 \pm 0.6) \times 10^{-4}$ K^{-1} for secondary structure elements, to $(2.1 \pm 0.3) \times 10^{-3}$ K^{-1} for the loop between β_5 and α_E, to $(3.6 \pm 0.1) \times 10^{-3}$ K^{-1} for residues 152-155 at the C-terminus. Thus, the characteristic energy scale for librations of the peptide backbone is nearly ten-fold greater for residues in secondary structural elements than for residues near the C-terminus.

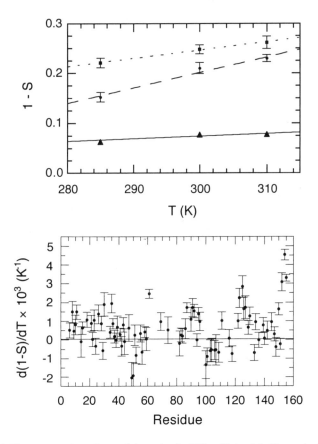

Figure 8. Temperature dependence of dynamics for RNase H. (top) 1–S is graphed versus T for (▲) the average values for secondary structure elements, (■) Gly 125, and (●) Gln 152. (bottom) Values of $d(1-S)/dT$ are shown as a function of sequence.

3. Microsecond-millisecond motions

Motions on microsecond-millisecond time scales that modulate the magnetic environment of a nucleus are recognized as an increase in the phenomenological R_2 in excess of the contributions from dipolar, chemical shift anisotropy, or quadrupolar relaxation. For two-site conformational exchange, the form of the exchange contribution depends on whether CPMG or $R_{1\rho}$ experiments are utilized [39, 40]:

$$R_{ex} = \Delta\omega^2 p_A p_B \tau_{ex} \left[1 - (2\tau_{ex} / \tau_{cp}) \tanh\left(\tau_{cp} / 2\tau_{ex}\right)\right] \quad (14)$$

$$R_{ex} = \Delta\omega^2 p_A p_B \tau_{ex} / \left(1 + \tau_{ex}^2 \omega_e^2\right) \quad (15)$$

respectively, in which p_i, ω_i, and σ_i are the populations, Larmor frequencies, and chemical shifts, respectively, for the spins in site i; $\Delta\omega = \omega_A - \omega_B = B_0(\sigma_A - \sigma_B)$ is the difference between the Larmor frequencies of the two sites; B_0 is the static magnetic field strength; τ_{cp} is the delay between 180° pulses in the CPMG spin echo sequence; $\omega_e = (\gamma_N B_1^2 + \Omega^2)^{1/2}$ is the effective field in the rotating reference frame; B_1 is the radiofrequency field amplitude; $\Omega = p_A \omega_A + p_B \omega_B - \omega_{rf}$ is the offset between the population weighted Larmor frequency and the B_1 carrier frequency, ω_{rf}; and $\tau_{ex} = (k_{A \rightarrow B} + k_{B \rightarrow A})^{-1}$ is the time constant for the exchange process.

3.1 EXPERIMENTAL TECHNIQUES

In the simplest approach, estimates of R_{ex} are obtained for selected residues when fitting R_1, R_2, and NOE data using the model-free formalism [6]. Alternatively, residues subject to exchange can be recognized from apparent increases in $J(0)$ determined from reduced spectral density mapping [20]. More precise characterization of chemical exchange can be obtained from data acquired at multiple static magnetic fields because R_{ex} varies a B_0^2 through its dependence on $\Delta\omega^2$ in Eqs. 14 and 15. Using Eqs. 9-11,

$$R_2 - 0.500 R_1 - 0.454 \sigma_{IS} = \frac{d^2}{2} J(0) + \left(\frac{2\gamma_N^2 \Delta\sigma^2}{3} J(0) + \frac{dR_{ex}}{d(B_0^2)}\right) B_0^2 \quad (16)$$

in which $dR_{ex}/d(B_0^2)$ is obtained from Eq. 14 or Eq. 15 depending on whether a CPMG or $R_{1\rho}$ experiment was used to measure R_2. Thus, assuming that the physical constants are known, $J(0)$ and $dR_{ex}/d(B_0^2)$ can be determined from the intercept and slope of a graph of the left hand side of Eq. 16 versus B_0^2. In many cases, $0.454\sigma_{IS} \ll 0.500 R_1$ and satisfactory results can be obtained by assuming $\sigma_{IS} = 0$ [41]. Nonetheless, the actual exchange constant τ_{ex} cannot be determined solely from the static magnetic field dependence of R_{ex}, because R_{ex} is a function of both $\Delta\omega^2 p_A p_B$ and τ_{ex}.

Exchange kinetics can be determined by measuring the phenomenological R_2 (Eq. 2) as a function of ω_e or τ_{cp} in $R_{1\rho}$ or CPMG experiments, respectively. The functional form of R_2 plotted versus ω_e or τ_{cp} is called a dispersion curve and is illustrated in Figure 9. Although some elegant applications of these methods to proteins have been reported in the literature [42, 43], significant dispersion is obtained only for relatively slow exchange processes due to experimental limitations on the maximum value of B_1 or $1/\tau_{cp}$.

3.2 ACTIVATION BARRIERS IN RIBONUCLEASE H

As discussed above, the microscopic exchange time constant τ_{ex} cannot be determined solely from R_{ex} measured for a single value of τ_{cp} or ω_e. However, the temperature dependence of R_{ex} measured using CPMG experiments does indicate whether τ_{ex} is greater or less than τ_{cp} and provides estimates of the activation barrier for the exchange process [7]. The temperature dependence of R_{ex} arises through the Arrhenius equation governing the exchange rate constant,

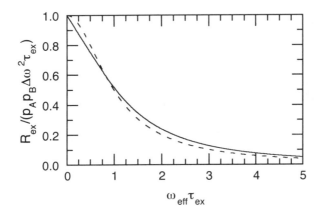

Figure 9. Chemical exchange rate constants. (———) Dependence of R_{ex} on $\omega_{eff} = 4/\tau_{cp}$ as given by Eq. 14 for a CPMG experiment. (- - - -) Dependence of R_{ex} on $\omega_{eff} = \omega_e$ as given by Eq. 15 for an $R_{1\rho}$ experiment. The CPMG curve is accurate for $(p_A p_B)^{1/2}|\Delta\omega|\tau_{ex} < 0.2$.

$$k_{A \to B} = k_0 \exp[-E_a/(k_B T)] \qquad (17)$$

and the Boltzmann distribution of the populations of the states A and B,

$$p_A = \{1 + \exp[-\Delta G/(k_B T)]\}^{-1} \qquad (18)$$

in which k_0 is a pre-exponential factor, E_a is the activation energy, and ΔG is the difference in free energy between conformational states A and B. Increasing T increases τ_{ex} because $E_a > \Delta G$ by construction. As shown in Figure 10, if $\tau_{ex} < \sim 0.3 \tau_{cp}$, then reducing the temperature, which increases τ_{ex}, will result in an increase in R_{ex}. If $\tau_{ex} > \sim 0.3 \tau_{cp}$, then reducing the temperature will lead to a decrease in R_{ex}. In general, $d\ln R_{ex}/d(1/T) < |E_a|$, both due to the dependence of p_A on T and due to the effect of the pulsing rate. Numerical calculations given in Figure 10 indicate that, for $\tau_{ex} < 0.1\tau_{cp}$ apparent activation barriers derived from the slope of plots of $\ln R_{ex}$ versus $1/T$ underestimate E_a by less than a factor of two.

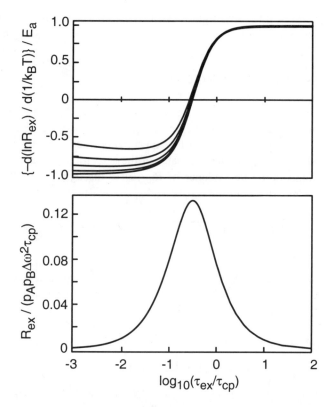

Figure 10. Chemical exchange in CPMG experiments. (bottom) The dependence of the exchange rate, R_{ex}, on the microscopic kinetic time constant, τ_{ex}, is illustrated for two-site exchange and unequal populations. τ_{cp} is the spacing between 180° pulses in the CPMG experiment. (top) The relative accuracy of apparent activation barriers derived from graphs of $\ln R_{ex}$ versus $1/T$ is shown. Curves are given for populations p_A of 0.5, 0.6, 0.7, 0.8 and 0.9 (from top to bottom).

As shown in Figure 11, the chemical exchange rate constants for Trp 90 and Lys 91 in RNase H decrease as the temperature increases; consequently τ_{ex} is less than 360 µs for τ_{cp} = 1.2 ms [7]. Assuming that $\Delta\omega$ < 2 ppm, the lower bound for τ_{ex} is ~10 µs. Arrhenius plots are shown in Figure 11 for residues Trp 90 and Lys 91, both of which are located in the handle region of RNase H. The apparent activation barriers are 52 ± 6 kJ/mol and 43 ± 7 kJ/mol, respectively. These barriers agree within experimental error, which suggests that both residues are subject to the same exchange process.

3.3 CONFORMATIONAL DYNAMICS IN FIBRONECTIN TYPE III DOMAINS

Although the $R_{1\rho}$ experiment has been applied to small molecules for 20 years [39], its applications to proteins have been limited due to experimental constraints. A new technique for measuring conformational exchange in proteins using off-resonance $R_{1\rho}$ spin-

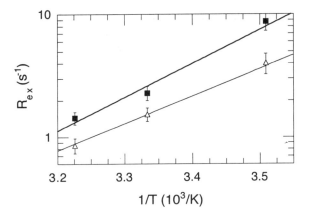

Figure 11. Temperature dependence of R_{ex} for (■) Trp 90 and (Δ) Lys 91 in RNase H.

locking techniques has been developed [44]. A pulse sequence for the new technique is illustrated in Figure 12. The experiment uses an off-resonance radiofrequency field to increase the effective magnetic field in the rotating reference frame. At the same time, the pulse sequence averages rotating frame and laboratory frame relaxation rate constants during a constant-relaxation-time period to simplify off-resonance effects. The effective relaxation rate constant during the experiment is given by,

$$R_{eff} = (R_2 - R_1 + R_{ex})\sin^2\theta \qquad (19)$$

in which $\theta = \arctan(\gamma_N B_1/\Omega)$ is the tilt angle of the effective field. Measuring $R_{eff}/\sin^2\theta$ as a function of the effective field ω_e permits determination of the three parameters Φ_{ex}

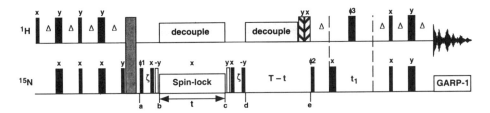

Figure 12. Off-resonance rotating-frame relaxation experiment. Narrow and wide solid bars depict 90° and 180° pulses. Narrow open bars depict pulses with a flip angle 90°–θ. The gray bar depicts a homospoil pulse. Hatched bars are high-power purge pulses. ^{1}H decoupling is applied during the relaxation delays. The spin-lock field is switched from the center of the spectrum to the off-resonance frequency at point b and switched back at point c. Elements between a and b and between c and d rotate coherences between the z-axis and the off-resonance rotating frame; ζ compensates for chemical shift offsets at large θ. The phase cycle is φ1 = y, –y; φ2 = y, y, –y, –y, y, y, –y, –y; φ3 = 4(x), 4(–x); receiver = x, –x, –x, x, –x, x, x, –x. Analogous experiments using adiabatic rotations have been described [45, 46].

= $\Delta\omega^2 p_A p_B$, $R_2 - R_1$, and τ_{ex} by non-linear curve-fitting. The precision of the fitted parameters obtained from relaxation dispersion curves can be improved by using the ^{15}N free precession linewidth, R_2^*/π, and longitudinal relaxation rate, R_1, to eliminate the unknown parameter $R_2 - R_1$ from Eq.19 to yield a new dispersion relation [9]:

$$R_2^* - R_1 - R_{eff}/\sin^2\theta = \Phi_{ex}\tau_{ex}^3\omega_e^2/(1 + \tau_{ex}^2\omega_e^2) + R_{ih} \qquad (20)$$

The contribution to the linewidth from magnetic field inhomogeneity broadening, R_{ih}, is estimated to be < 1 s^{-1} from comparison of R_2^* and the values of R_2 obtained from CPMG experiments performed on calbindin D$_{9k}$.

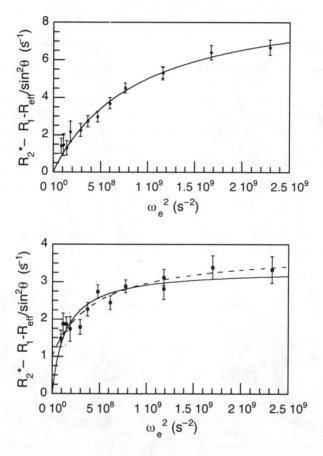

Figure 13. Relaxation dispersion curves described by Eq. 20. (*top*) Data for residue Arg 45, representative of residues that exhibit large exchange contributions to the ^{15}N transverse relaxation rates in CPMG experiments ($R_{ex} \sim$ 4–17 s^{-1}) [8]. (*bottom*) Data for residue Ser 58, representative of residues that do not exhibit obvious exchange contributions to the ^{15}N CPMG transverse relaxation rates. The solid lines are the fit of Eq. 20 to the data assuming that $R_{ih} = 0$ s^{-1}. The dotted line illustrates the effect of fitting the data for Ser 58 assuming $R_{ih} = 1.0$ s^{-1}

The new experiment was applied to the fibronectin type III domain from TNfn3 [9]. Figure 13 shows relaxation dispersion curves for residue Arg 45, which exhibits large dispersion amplitudes, with $\tau_{ex} = 32 \pm 3$ μs and $\Phi_{ex} = (3.0 \pm 0.6) \cdot 10^5$ s^{-2}. Figure 13 also displays relaxation dispersion curves for Ser 58, which exemplifies the data obtained for residues with smaller dispersion amplitudes, with $\tau_{ex} = 62 \pm 8$ μs and $\Phi_{ex} = (0.6 \pm 0.1) \cdot 10^5$ s^{-2}. Residue Arg 45 is located at the end of a β-strand, with a crystallographic B-factor of 9.2 Å^2 for the backbone nitrogen atoms, while Ser 58 is located in the middle of a well-defined β-strand, with a B-factor < 9 Å^2 [34].

Correlation times, τ_{ex}, and dispersion amplitudes, Φ_{ex}, were measured for 47 individual ^{15}N spins in TNfn3 that experience distinct magnetic environments due to transitions between molecular conformations. Values of τ_{ex} and Φ_{ex} are coded onto the

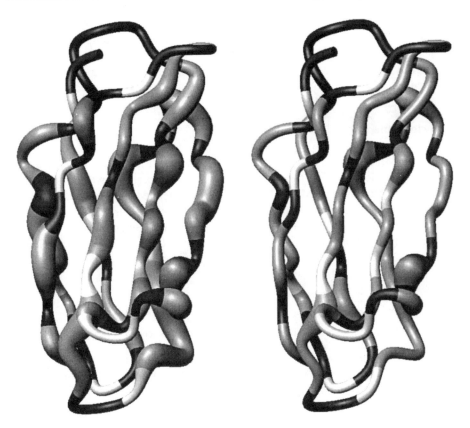

Figure 14. Location in the structure of residues exhibiting conformational fluctuations on sub-millisecond time scales. Values of (left) τ_{ex} and (right) Φ_{ex} are width-coded onto a Cα chain trace of structure PDB file 1ten. Widths were interpolated on a continuous scale from narrow to wide corresponding to the range of $\tau_{ex} = 30$–120 μs and 2.0×10^4 s^{-2} to 3.0×10^5 s^{-2}. Prolines and residues for which no data were obtained due to spectral overlap are colored dark gray, while residues that do not show significant relaxation dispersion are colored white. The RGD loop is the frontmost loop at the upper right side of the molecule.

structure in Figure 14. Conformational fluctuations are observed throughout the TNfn3 domain. The slower motions in TNfn3 tend to occur in loops, turns and near the ends of β-strands. Similarly, the largest dispersion amplitude parameters are observed for residues predominantly located in, or surrounding, loops and turns. Possible mechanisms of the conformational exchange observed in TNfn3 may thus include "breathing" motions of the β-sandwich, such as transient twisting or buckling of the β-sheets.

4. Conclusions

The investigations of spin relaxation in calbindin D_{9k}, ribonuclease H, and the fibronectin type III domains of fibronectin and tenascin described above, as well as the numerous studies conducted by other laboratories [2], illustrate the power of NMR spectroscopy to delineate detailed aspects of protein dynamics on multiple time scales. Nonetheless, additional experiments, both spectroscopic and biochemical, will be required to firmly establish the importance of protein dynamics for biological function in these molecules.

5. Acknowledgements

We thank Drs. Mikael Akke, Peter A. Carr, and Arthur M. Mandel for their contributions to the work described in this article. Fruitful collaborations with Drs. Walter Chazin, John Cavanagh, and Mark Rance are acknowledged gratefully. A.G.P. acknowledges financial support from the National Institutes of Health through grant GM-50291, a Searle Scholar Award and a Irma T. Hirschl Career Scientist Award. W.C.B was supported by a National Institutes of Health National Research Service Award GM-17562.

6. References

1. Palmer, A.G., Williams, J., and McDermott, A. (1996) Nuclear magnetic resonance studies of biopolymer dynamics, *J. Phys. Chem.* **100**, 13293-13310.
2. Palmer, A.G. (1997) Probing molecular motion by NMR, *Curr. Opin. Struct. Biol.* **7**, 732-737.
3. Akke, M., Brüschweiler, R., and Palmer, A.G. (1993) NMR order parameters and free energy: An analytic approach and application to cooperative Ca^{2+} binding by calbindin D_{9k}, *J. Am. Chem. Soc.* **115**, 9832-9833.
4. Akke, M., Skelton, N.J., Kördel, J., Palmer, A.G., and Chazin, W.J. (1993) Effects of ion binding on the backbone dynamics in calbindin D_{9k} determined by ^{15}N NMR relaxation, *Biochemistry* **32**, 9832-9843.
5. Kördel, J., Skelton, N.J., Akke, M., Palmer, A.G., and Chazin, W.J. (1992) Backbone dynamics of calcium-loaded calbindin D_{9k} studied by two-dimensional proton-detected NMR spectroscopy, *Biochemistry* **31**, 4856-4866.
6. Mandel, A.M., Akke, M., and Palmer, A.G. (1995) Backbone dynamics of *Escherichia coli* ribonuclease HI: Correlations with structure and function in an active enzyme, *J. Mol. Biol.* **246**, 144-163.
7. Mandel, A.M., Akke, M., and Palmer, A.G. (1996) Dynamics of ribonuclease H: Temperature dependence of motions on multiple time scales, *Biochemistry* **35**, 16009-16023.

8. Carr, P.A., Erickson, H.P., and Palmer, A.G. (1997) Backbone dynamics of homologous fibronectin type III cell adhesion domains from fibronectin and tenascin, *Structure* **5**, 949-959.
9. Akke, M., Liu, J., Cavanagh, J., Erickson, H.P., and Palmer, A.G. (1998) Pervasive conformational fluctuations on microsecond time scales in a fibronectin type III domain, *Nat. Struct. Biol.* **5**, 55-59.
10. Peng, J.W., Thanabal, V., and Wagner, G. (1991) 2D heteronuclear NMR measurements of spin-lattice relaxation times in the rotating frame of X nuclei in heteronuclear HX spin systems, *J. Magn. Reson.* **94**, 82-100.
11. Skelton, N.J., Palmer, A.G., Akke, M., Kördel, J., Rance, M., and Chazin, W.J. (1993) Practical aspects of two-dimensional proton-detected ^{15}N spin relaxation measurements, *J. Magn. Reson., Ser. B* **102**, 253-264.
12. Farrow, N.A., Muhandiram, R., Singer, A.U., Pascal, S.M., Kay, C.M., Gish, G., Shoelson, S.E., Pawson, T., Forman-Kay, J.D., and Kay, L.E. (1994) Backbone dynamics of a free and a phosphopeptide-complexed Src homology 2 domain studied by ^{15}N NMR relaxation, *Biochemistry* **33**, 5984-6003.
13. Abragam, A. (1961) *Principles of Nuclear Magnetism*. Clarendon Press, Oxford
14. Hiyama, Y., Niu, C.-H., Silverton, J.V., Bavoso, A., and Torchia, D.A. (1988) Determination of ^{15}N chemical shift tensor via ^{15}N-^{2}H dipolar coupling in Boc-glycylglycyl [^{15}N] glycine benzyl ester, *J. Am. Chem. Soc.* **110**, 2378-2383.
15. Tjandra, N., Szabo, A., and Bax, A. (1996) Protein backbone dynamics and ^{15}N chemical shift anisotropy from quantitative measurement of relaxation interference effects, *J. Am. Chem. Soc.* **118**, 6986-6991.
16. Halle, B., Wennerström, H. (1981) Interpretation of magnetic resonance data from water nuclei in heterogeneous systems, *J. Chem. Phys.* **75**, 1928-1943.
17. Lipari, G., Szabo, A. (1982) Model-free approach to the interpretation of nuclear magnetic resonance relaxation in macromolecules. 1. Theory and range of validity, *J. Am. Chem. Soc.* **104**, 4546-4559.
18. Lipari, G., Szabo, A. (1982) Model-free approach to the interpretation of nuclear magnetic resonance relaxation in macromolecules. 2. Analysis of experimental results, *J. Am. Chem. Soc.* **104**, 4559-4570.
19. Ishima, R., Nagayama, K. (1995) Quasi-spectral density function analysis for nitrogen-15 nuclei in proteins, *J. Magn. Reson., Ser. B* **108**, 73-76.
20. Farrow, N.A., Zhang, O., Szabo, A., Torchia, D.A., and Kay, L.E. (1995) Spectral density function mapping using ^{15}N relaxation data exclusively, *J. Biomol. NMR* **6**, 153-162.
21. Peng, J., Wagner, G. (1995) Frequency spectrum of NH bonds in eglin c from spectral density mapping at multiple fields, *Biochemistry* **34**, 16733-16752.
22. Lefevre, J.F., Dayie, K.T., Peng, J.W., and Wagner, G. (1996) Internal mobility in the partially folded DNA binding and dimerization domains of GAL4: NMR analysis of the N-H spectral density functions, *Biochemistry* **35**, 2674-86.
23. Woessner, D.E. (1962) Nuclear spin relaxation in ellipsoids undergoing rotational Brownian motion, *J. Chem. Phys.* **37**, 647-654.
24. Schurr, J.M., Babcock, H.P., and Fujimoto, B.S. (1994) A test of the model-free formulas. Effects of anisotropic rotational diffusion and dimerization, *J. Magn. Reson., Ser. B* **105**, 211-224.
25. Clore, G.M., Szabo, A., Bax, A., Kay, L.E., Driscoll, P.C., and Gronenborn, A.M. (1990) Deviations from the simple two-parameter model-free approach to the interpretation of nitrogen-15 nuclear magnetic relaxation of proteins, *J. Am. Chem. Soc.* **112**, 4989-4991.
26. Brüschweiler, R., Wright, P.E. (1994) NMR order parameters of biomolecules: A new analytical representation and application to the Gaussian axial fluctuation model, *J. Am. Chem. Soc.* **116**, 8426-8427.
27. Bremi, T., Brüschweiler, R., and Ernst, R.R. (1997) A protocol for the interpretation of side-chain dynamics based on NMR relaxation: Application to phenylalanines in antamanide, *J. Am. Chem. Soc.* **119**, 4272-4284.
28. Heizmann, C.W., Hunziker, W. (1991) Intracellular calcium-binding proteins: more sites than insights, *Trends Biol. Sci.* **16**, 98-103.
29. Yang, D., Kay, L.E. (1996) Contributions to conformational entropy arising from bond vector fluctuations measured from NMR-derived order parameters: application to protein folding, *J. Mol. Biol.* **263**, 369-382.

30. Li, Z., Raychaudhuri, S., and Wand, A.J. (1996) Insights into the local residual entropy of proteins provided by NMR relaxation, *Protein Sci.* **5**, 2647-2650.
31. Linse, S., Johansson, C., Brodin, P., Grundström, T., Drakenberg, T., and Forsén, S. (1991) Electrostatic contributions to the binding of Ca^{2+} in calbindin D_{9k}, *Biochemistry* **30**, 154-162.
32. Hynes, R.O. (1990) *Fibronectins*. Springer-Verlag, New York, NY, USA
33. Erickson, H.P. (1993) Tenascin-C, tenascin-R and tenascin-X: a family of talented proteins in search of functions. [Review], *Current Opinion in Cell Biology* **5**, 869-76.
34. Leahy, D., Hendrickson, W.A., Aukhil, I., and Erickson, H.P. (1992) Structure of a fibronectin type III domain from tenascin phased by MAD analysis of the selenomethionyl protein, *Science* **258**, 987-991.
35. Leahy, D.J., Aukhil, I., and Erickson, H.P. (1996) 2.0 Å crystal structure of a four-domain segment of human fibronectin encompassing the RGD loop and synergy region, *Cell* **84**, 155-164.
36. Main, A.L., Harvey, T.S., Baron, M., Boyd, J., and Campbell, I.D. (1992) The three-dimensional structure of the tenth type III module of fibronectin: An insight into RGD-mediated interactions, *Cell* **71**, 671-678.
37. Dickinson, C.D., Veerapandian, B., Dai, X.-P., Hamlin, R.C., Xuong, N., Ruoslahti, E., and Ely, K.R. (1994) Crystal structure of the tenth type III cell adhesion module of human fibronectin, *J. Mol. Biol.* **236**, 1079-1092.
38. Crouch, R.J. (1990) Ribonuclease H: From discovery to 3D structure, *New Biol.* **2**, 771-777.
39. Deverell, C., Morgan, R.E., and Strange, J.H. (1970) Studies of chemical exchange by nuclear magnetization relaxation in the rotating frame, *Mol. Phys.* **18**, 553-559.
40. Allerhand, A., Gutowsky, H.S. (1965) Spin-echo studies of chemical exchange. II. Closed formulas for two sites, *J. Chem. Phys.* **42**, 1587-1599.
41. Phan, I.Q.H., Boyd, J., and Campbell, I.D. (1996) Dynamic studies of a fibronectin type I module pair at three frequencies: Anisotropic modeling and direct determination of conformational exchange, *J. Biomol. NMR* **8**, 369-378.
42. Szyperski, T., Luginbühl, P., Otting, G., Güntert, P., and Wüthrich, K. (1993) Protein dynamics studied by rotating frame ^{15}N spin relaxation times, *J. Biomol. NMR* **3**, 151-164.
43. Orekhov, V.Y., Pervushin, K.V., and Arseniev, A.S. (1994) Backbone dynamics of (1-71)bacterioopsin studied by two-dimensional ^{1}H-^{15}N NMR spectroscopy, *Eur. J. Biochem.* **219**, 887-896.
44. Akke, M., Palmer, A.G. (1996) Monitoring macromolecular motions on microsecond–millisecond time scales by $R_{1\rho}$–R_1 constant-relaxation-time NMR spectroscopy, *J. Am. Chem. Soc.* **118**, 911-912.
45. Mulder, F.A.A., de Graaf, R.A., Kaptein, R., and Boelens, R. (1998) An off-resonance rotating frame relaxation experiment for the investigation of macromolecular dynamics using adiabatic rotations, *J. Magn. Reson.* **131**, 351-357.
46. Zinn-Justin, S., Berthault, P., Guenneugues, M., and Desvau, H. (1997) Off-resonance rf fields in heteronuclear NMR: Application to the study of slow motions, *J. Biomol. NMR* **10**, 363-372.

STRUCTURAL DIVERSITY OF THE OSMOREGULATED PERIPLASMIC GLUCANS OF GRAM-NEGATIVE BACTERIA BY A COMBINED GENETICS AND NUCLEAR MAGNETIC RESONANCE APPROACH

GUY LIPPENS[a]* AND JEAN-PIERRE BOHIN[b]*

[a]*CNRS URA 1309, Pasteur Institute of Lille, 1 rue du Professeur Calmette, 59019 Lille, France*

[b]*CNRS UMR 111, Université des Sciences et Technologies de Lille, 59655 Villeneuve d'Ascq Cedex, France,*

*Laboratoire Européen Associé "Analyse structure-fonction des biomolécules" (CNRS, France and FNRS, Belgium).

1. Abstract

The periplasmic space is a cellular compartment lying between the inner and the outer membranes of gram-negative bacteria. Located in this region are the peptidoglycan layer, the osmoregulated periplasmic glucans (OPGs), and a number of proteins implicated in the detection, the processing and the transport into the cell of nutrients, the detoxification of deleterious substances, and the biogenesis of the envelope. OPGs exhibit structural features that are characteristic of a bacterial genus. However, they share several major traits: glucose is the only monosaccharide; they are oligosaccharides made of a limited number of units (c.a. 10-20); they are accumulated in the periplasmic space in a higher amount in response to a lower osmolarity of the medium.

NMR has been extensively used in our laboratories to establish the primary sequences of different OPGs. Homonuclear COSY, and relayed COSY or TOCSY experiments are used to assign all protons in every individual glucose ring, and are combined with HSQC and HSQC-TOCSY experiments for complete carbon assignments. NOESY and HMBC spectra, exploiting respectively the short distance between protons on both sides of the glycosidic linkage and the heteronuclear coupling over the linkage, are used to establish the sequential assignments, but suffer from poor separation in all but the anomeric proton signals. We exploited the wider spread in carbon chemical shifts by implementing the NOESY experiment in its HSQC-NOESY version. Alternatively, the assignment can be done solely on the basis of the anomeric proton region by the NOESY-TOCSY and/or TOCSY-NOESY pulse sequences.

Determination of the 3D structures of the OPGs has been addressed in our laboratories after the discovery of several cyclic OPGs that contain one single α–(1-6) linkage next to all β–(1-2) linkages. This single α–(1-6) linkage introduces conformational strains to such an extent that all anomeric protons can be observed separately. We have focused on the cyclic OPG of *Ralstonia solanacearum*, that because of its small size (DP13) makes an ideal model to address the problems of potential cavities or other structural features that might be relevant for the function of these intriguing molecules. In contrast to protein NMR, the number of experimental parameters that can be used to derive a structural model are very limited. They basically are the same as the parameters used for the sequential assignment, i.e. the NOE between the H1-H2' proton pair and the ^3J H1-C2' and C1-H2' long-range coupling constants. A last parameter that has been correlated to the glycosidic conformation is the direct ^1J (H1-C1) coupling constant. The limited number of experimental structures, however, leads to a probably imperfect parametrization of the Karplus relationships that relate coupling constants to structure in oligosaccharides.

Model building led to different structures, but none was compatible with all experimental observations. Further indication of dynamical equilibrium came from the line broadening as observed on the anomeric proton of the b unit. This was confirmed by a detailed relaxation study, and we concluded that the molecule undergoes slow dynamics on the microsecond time scale. The biological relevance of this dynamics is currently under investigation in our groups.

In order to address question of the molecular mobility of the OPGs in the periplasm, we have started *in vivo* experiments, with bacteria containing isotopically enriched OPGs. The obtaining of workable NMR spectra indicates that at least a majority of the molecules maintain a high degree of diffusional mobility.

2. Diversity of the OPGs

The periplasmic space lies between the inner and the outer membranes of gram-negative bacteria [1]. A number of processes that are vital to the growth and viability of the cell occur within this compartment. Proteins residing in the periplasm fulfill important functions in the detection and processing of essential nutrients and their transport into the cell. The contents of the periplasmic space provide a microenvironment of small and middle-sized molecules that buffer the cell from changes that occur in its local surroundings. In fulfilling these essential roles, the periplasm is not static but dynamic, capable of regulation to accommodate changes in the external and internal environments that surround it. While an overall understanding of the structure and contents of the periplasmic space has been obtained, the periplasm still remains the most controversial and poorly defined compartment in the gram-negative bacterial cell. The exact size and internal architecture of the periplasm remain area of dispute, and the fine structure that leads to low mobility of periplasmic proteins needs further clarification [1]. Osmoregulated periplasmic glucans (OPG) are general components of the periplasm of gram-negative bacteria [2] and share the following features: glucose as the sole sugar, glucose units linked, at least partially, by β–glycosidic bonds, and synthesis under osmotic control and that is inversely correlated to the osmolarity of the growth medium of the cells [3]. Recent investigations have revealed that beyond these common features OPGs from various bacterial species show an unexpected structural diversity. Four families can be distinguished (Figure 1).

Family 1. OPGs of *E. coli*, also known as MDOs (Membrane-Derived Oligosaccharides), appear to range from 5 to 12 glucose residues, with the principal species containing 8 or 9 glucose residues (4-5)]. The structure is highly branched, the backbone consisting of β–(1-2)-linked glucose units to which the branches are attached by β–(1-6) linkages. Similar structures were found for the OPGs of the closely related species *Erwinia chrysanthemi* [6] and also for the OPGs of the more distantly related species *Pseudomonas syringae* [7].

Family 2. Among members of the family *Rhizobiaceae*, periplasmic glucans are cyclic [8]. Studies have demonstrated that *Agrobacterium* and *Rhizobium* species synthesize periplasmic glucans with similar structure [9, 10, 11]. In both genera, periplasmic glucans are composed of a cyclic β–(1-2)-glucan backbone containing 17 to 24 glucose residues. Bundle et al. [12] reported that *Brucella* spp. synthesize cyclic β–(1-2)-glucans essentially identical to those synthesized by members of the *Rhizobiaceae*.

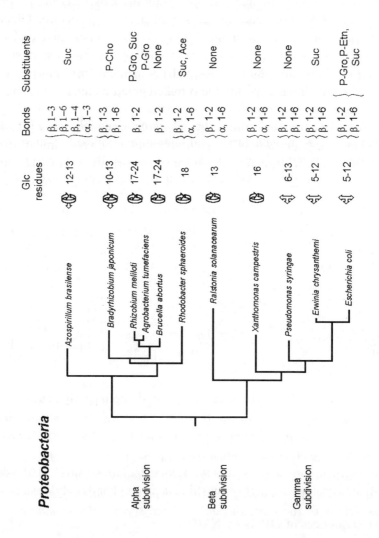

Figure 1. OPG structures in relation to the phylogenetic tree of representative species from three subdivisions of the purple bacteria (Proteobacteria); see the text for details. (Ace= acetate, Suc= succinate, P-Cho= phosphocholine, P-Gro= phosphoglycerol, P-Etn= phosphoethanolamine).

Family 3. Extracts of *Bradyrhizobium* spp. revealed the presence of β–(1-6)- and β–(1-3) cyclic glucans containing 10 to 13 glucose units per ring [13, 14] Three distinct glucans (I, II, and III) are synthesized by cells of *Azospirillum brasilense* [15]. The three glucans consist of a cyclic structure. Glucan I is made of 12 glucose units linked by 3 β–(1-3), 8 β–(1-6) and one β–(1-4) linkages. Glucan II is derived from the glucan I by the addition of a glucose linked by an α–(1-3) linkage, and glucan III is derived from glucan II by the addition of a 2-O-methyl group onto the α linked glucose unit.

Family 4. *Ralstonia solanacearum* [16], *Xanthomonas campestris* [16, 17, 18] and *Rhodobacter sphaeroides* [19] synthesize OPGs of very similar structural features. OPGs have a unique degree of polymerization [13, 16, and 18, respectively]. All the glucose residues are linked by β–(1-2) linkage with the exception of one α–(1-6) linkage.

OPG substitutions. In *E. coli*, the OPGs are highly substituted with *sn*-1-phosphoglycerol, phosphoethanolamine, and succinyl ester residues [5]. In *E. chrysanthemi*, succinyl residues are the sole substituent [6] while OPGs of *P. syringae* are not substituted [7].
In *A. tumefaciens*, one or more phosphoglycerol moieties are substituted for approximately 50% of the total periplasmic glucans [20]. In *R. meliloti*, anionic moieties may be substituted for as much as 90% of the periplasmic cyclic β–(1-2) glucans and the predominant anionic substituent present is phosphoglycerol [21].
Periplasmic glucans of *Bradyrhizobium* spp. are predominantly neutral [13] but Rolin and co-workers [14] have purified a phosphocholine substituted β–(1-6)- and β–(1-3) glucan from *B. japonicum* USDA 110. The OPGs of *A. brasilense* can be substituted by one succinyl residue [15].
The OPGs of *R. solanacearum* and *X. campestris* are neutral [16] while the OPGs of R. sphaeroides are highly substituted succinyl and acetyl residues [19].

3. Primary sequences of OPGs by NMR

3.1 INTRA-RESIDUE ASSIGNMENTS (COSY, RELAYED COSY, TOCSY, HSQC)

Intra-residue proton assignments are obtained through the conventional use of COSY, relayed COSY and TOCSY experiments. Heteronuclear assignments are obtained from the correlation of proton and carbon resonances in a HMQC or

HSQC experiment, and can be completed by a HSQC-TOCSY experiment. The following Figure 2 shows the fully assigned ^{1}H-^{13}C spectrum of the DP13 of *R. solanacearum*.

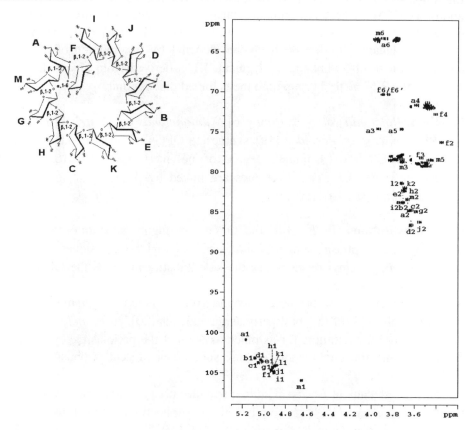

Figure 2. Heteronuclear Single Quantum Experiment correlating the 1H and 13C resonances of the non-enriched DP13 molecule of *R. solanacearum* at 301K. The numbering of the units goes from a for the unit in α–(1-6) to m, as depicted on the macrocycle on the left.

3.2 INTER-RESIDUE ASSIGNMENTS

The NOE between the two protons flanking the linkage forms a self-evident and easily accessible probe to establish the linkage between the different units. Unfortunately, the limited chemical shift dispersion for all but the anomeric protons makes this method less obvious in the case of the cyclic OPGs. However, homonuclear techniques can still be used to establish the connectivities between the units, through the linking of the anomeric resonances in a NOESY-TOCSY and/or TOCSY-NOESY experiment.

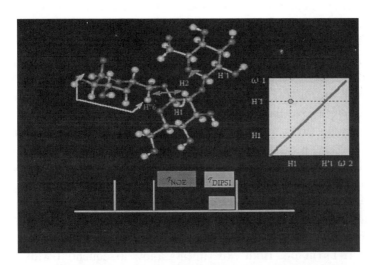

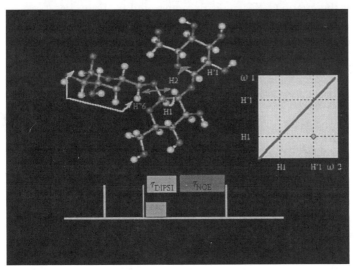

Figure 3. Homonuclear NOESY-TOCSY (Top) and TOCSY-NOESY (Bottom) pulse sequences and the resulting anomeric proton spectral zones, showing the asymmetry of the cross peaks. The magnetization transfer is shown on a trisaccharide where two units are linked in β-(1-2), and two in α-(1-6). Only the cross-peak corresponding to the β-(1-2) linkage is shown

The sequences are shown in the Figure 3. They can easily be understood as a merging of the regular NOESY mixing period, during which the dipolar cross-relaxation will establish a transfer of magnetization between the protons flanking the glycosidic linkage, combined with a z-TOCSY period, where the magnetization is again transferred to the anomeric proton.

Schematically, the magnetization transfer for a typical disaccharide linked in β-(1-2) can be seen in the same Figure 3: during the t_1 period, magnetization at the anomeric frequency of unit i evolves and gives rise to a peak at $\omega_{ano}(i)$. In a NOESY-TOCSY experiment, it is first transferred during the NOESY period to the H2 proton of unit i+1, and then by scalar coupling to the anomeric proton of this same unit i+1. We therefore will observe a cross peak between the anomeric protons of units i (in ω_1) and (i+1) (in ω_2). Its symmetric counterpart is missing in this experiment, because of the order of the NOESY and TOCSY transfer. In the equivalent TOCSY-NOESY experiment, however, the other peak will appear, because magnetization labeled in ω_1 by the frequency of the anomeric proton of residue i will first be transferred to the H2 proton of the same residue, and then transfer through cross-relaxation to the anomeric proton of the preceding unit. We therefore will observe a cross peak between the anomeric protons of units i (in ω_1) and (i-1) (in ω_2). Both experiments allow a sequential walk through the sequence of the cyclic OPGs, as is demonstrated in Figure 4.

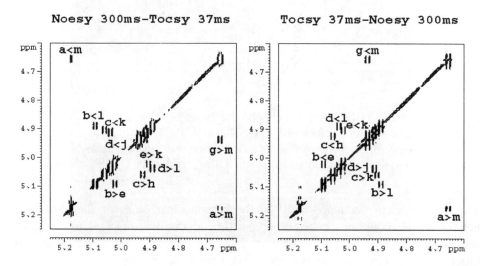

Figure 4. Homonuclear NOESY-TOCSY (Left) and TOCSY-NOESY (Right) experiments on the DP13 molecule of *R. solanacearum* at 301K.

In the case of a β–(1-2) linkage as in the above cited example, short mixing periods for both NOE and scalar coupling transfer will be sufficient, because of the short distance between the protons flanking the linkage and the large ^3J coupling constant between H1 and H2 protons. Special care should be exerted when one expects a (1-6) linkage, where especially the z-TOCSY delay should be considerably increased. An example is shown in Figure 5 , where the sequential peak between the c and (k or l) units linked by an β–(1-6) linkage is not present when using a 80 ms TOCSY period, but becomes visible upon applying a 160 ms TOCSY mixing period.

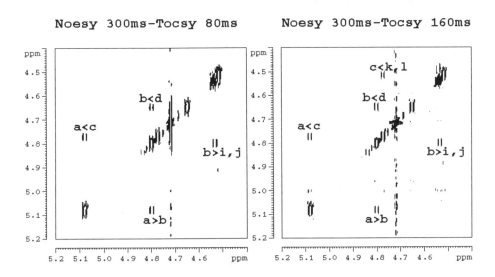

Figure 5. NOESY-TOCSY experiments on the cyclic OPG of *Azospirillum brasilense* using a shorter (Left) and longer (Right) TOCSY mixing time. The contact between the anomeric proton of k or l (same resonance position) linked in β–(1-6) to the c unit is only visible when using the longer TOCSY transfer time.

The heteronuclear sequences rely on the homonuclear NOE or the long-range heteronuclear coupling constant over the glycosidic linkage to establish sequential contacts. The NOE between flanking protons H1 and H2', where the latter is encoded by its corresponding carbon frequency to improve chemical shift dispersion, is at the basis of the HSQC-NOESY experiment. The ^3J (H1-C2') and ^3J(C1-H2') coupling constants are used in the HMBC experiment. The delay during which the long-range coupling creates antiphase was optimized for a ^3J coupling constant of 5 Hz in our HMBC experiment on the DP13 molecule.

4. Beyond the primary sequence – towards a tertiary structure

4.1 INTRODUCTION

The structure of cyclic osmoregulated periplasmic glucans (OPG) synthesized by different Gram-negative bacteria might shed light on the role that these molecules play in the adaptation to osmotic stress and in the phytopathogenicity of these bacteria [11]. The cyclic nature of these molecules could be essential for these two aspects, but a precise structural model is needed to assess this point. The glucans produced by members of the *Rhizobiaceae* family are cyclic molecules composed of 17 to 40 glucose units, all linked in β–(1-2) ; their structure determination presents an enormous challenge, as the NMR spectrum of a purified subspecies is completely degenerate and equivalent to that of a single D-glucopyranosyl unit [22, 23, 24, 25]. Whereas this might be the indication of a completely symmetric molecule, with magnetic equivalence for every glucose unit, recent work has shown that no rigid cyclic structure can be formed with identical (Φ, Ψ) angles for all linkages [26]. The proposed origin for the equivalence of all NMR signals is the high flexibility of the molecule, where each glucose unit rapidly samples a number of conformations. The NMR parameters are therefore averaged twice, over all equivalent units and over time.

Beyond this problem of time-averaging, the scarceness of experimental constraints in oligosaccharides makes an accurate structure determination extremely challenging. In proteins, sequential assignment can be obtained from NOE contacts between neighbouring residues or on the basis of J-coupling in the so-called triple-resonance experiments, but the 3D structure is calculated on the basis of a large number of long-range NOE constraints. For the cyclic glucan under study, the experimental parameters used for the sequential assignment are the only ones available for the complete structure determination.

4.2 ONE A-(1-6) LINKAGE CHANGES IT ALL.

Recently, the structures of three fundamentally different cyclic glucans extracted from cells of *R. solanacearum* [16, 27], *X. campestris* [18, 16, 27] and *R. sphaeroides* (19) were described. Although they share the cyclic nature with the OPG of members of the *Rhizobiaceae* family, they can be distinguished by the unique degree of polymerization (DP ; 13, 16 or 18 units) and one α–(1-6) linkage, whereas all other glucose residues are linked by β–(1-2) linkages. The presence of this α–(1-6) linkage induces structural constraints in this molecule to such extent that all individual anomeric proton resonances can be distinguished (Figure 6).

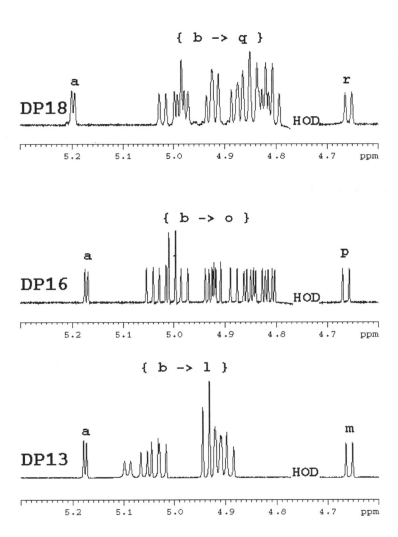

Figure 6. Anomeric proton regions of the DP18 (top), DP16 (middle) and DP13 (bottom) molecules. The two groups of chemical shifts discussed in the text for the DP13 molecule are present as well in the two other molecules, confirming their belonging to the same structural family.

Its minimal size and the complete resonance assignment of the cyclic OPG from *R. solanacearum* make this molecule an ideal candidate for a detailed structural study, and we will focus on the results we have recently obtained on this molecule.

4.3 CHEMICAL SHIFT PATTERN ON THE DP13 AND DP16

While the chemical shift remains one of the most easily measured parameter, it simultaneously is one of the most difficult to calculate from a structure. Still, it is a parameter which is extremely sensitive to the precise three-dimensional environment of the nucleus. The observation by both our group and Dr. W. York (CCRC, Athens, USA) that the chemical shifts of the H_1 and C_2 atoms flanking the glycosidic β–(1-2) linkage show an alternating pattern (Figure 7), suggests therefore a non-uniform location of the interglycosidic linkages in the (Φ, Ψ) map. This is in general agreement with the observed alternating pattern for the ^1J coupling constants, that have been related in different theoretical studies [28] to the conformation of the linkage.

Residue	H_1	C_1	H_2	C_2	$^1J_{H_1 C_1}$
a	5.18	99.48	3.66	83.43	175.0
f	4.94	103.3	3.36	74.86	165.0
i	4.91	103.4	3.56	84.80	167.0
j	4.94	103.5	3.75	82.39	164.6
d	5.04	102.0	3.64	85.17	167.0
l	4.89	102.6	3.72	79.96	165.6
b	5.09	101.8	3.72	82.36	167.0
e	5.02	102.2	3.72	80.91	167.6
k	4.90	102.7	3.70	80.50	164.6
c	5.06	102.3	3.63	83.36	167.0
h	4.93	102.9	3.70	80.90	165.6
g	4.94	103.3	3.60	83.50	165.0
m	4.66	104.5	3.67	81.97	162.7

Figure 7. Alternating chemical shift values and $^1J(^1H-^{13}C)$ coupling constant values for the different glucose units of the DP13 molecule.

4.4 LABELING THE MOLECULE

Chemical shift dispersion in the carbon dimension being far superior to the one in the proton dimension, especially for the C2 carbons, both distance and coupling constant measurements could greatly benefit of a heteronuclear detection scheme. For this purpose, we decided to produce a ^{13}C enriched sample. *R. solanacearum* T11 was grown at 24°C with agitation in a low osmolarity medium which contained K_2HPO_4 (1 mM), $(NH_4)_2SO_4$ (1.5 mM), $MgCl_2$ (0.08 mM), $FeSO_4$ (0.5 mg l^{-1}), thiamin (2 mg l^{-1}), and casein hydrolysate (4 g l^{-1} ; casamino acids, vitamin free, Difco Laboratories). This medium was adjusted to pH 7.0 with Tris-free base. ^{13}C enrichment was obtained by replacing one fourth of the casein hydrolysate by an equal amount of a mixture of uniformly enriched amino acids and glucose ("sugar mix ", EMBL, Heidelberg, Germany). The cyclic glucan was purified from cell extract (16). Mass spectrometry spectra on both the native and ^{13}C enriched DP13 samples indicated an average mass increase of 37.5 Da due to the ^{13}C labeling. Taking into account that the DP13 molecule contains 78 carbon atoms, this gives a mean enrichment of 48%. A similar result close to 50% was found on the basis of integrations made on the non-decoupled ^{1}H spectrum.

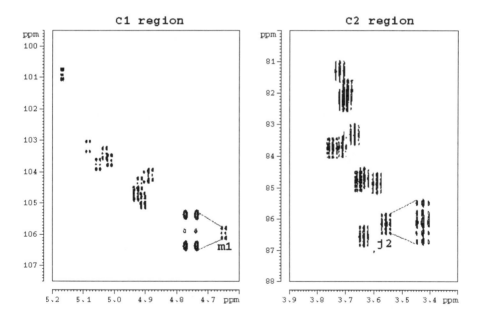

Figure 8. Heteronuclear Single Quantum Experiment correlating the ^{1}H and ^{13}C resonances of the ^{13}C enriched DP13 molecule of *R. solanacearum* at 301K. The doublet and quintuplet ^{13}C line shapes of units m and j, respectively, are magnified.

The regular HSQC spectrum, where the ^{13}C-^{13}C coupling constants are not refocused in a constant-time period [29] indicates that the labeling is not random. Indeed, we observed a doublet for the C_1 resonances (Figure 8, left panel), but a quintuplet for the C_2 correlation peaks (Figure 8, right panel). ^{13}C enrichment of the C_1 position implying ^{13}C enrichment of the C_2 position leads to the doublet structure for C_1, and the quintuplet structure for the C_2 resonances can be explained by the simultaneous presence of $^{13}C_1$ - $^{13}C_2$ - $^{12}C_3$ triplets, leading to a doublet for C_2, and $^{13}C_1$ - $^{13}C_2$ - $^{13}C_3$ triplets on other molecules leading to a triplet structure. Considering the usual bacterial glucose synthesis pathway [30], this implies that the ^{13}C enriched amino acids have been broken down to the level of a two-carbon containing molecule (probably acetyl CoA or acetate), and ^{13}C has been incorporated in a pair-wise fashion.

4.5 A GEOMETRIC INTERPRETATION OF THIS ALTERNATING CHARACTER

A limited number of studies [31, 32] have attempted to examine the conformational behavior of glycosidic linkages on the basis of precise NOE curves. However, whereas the small number of distance restraints between the two pyranosyl units requires a precise determination of the latter, the flexibility of the glycosidic linkage may lead to wrong distances due to a difference in dynamics with the proton pair considered as the internal distance reference. To take into account the possible variations of internal flexibility, at least two independent measures should be obtained for each pair of protons. The off-resonance ROESY technique [33] allows the measurement of the dipolar cross-relaxation rates along an effective field axis in the rotating frame tilted by an angle θ with respect to the static magnetic field. By varying this angle from zero to higher values, the longitudinal (σ) and transverse (μ) dipolar cross-relaxation rates can be determined with a high accuracy since they result from an over-determined fitting procedure. Finally these two values can be exploited simultaneously to obtain distances without the requirement of any internal reference [34].
Build-up rates for the normalized cross-peaks in HSQC-NOESY spectra were determined on the basis of four mixing times between 100 and 400 ms with a good linear behavior up to at least 300 ms. The off-resonance ROESY spectra were recorded as a series of 2D spectra with a constant spin-lock field strength $\gamma B_1 = 7$ kHz but with varying offset Δ of the spin-lock field. Eleven values of $\theta =$ arc tangent($\gamma B_1/\Delta$) were sampled between 5° and 55°, with two

mixing times of 150 ms and 300 ms for every angle. Build-up curves were determined independently for every angle, using the cross-peak volumes normalized to the diagonal peak ones in order to extend the validity of the initial slope approximation. A least square fit of the resulting slopes as a function of the angle θ to the theoretical expression $\sigma' = \sigma\cos^2(\theta) + \mu\sin^2(\theta)$ yields both relaxation rates.

We observe excellent agreement between the σ rates determined by the classical NOE build up rates and those obtained by the off-resonance ROESY fit (Table I). The independent determination of σ and μ allows extraction of pairwise structural (r) and dynamic (τ_{cp}) parameters (Table I, columns 5 and 6).

	HSQC-NOE	HSQC-(off-resonance ROESY)			
	$\sigma(s^{-1})$	$\sigma(s^{-1})$	$\mu\ (s^{-1})$	r(Å)	τ_{cp} (ns)
f1-i2	ND	ND	ND	ND	ND
i1-j2	ND	ND	ND	ND	ND
j1-d2	-0.35	-0.37	1.0	2.19	0.9
d1-l2	-0.20	-0.20	0.52	2.47	0.9
l1-b2	-0.40	-0.40	1.43	2.06	0.7
b1-e2	-0.38	-0.38	0.93	2.25	1.0
e1-k2	-0.19	-0.18	0.49	2.46	0.8
k1-c2	-0.33	-0.32	1.10	2.14	0.7
c1-h2	-0.21	-0.20	0.58	2.43	0.8
h1-g2	-0.42	-0.37	1.15	2.09	0.7
g1-m2	-0.39	-0.46	1.22	2.14	0.9
m1-a2	-0.48	-0.59	1.70	2.00	0.8

Table I : Longitudinal cross-relaxation rates as determined from the HSQC-NOE build-up curves (column 2) and the HSQC-(off resonance ROESY) fitting procedure (column 3), transverse cross-relaxation rates by the same procedure (column 4), and resulting pairwise distances r and dipolar correlation times τ_{cp} (columns 5 and 6). ND = not determined due to overlap.

Two classes of interglycosidic distances can be distinguished, those linking a H1 proton from group 1 [b-e] to the H2 proton from group 2 [f-l], with values around 2.45 Å, and those where a unit from group 2 substitutes a unit from group 1, with significantly shorter distances (around 2.15 Å). An averaged distance between these both extremes is found for the pair b1-e2. However, the anomeric proton of residue b is appreciably broadened (around 0.5 Hz at 301 K) compared to the other anomeric resonances. At lower temperatures (288K), the splitting due to the homonuclear coupling constant disappeared completely due to the line broadening, proving that the glycosidic linkage exhibits some low frequency jumps between two (or more) positions. The longer correlation time observed for this NOE contact indicates that, in the nanosecond range, this glycosidic bond is more rigidly fixed than are the other glucose units. The b–e glycosidic bond is diametrically opposed to the α–(1-6) linkage between units a and f, and it links two units within group 1 ; these two observations might be the basis of its particular behavior.

Relaxed energy maps for β–sophorose [35] show the main low-energy region at $20° < \Phi < 80°$ and $-60° < \Psi < 60°$. A comparison of the calculated energy map with a map showing the H1-H2' inter-proton distance of β–sophorose as a function of the interglycosidic angles together with the observed strong NOEs indicates that all β–(1-2) bonds of the cyclic OPG of *B. solanacearum* are situated in this main low-energy zone. However, the calculated precision of the interglycosidic distances (Table I) allows us to discern a pattern of alternating shorter and longer H1'-H2 distances along the primary sequence, following closely the alternating pattern of chemical shift values. The deviating distance found for the b1-e2 contact is accompanied by a similar perturbation for the H1 and C2 chemical shift values of this particular linkage.

4.6 MEASUREMENTS OF LONG-DISTANCE ^{3}J COUPLING CONSTANTS.

The delay in the HMBC during which the long-range H-C correlations were established for the sequential assignment was optimized for a hypothetical coupling constant of 5 Hz. The resulting spectrum showed uniformly intense cross-peaks in the region connecting C1 and H2' resonances, whereas the C2'-H1 region showed important differences in intensity for the various cross peaks [36]. This observation suggests uniform $^3J(C_1\text{-}H_2')$ coupling constants with values close to 5 Hz, but a less uniform distribution for the $^3J(H_1\text{-}C_2')$ coupling constants. However, differential relaxation rates as well as any other long range carbon coupling might in part contribute to the latter intensity difference, and the absolute value character of the cross peaks makes a quantitative evaluation of the coupling constants far from straightforward. Numerous other techniques for measuring the ^{3}J coupling constants have been described in the literature [37,

38]. Poppe *et al.* (23) described a ^{13}C filtered (^{1}H, ^{1}H) ROESY experiment based on the general ^{1}J-resolved E.cosy approach (39). This experiment resulted in accurate ^{3}J values on the homogeneous cyclic β–(1-2)-glucan icosamer isolated from *Rhizobium trifolii*. The experiment was performed on a 5mM natural abundance sample, but the high flexibility of this molecule leads to a completely degenerate ^{1}H and ^{13}C spectrum, equivalent to that of one single unit [22, 23, 24, 25]. This reduces the information content of the measured coupling constants, as they represent time-averaged values, but enhances simultaneously the feasibility of the experiment, as the 5 mM sample results in an effective glucose concentration of 100 mM. Despite this high concentration, the resulting experiment took 54 hours. A similar experiment on a natural-abundance DP13 sample, where no degeneracy exists, would not be feasible, making ^{13}C labeling essential.

The original experiment as proposed by Poppe et al. [23] exploits the presence of $^{13}C_1$-$^{12}C'_2$ and $^{12}C_1$-$^{13}C'_2$ carbon pairs at natural abundance. The 50% incorporation of ^{13}C in the present DP13 molecule optimized the number of such carbon pairs, but forced us to introduce an additional step of ^{12}C filtering. The complete sequence is depicted in Figure 9, and can easily be broken down into three modules : first, a ^{13}C filter is applied in order to select for the t_1 evolution only those protons that are ^{13}C bound. We used a gradient-enhanced scheme (Figure 9A) that is quite similar to the solvent suppression scheme used in the ^{1}H-^{15}N HSQC experiment [40]. When the first gradient is applied, ^{13}C bound proton magnetization is in the IzSz state, whereas all other magnetization components are in the *xy* plane and will be effectively scrambled by G_1. After refocusing of the proton magnetization, it is stored temporarily along the z-axis, and the second gradient removes effectively all residual terms.

In contrast to the ^{13}C half-filter of Poppe et al. [23], our sequence selects no antiphase proton magnetization with respect to the $^1J_{CH}$ coupling but pure in-phase proton magnetization (term Ix). This magnetization will evolve during t_1 under the influence of both the chemical shift and the $^1J_{CH}$ scalar coupling term, and terms of the form Iz $\cos(\omega_1 t_1) \cos(\pi J\, t_1)$ will enter the NOESY mixing time (Figure 9B). The length of the mixing time should be a compromise between two requirements : it should be long enough to allow an efficient transfer of proton magnetization between the H_1 and H_2' positions, but it should not be too long such that ^{13}C T_1 relaxation will destroy the E.cosy type pattern that was constructed during t_1. Intuitively, the second requirement can be understood as follows: during t_1, two populations of proton spins can be distinguished, depending on the state of the attached ^{13}C spin.

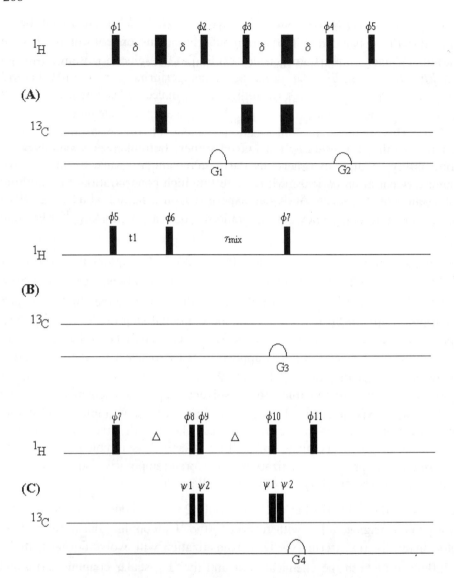

Figure 9. Pulse sequence for the double half-filtered NOESY spectrum. The sequence has been divided in three parts, the ^{13}C filter (A), the t_1 evolution and subsequent NOE mixing time (B), and the final ^{12}C filter (C). Phases used are: Φ_1=y, Φ_2=(x, -x), Φ_3=(8*x,8*(-x)), Φ_4=(x, x, -x, -x, y, y, -y, -y), Φ_5=(4*x, 4*y), Φ_6=(-x, -x, x, x, -y, -y, y, y), Ψ_1=x, Ψ_2=x or -x (according to the choice of the subspectra), acquisition phase = (x, -x, -x, x, y, -y, -y, y, -x, x, x, -x, -y, y, y, -y).

During the NOE mixing time, we want this relationship to be conserved, but it will tend to disappear due to T_1 relaxation. If the mixing time is so long that the correlation between initial and final ^{13}C spin state has completely disappeared, no displacement in ω_2 representative for the ^3J value will be observed [41]. In the case of the cyclic glucan, both requirements can easily be fulfilled by a short mixing time of 100 ms, as the H_1-H_2' distances are all very short [42].

After the mixing time, we apply a ^{12}C filter that is similar to the one used for distinguishing NOEs in macromolecular complexes in which one of the components has been enriched with a stable isotope [43, 44, 45] or intra- from inter-subunit NOE cross peaks in symmetrical homodimers [46, 47]. The same two ^{13}C 90° pulses that are used for ^{12}C selection (once with the phases Ψ_1 and phases Ψ_2 equal and once opposite) have to be applied at the end of the filter, in order to maintain the ^{13}C spin state (Figure 9C). Failure of applying those pulses leads to E.cosy type cross peaks that for the two spectra are oppositely shifted in ω_2 with respect to the ^3J coupling value, making the addition of the two subspectra in order to produce a ^{12}C filtered spectrum impossible.

In the resulting 2D spectrum, suppression of the diagonal peaks at the level of the anomeric proton resonances was optimized by introducing a correction factor of 1.08 when adding both subspectra.

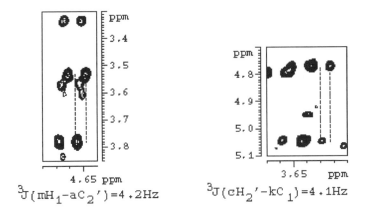

Figure 10. Final 2D spectrum obtained by adding the two subspectra obtained with the sequence of Figure 10. The zooms show the aH'$_2$-mH$_1$ cross peak and the extraction of the corresponding ^3J (mH$_1$ - aC'$_2$) coupling constant (Left), and similar for the kH$_1$-cH$_2$' cross peak and corresponding ^3J (kC$_1$ - cH$_2$') coupling constant (Right).

The left panel of Figure 10 shows an enlargement of the aH_2'-mH_1 cross peak: in ω_1, the central frequency of the doublet is that of the aH_2' proton, and we observe the doublet with respect to the $^1J_{HC}$ coupling constant of 145.8 Hz. In ω_2, both lines correspond to a ^{12}C bound mH_1 proton, but the upper line of the ω_1 doublet is displaced with respect to the lower one by 4.2 Hz, which is the value for the corresponding ^3J (mH_1 - aC_2') coupling constant. In the other quadrant of the spectrum, we will observe H_2 magnetization, and the right panel shows a zoom of the kH_1-cH_2' cross peak. The upper triplet (corresponding to cH_2' magnetization coupled through a long-range ^3J constant to a k $^{13}C_1$ nucleus in a well-defined state) is displaced by 4.1 Hz with respect to the lower one, corresponding to the c H_2' proton bound to a k $^{13}C_1$ is in the other state. In both cases the fine structure of the observed cross peak (a doublet for the anomeric protons, and a triplet for the H_2 protons) increases the level of confidence for the measurement of the coupling constant : indeed, we can measure the displacement of every component of the multiplet at $\omega_1+{}^1J/2$ with respect to the corresponding component in the multiplet at $\omega_1-{}^1J/2$. From this multiple measurement, we estimated error bars of ± 0.2 Hz on the reported ^3J values reported in Table II.

	$^3J(H_1-C'_2)$	Φ	$^3J(C_1-H'_2)$	Ψ
f1-i2	ND	ND	ND	ND
i1-j2	ND	ND	ND	ND
j1-d2	4.1	32	4.0	327
d1-l2	2.2	54	4.2	329
l1-b2	3.0	44	4.7	336
b1-e2	3.4	39	3.8	324
e1-k2	2.9	45	3.9	326
k1-c2	4.1	32	4.1	328
c1-h2	2.9	45	4.1	328
h1-g2	4.0	33	4.2	329
g1-m2	3.0	44	4.3	330
m1-a2	4.2	30	4.0	327

Table II : Long-range ^1H-^{13}C coupling constants obtained for the cyclic glucan *of Ralstonia solanacearum*. The angles were derived according to the Karplus relationship [48] and assuming that the linkages are in the main low-energy zone. The first two values were not determined due to overlap of the corresponding cross peaks.

The presence of intra-residue H5-H1 cross peaks despite the combined ^{13}C and ^{12}C filters stems from glucose residues that contain a ^{13}C$_1$ but a ^{12}C$_5$ nucleus, and is compatible with the pairwise incorporation of ^{13}C nuclei (*vide supra*). From these cross peaks, coupling constants for H$_1$-C$_5$ smaller than 0.5 Hz were measured, in agreement with the very weak cross peaks observed in the HMBC.

Using both quadrants, 10 values of both ^{3}J values could be extracted (Table II). The missing values could not be determined because of heavy overlap of the corresponding cross peaks. Finally, it should be noted that the single α–(1-6) linkage can equally be characterized by the same procedure, as we could measure the coupling constants between H$_6$-C$_1$ and H$_6$'-C$_1$ (the notation H$_6$-H$_6$' being at this moment arbitrary). The values found are ^{3}J(aC$_1$-fH$_6$) = 2.4Hz, and ^{3}J(aC$_1$--f H$_6$') = 3.0 Hz.

4.7 A UNIQUE STRUCTURE OF THE MOLECULE?

The techniques of Distance Geometry and/or Simulated Annealing are commonly used to generate families of protein structures compatible with the majority of the experimental data. However, it has been amply demonstrated that the reliability of both techniques depends to a large extent on the number of experimental constraints that can be imposed rather than on the accuracy of these constraints. In the case of our cyclic oligosaccharide molecule, however, this number of constraints is limited to the distance measurements between protons flanking the glycosidic linkages, and the long-range coupling constants. An additional problem with the latter is that the empirical Karplus relationship, well parametrized for protein structures due to the large number of crystal structures and coupling constants, is not as reliable for oligosaccharides. Indeed, the majority of experimental points that were used for the actual parametrization of the Karplus relationship concern a series of conformationally rigid carbohydrate derivatives with dihedral angles between 90° and 270° [48]. As the authors had not yet prepared rigid compounds containing fixed C-O-C-H segments with dihedral angles outside this region, they included three experimental values for compounds where the dihedral angles were not directly determined by X-ray crystallography. Therefore, we believe that for our compound characterized mainly by dihedral angles outside of the (90°, 270°) region, the current parametrization (^{3}J$_{C-H}$ = 5.7 cos^2(Φ) - 0.6 cos(Φ) + 0.5) and thus the angles derived from the coupling constant values have to be used with caution.

For the reasons mentioned above, we developed a genetic algorithm approach to generate different structures [49]. The solvent contribution, which plays an important role in oligosaccharide stability, is taken implicitly into account by the boundary element method [50, 51], which averages over all possible water configurations. The distance results stemming from the off-resonance ROESY experiment were included as a harmonic energy term.

The algorithm yielded several interesting structures, characterized by an overall low energy. These structures consistently contain fragments that adopt the ribbon-like right-handed helix as was described for the DP16 molecule [18]. However, we found no single structure that was of low energy and compatible with all experimental data simultaneously. For this reason, we decided to first look at the dynamics of the molecule in solution. The result of this study is described in the paragraph below.

5. Complex dynamics of the DP13 molecule

5.1 IMPORTANCE OF DYNAMICS

The dynamics of biomolecules has been an intensive field of research over the last years [for a review, see 52]. In the field of protein dynamics, numerous studies have probed the rapid motions on the pico- to nanosecond time scales, with relatively small amplitudes. These motions are generally considered as librations or hindered motions in an energetic well, and contribute by influencing barriers for slower motions and by modulating entropic changes that might be associated with functional processes. Recent studies in the field of oligosaccharide NMR have also focused on this rapid dynamics. Slower domain motions, however, have proved to be more difficult to study, despite their enormous biological relevance, as they are close to the time scales on which docking, catalysis and ligand release take place. Examples are long-range loop-flip motions in proteins, or the "flipped-out" substrate base in the crystal structure of DNA cytosine-5-methyltransferase. However, to our knowledge, no such slow-time scale motions have been reported for isolated oligosaccharides. As we will show further, the DP13 molecule of *R. solanacearum* does exhibit fluctuations of probably large amplitude on the μs time-scale, adding to its intriguing function *in vivo* the question of the functional relevance of this complex dynamics.

5.2 RESULTS OF HOMONUCLEAR EXPERIMENTS

A side result of the off-resonance ROESY experiments as described above is the set of τ_c values describing the individual correlation functions describing the pairwise dipolar interaction between H1 and H2' protons. These values were found to be fairly homogeneous, with an average value of 0.9 ns. At first, we attributed them to the overall tumbling of the hypothetical rigid molecule, but fast internal dynamics could also contribute to the modulations of the dipolar terms. However, a separation of overall and fast internal dynamics, as is often done following heteronuclear relaxation studies, is not of much use here, as every individual dipolar interaction is described by its own τ_c.

Indication of slow dynamics came from the observation that the b anomeric proton resonance is appreciably broadened compared to all other anomeric resonances. We therefore decided to start a detailed heteronuclear relaxation study to better understand the complex dynamics of the DP13 molecule.

5.3 HETERONUCLEAR RELAXATION EXPERIMENTS INDICATE A SLOW CONFORMATIONAL EXCHANGE

A nowadays well-established approach to the investigation of dynamical properties in carbohydrates (and other biomolecules) is by means of heteronuclear relaxation measurements. A "model-free" analysis of the relaxation rates has been proposed, decoupling the overall tumbling motion and the local motion where the extent and dynamics of the latter are characterized by an order parameter S^2 and a local correlation time [53]. Moreover, in several studies, especially in proteins, it has been necessary to extend the parameter set to obtain consistency between the different relaxation data. An example is the *ad hoc* inclusion of an exchange term into the expression of the T_2 relaxation rate [54], or, recently, the anisotropy of the rotational diffusion as a supplementary parameter [55].

A complete heteronuclear relaxation study on the cyclic glucan of *R. solanacearum* has allowed an experimental verification of the agreement between the correlation times stemming from the off-resonance ROESY experiment (that describe the motion of the proton-proton vector) and the ones from the heteronuclear relaxation experiments, describing the motion of the ^{1}H-^{13}C vector. Secondly, the heteronuclear relaxation data yield very important information about the slow dynamics of the molecule, with exchange rates of the order of several microseconds.

The T_1 values of all carbon C1 resonances, when fitting the magnetization curves measured on a ^{13}C labeled sample to a mono-exponential are fairly homogeneous, with values centered around 276 ms (Table III). Error bars on the individual T_1 times were estimated to ±3 ms. When measuring the T_1 relaxation rates on a non-enriched sample, we systematically found longer values. We checked for a possible influence of the sample concentration by comparing one-dimensional T_1 data on a native and ^{13}C enriched sample at the same concentration, but the difference consistently showed up. The average decrease of 14 ms for the C1 relaxation times upon ^{13}C enrichment was therefore attributed to the contribution of the carbon-carbon dipolar interaction to the relaxation pathway, and is in good agreement with the 4.6% variation in the Cα T_1 time when going from a singly towards a uniformly ^{13}C-labeled alanine residue dissolved in perdeuterated glycerol at 40°C where its correlation time τ_c is equal to 1ns [56]. The magnitude of the contribution of the $^{13}C2$-$^{13}C1$ homonuclear dipolar interaction being a monotonic function of correlation time, the observed difference is an independent estimate of a correlation time of the order of 1ns.

Residue	T_1 (C1) (ms)	T_1(C2) (ms)	NOE (C1) %	NOE (C2) %
a	269	257	35	33
f	283@	296	36@	42
i	274&	257	35&	34
j	274&	261	35&	33
d	281	262	34	26
l	280§	278	34§	29
b	268	284	34	36
e	274	281#	34	32°
k	280§	268	34§	31
c	274	280	32	32
h	269	281#	35	32°
g	283@	279	36@	34
m	281	273	35	38

Carbon	T_1 (C1) (ms)	NOE (%)
aC3	274	34
aC5	272	35
fC3	308	36
fC4	289	35
mC3	279	33
mC5	294	38

Table III. (a) T_1 and NOE values for the C1 and C2 carbons of the cyclic OPG of *B. solanacearum*, as measured on a ^{13}C enriched sample at 301K. Residues are labeled as in Figure 2. Overlapping resonances were grouped in pairs, indicated by special symbols. The average value is reported. (b) T_1 and NOE values for other selected carbon resonances.

The good chemical dispersion around the α–(1-6) linkage both at the ^{1}H and ^{13}C level allowed the determination of the T_1 times of some other carbon nuclei. Therefore, we could verify whether the ring carbons in each ring are dynamically equivalent. For both the a and m units, the values for all carbons in the hexacycle are very uniform (Table III), confirming that this unit moves as a rigid body. For the f unit, we see some more variation, which might point to an enhanced mobility for this residue implicated in the α–(1-6) linkage.

The NOE values are also in good agreement with the theoretical value calculated for a τ_c=0.9 ns correlation time. As we do not expect any influence of the carbon-carbon dipolar coupling, the measurements were only performed on the enriched sample. The theoretical expression for the NOE value contains spectral density terms in both the numerator and enumerator. Therefore, this parameter shows no explicit dependency on the order parameter S^2, leaving us with a pure estimation of τ_c. The average value of 34% is in excellent agreement with the previously determined value of 0.9 ns for τ_c.

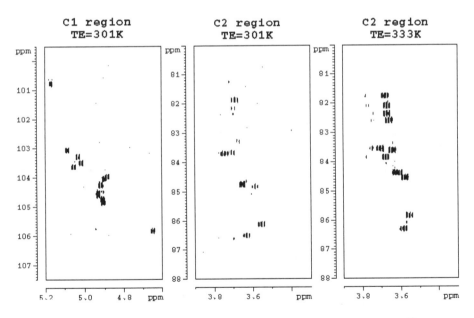

Figure 11. Selected regions of a constant-time HSQC spectrum of the ^{13}C enriched DP13. a) H1-C1 region at 301K; b) H2-C2 region at 301K; c) H2-C2 region at 333K. The intensities were normalized with respect to the f H2-C2 correlation peak.

Based on the same estimates of the correlation times τ_c=0.9 ns and order parameter $S^2 = 0.85$, we expect T_2 relaxation times of the order of 190 ms. A first indication of a discrepancy between these theoretical value and the experimental ones came from the observation of the constant-time HSQC on a ^{13}C enriched sample. The constant-time period for the evolution of the chemical shift was set to 44 ms, allowing a correct refocusing of the average ^{13}C-^{13}C one-bond J-coupling constant. In the spectrum of Figure 11 (left) recorded at 301K, all H1-C1 direct correlation peaks are visible, be it with a weaker intensity for the bC1 resonance. In the same spectrum, all H2-C2 correlation peaks are substantially weaker, and certain resonances such as the l unit almost disappear (Figure 12, middle).

A strong indication that the additional line broadening is due to chemical exchange comes from the temperature dependence of the line broadening. Already for the anomeric proton resonance of unit b, we had found an increased broadening at lower temperatures. At higher temperatures, the chemical exchange rate k_{ex} should increase, and hence contribute less to the exchange broadening. We tested this by recording constant-time HSQC spectra at different temperatures ranging from 20 °C to 60 °C. As can be seen in Figure 11 (right), the C2 resonances gain considerably in intensity at the higher temperature, in agreement with the hypothesis of exchange broadening. The fC2 resonance, that is not involved in a linkage, provides a reference as it hardly changes in intensity upon increasing temperatures.

More precise measurements of the T_2 values were obtained with the 2D extension of the Carr-Purcell-Meiboom-Gill (CPMG) sequence [57, 58]. The large distribution of expected T_2 times (inferior to 40 ms for certain C2 resonances, based on the results of the constant-time HSQC experiment, and of the order of 150 ms for the a and m C1 resonance, based on preliminary 1D relaxation experiments) led to the delicate problem of choosing the CPMG times that should be used in the series of 2D experiments. The CPMG times were selected according to the results of a recent theoretical study where the Cramér-Rhao theory was used to define an optimal sampling strategy [59]. The results of the T_2 measurements are summarized in Table IV. The series was repeated twice at a two-months interval, and the results proved reproducible to within ± 2ms.

Residue	T_2 (C1) (ms)	T_2(C2) (ms)	Δ_{ex} C2 (Hz)
a	130	117	3
f	93@	151	-
i	81#	64	11
j	81#	53	14
d	107	33	25
l	92§	19	54
b	47	27	33
e	89	18#	50
k	92§	45	15
c	81	21	38
h	124	18#	50
g	93@	29	28
m	136	24	35

Table IV. ^{13}C T_2 values measured on the non-enriched OPG of *R. solanacearum* at 301K and 150 MHz. We used the CPMG sequence with a delay δ=500 μs between the π pulses, and 11 CPMG times ranging from 1ms to 240 ms (see text). Overlapping resonances were grouped in pairs, indicated by special symbols. The average value is reported. Exchange contributions were calculated from ($1/T_2^{obs} = 1/T_2^{dip} + \Delta_{ex}$) with a T_2^{dip}= 190 ms.

Whereas the values of all C1 resonances except for the b unit experience relatively little exchange broadening, the bC1 and even more all C2 resonances are dramatically broadened. With a purely dipolar T_2 value of 190 ms based on the estimates of 0.9 ns for the correlation time and 0.85 for the order parameter S^2, the exchange contribution to the bC1 linewidth can be estimated to 14 Hz, whereas for certain C2 resonances, it obtains values as large as 54 Hz (Table IV). Alternatively, we could consider the fC2 T_2 relaxation time as the basis for the non-exchange broadened T_2 value, but this leads to no major differences for the exchange term.

In the constant time HSQC experiment, only one 180° ^{13}C pulse is applied during the delay of 44 ms, therefore no refocusing of the chemical exchange can happen (*vide infra*), and the exchange contribution is given by (60)

[1] $\Delta_{ex} = \Omega_{ex}^2 / 8 / k_{ex}$

The quadratic dependence on Ω_{ex}, the chemical shift separation (in radial units) of the different chemical species, emphasizes that physically an exchange process is similar to the molecular self-diffusion process, here in one dimension.

Furthermore, the analytical dependence of the exchange contribution on the quadratic difference in chemical shift for the co-existing conformations suggests a straightforward way of verifying the nature of the broadening mechanism. We recorded at 301K a series of T_2 experiments at the lower field of 7.1 T (or 75 MHz for ^{13}C). The disadvantage of the lower field strength was the decrease in sensitivity and the increased spectral overlap, leading to considerably higher uncertainties in the T_2 values. Still, the good separation of the H2-C2 correlation peaks allowed us to extract a certain number of them (Table V). If again we estimate the dipolar contribution to the T_2 relaxation time based on a correlation time $\tau_c = 0.9$ ns and an average order parameter $S^2 = 0.85$, we obtain a value of 133 ms. Subtracting this dipolar

Residue	T_2 (C2) (ms)	Δ_{ex} C2 (Hz)	Δ_{ex} (14T)/ Δ_{ex} (7T)
a	ND	ND	ND
f	122±5	0	-
i	84±7	4.4	3.2
j	84±6	4.4	3.2
d	75±6	5.8	4.3
l	59±10	9.4	5.7
b	64±6	8.1	4.1
e	ND	ND	ND
k	ND	ND	ND
c	ND	ND	ND
h	ND	ND	ND
g	73±10	6.2	4.5
m	57±7	10.0	3.5

Table V. ^{13}C T_2 values measured on the non-enriched OPG of *R. solanacearum* at 301K and 75 MHz. Exchange contributions were calculated from (1/ T_2^{obs} = 1/ T_2^{dip} + Δ_{ex}) with T_2^{dip}= 133 ms.

contribution from the observed relaxation rates, we obtain the exchange contribution at 75 MHz (Table V), and can evaluate the ratio of the exchange contributions at 14.1T and 7.1T. Taking into account the large uncertainties on the T2 values at the lower field of 7.1T, the obtained ratios are reasonably close to the theoretical value of 4 as expected from Eq. 1.

5.4 STRUCTURAL IMPLICATIONS OF THE RELAXATION DATA

The above described results indicate unambiguously that the oligosaccharide passes through different conformations that are in dynamic exchange on the microsecond time scale [61]. One major problem of any NMR study on a system that shows rapid chemical exchange is the average nature of all obtained structural parameters. A very illustrative example is formed by the cyclic decapeptide antanamide, for which the available NOE and J coupling data could not be satisfied by one single conformation [62]. The simplest model compatible with all NMR parameters required two different conformations. Another extreme example of the average nature of the NMR parameters is formed by the all β–(1-2) cyclic glucans of the *Rhizobiaceae*. Both the chemical shift values and the ^3J coupling constants measured over the glycosidic linkage reflect an average behaviour over both the individual monomers and over the different conformations adopted by the macrocycle over time.

Our initial hope when starting the structural studies on the OPG of *R. solanacearum* was that the single α–(1-6) linkage would induce structural constraints to such an extent that both mechanisms of averaging would be greatly reduced. Whereas the first averaging evidently has disappeared - we observe all anomeric protons resonances individually - the above described relaxation data show that our structural data still result from an average over two or more conformations.

It is illustrative to map the exchange contribution on the macrocycle (Figure 12). With the exception of the b unit, only the C2 carbon T_2 values are appreciably affected by the chemical exchange, implying the most important variation for the Ψ angle between the different conformations. We further observe that chemical exchange is not uniformly distributed over the macrocycle, as the residues around the α–(1-6) linkage are relatively little affected, whereas the main exchange phenomenon takes place around the b unit, diametrically opposed to the α–f linkage. For the latter b residue, both C1 carbon and C2 carbon are considerably broadened. In the case of the DP16 molecule of *Xanthomonas campestris*, York presented recently (18) a structural model, and proposed a topological reversal point between the two helical domains to occur at the level of the flexible α–(1-6) linkage and at the residue where a frameshift in the alternating chemical shift pattern was observed. Our current hypothesis is that this point of asymmetry in the DP13 molecule of *R. solanacearum* occurs at the level of the b residue, but not in a static fashion - i.e. that the point hops between both linkages preceding or following this b residue.

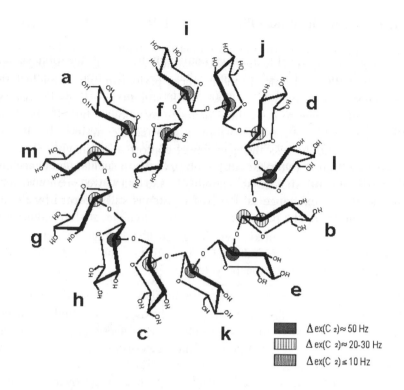

Figure 12. The macrocycle of the DP13 OPG of *R. solanacearum*. The circles on the C2 carbons graphically represent the exchange contribution to the T_2 relaxation of the C2 nucleus. All C1 carbons except for the bC1 have values of Δ_{ex} inferior to 10Hz.

In the ^{13}C NMR spectra of oligo- and poly-saccharides, the chemical shifts of the carbon atoms on either side of the glycosidic linkage have been found to vary over a range of up to 12 ppm, depending on the conformation of the linkage [63]. Excluding the fC2 carbon that is not implicated in a linkage, the variations for the C2 resonance positions have to be very large to lead to the observed line broadenings, indicating a large conformational variety. The combination of the explicit dependence of the chemical shift on the conformation of the glycosidic linkage, at least in a β–(1-2) linked disaccharide, with the previously determined short distances and ^3J coupling constants will probably allow to alleviate this ambiguity.

6. *In vivo* observation of the OPGs

The functional relevance of the complex internal dynamics of the OPG in vitro is not clear at this moment. However, NMR allows equally to probe the motional freedom of the OPG molecules in the bacterial periplasm, which might be more directly related to its function. In vivo NMR studies have mostly focused on small molecules, aiming to detect and quantify metabolites by ^{31}P NMR or ^{13}C NMR, but have not yet been used to investigate more complex biomolecules in the organism, and to address the question of conformation in vivo. The aim of our study was therefore twofold : can we observe the molecule, and hence deduce a certain degree of molecular mobility? and can we see whether the conformational slow exchange as observed in vitro is still present in the complex environment that forms the periplasm?

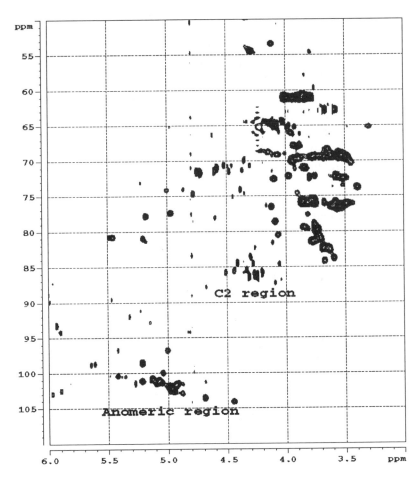

Figure 13. ^{1}H-^{13}C HSQC spectrum of a cellular suspension of *R. solanacearum* in D$_2$O.

One of the prerequisites of high-resolution NMR is that the external magnetic field is extremely homogeneous over the volume of the detection coil. Modern techniques of magnet and coil design have led to the impressive achievement of a homogeneity smaller than one Hz, on a total frequency that approaches the GHz. However, when bodies with a different magnetic susceptibility are introduced in the sample, this resolution is partly lost, due to the macroscopic dipolar fields that are generated at the interface of differential magnetic susceptibility. Recent efforts aimed at minimizing such effects have used the technique of Magic Angle Spinning [64, 65], borrowed from solid state NMR, to minimize macroscopic inhomogeneities, and obtain high-resolution spectra on heterogeneous samples such as polystyrene resin-bound organic molecules or peptides [66, 67].

We performed a first experiment with a 60ml solution of concentrated cells in a 4mm rotor, but the spinning at 4kHz altered the hydration of the cells, and no workable NMR spectra could be obtained. In a second attempt, we recorded a spectrum of the same solution in a regular non-spinning 5mm tube, and here, we were able to obtain a good spectrum, where both the anomeric and C2 region could be clearly distinguished (Figure 13). We ascribe the success of this experiment to the small difference in magnetic susceptibility between the cells and the aqueous solvent, as had been observed earlier for entire embryos [68]. Future work of our groups will therefore be directed to selective labeling of the OPG by using C1 or C2 labeled glucose as sole carbon source, and further investigate the structure and dynamics of the OPG in the bacterial periplasm.

7. Acknowledgments

Dr. J.-M. Wieruszeski has contributed greatly to all NMR results presented in this lecture. Dr. P. Talaga was responsible for all biochemistry work, and Dr. D. Horvath performed all modeling studies. Without these three persons, we would not have been able to present this lecture, and we want to express our warmest thanks. We thank Dr. H. Desvaux (CEA, France) for fruitful collaboration in the off-resonance ROESY experiment, Prof. P. Albersheim (Complex Carbohydrate Research Center, University of Georgia) for access to the mass spectrometer and Dr. W. S. York for performing the MALD-TOF MS analysis. The 600 MHz facility used in this study was funded by the European funds of the FEDER attributed to the Région Nord - Pas de Calais (France), by the CNRS and by the Institut Pasteur de Lille.

8. References

1 – Oliver, D. B. (1996) in *Escherichia coli* and *Salmonella*. Cellular and Molecular Biology, Neidhardt, F.C. (ed.), American Society for Microbiology, Washington, D.C., pp. 88-103.

2 – Schulman, H., and Kennedy, E. P.(1979) *J. Bacteriol.* **137**, 686-688.

3 – Miller, K. J., Kennedy, E. P., and Reinhold, V. N. (1986) *Science* **231**, 48-51.

4 – Van Golde, L. M. G., Schulman, H., and Kennedy, E. P. (1973) *Proc. Natl. Acad. Sci. USA* **70**, 1368-1372.

5 – Kennedy, E. P. (1996) in *Escherichia coli* and *Salmonella*. Cellular and Molecular Biology, Neidhardt, F.C. (ed.), American Society for Microbiology, Washington, D.C., pp. 1064-1071;

6 – Talaga, P., and Bohin, J.-P. (unpublished).

7 – Talaga, P., Fournet, B., and Bohin, J.-P.(1994) *J. Bacteriol.* **176**, 6538-6544.

8 – Dell, A., York, W. S., McNeil, M., Darvill, A. G., and Albersheim, P. (1983) *Carbohydr. Res.* **117**,185-200.

9 – Koizumi, K., Okada,Y., Horiyama, S., and Utamura, T.(1983) *J. Chromat.* **265**, 89-96.

10 – Hisamatsu, M. (1992) *Carbohydr. Res.* **231**, 137-146.

11 – Breedveld, M. W., and Miller, K. J. (1994) *Microbiol. Rev.* **58**, 145-161

12 – Bundle, D. R., Cherwonogrodzky, J. W., and Perry, M. B. (1988) *Infect. Immun.* **56**, 1101-1106.

13 – Miller, K. J., Gore, R. S., Johnson, R., Benesi, A. J., and Reinhold, V. N. (1990) *J. Bacteriol.* **172**, 136-142.

14 – Rolin, D. B., Pfeffer, P. E., Osman, S. F.,Szwergold, B. S., Kappler, F., and Benesi, A. J. (1992) *Biochim. Biophys. Acta* **1116**, 215-225.

15 – Altabe, S. G., Talaga, P., Wieruszeski, J.-M., Lippens, G., Ugalde, R., and Bohin, J.-P. (1998) in Biological Nitrogen Fixation for the 21st Century, C. Emelrich et al. (eds), p 390.

16 – Talaga, P., Stahl, B., Wieruszeski, J.-M., Hillenkamp, F., Tsuyumu, S., Lippens, G., and Bohin, J.-P. (1996) *J. Bacteriol.* **178**, 2263-2271.

17 – Amerura, A., and Cabrera-Crespo, J. (1986) *J. Gen. Microbiol.* **132**, 2443-2452.

18 – York, W. S. (1995) *Carbohydr. Res.* **278**, 205-225

19 – Talaga, P, Wieruszeski, J.-M., Cogez, V., Bohin, A., Lippens, G., and Bohin, J.-P. (in preparation)

20 – Miller, K. J., Reinhold, V. N., Weissborn, A. C., and Kennedy, E. P. (1987) *Biochim. Biophys. Acta* **901**, 112-118.

21 – Miller, K. J., Gore, R. S., and Benesi, A. J. (1988) *J. Bacteriol.* **170**, 4569-4575.

22 – Serrano, G., Franco-Rodriguez, G., Gonzalez-Jimenez, I., Tejero-Mateo, P., Molina Molina, J., Dobado, J. A. , Mégias, M. and Romero, M. J. (1993) *J. Mol. Struct.* **301**, 211

23 – Poppe, L., York, W. S. and van Halbeek, H. (1993) *J. Biomol. NMR* **3**, 81- 89

24 – André, I., Mazeau, K., Taravel, F. R., and Tvaroska, I. (1995a) *Int. J. Biol. Macromol.* **17**, 189.

25 – André, I., Mazeau, K., Taravel, F. R., and Tvaroska, I. (1995b) *New J. Chem.* **19**, 331.

26 – York, W. S., Thomsen, J. U., and Meyer, B. (1993) *Carbohydr. Res.* **248**, 55-80

27 – Lippens, G., Talaga, P., Wieruszeski, J.-M., and Bohin, J.-P. (1995), in *Spectroscopy of Biological Molecules*, Merlin, J. C., Turrell, S., and Huvenne, J.-P. (ed.), Kluwer Academic Publishers, Dordrecht, the Netherlands.

28 – Tvaroska, I., and Taravel, F. R. (1992) *J. Biomol. NMR* **2**, 421-430

29 – Vuister, G. and Bax, A. (1992) *J. Magn. Res.* **98**, 428-435

30 – Gottschalk, G. (1986) *Bacterial Metabolism*, Springer, New York, NY

31 – Widmalm, G., Byrd, R. A. , and Egan, W. (1992) *Carbohydr. Res.* **229**, 195.

32 – Peters, T., and Weimar, T.(1994*) J. Biomol. NMR*, **4**, 97

33 – Desvaux, H., Berthault, P., Birlirakis, N. Goldman, M. (1994) *J. Magn. Res. A* , **108**, 219.

34 – Desvaux, H., Berthault, P., and Birlirakis, N. (1995) *Chem. Phys. Letters* **233**, 545.

35 – Dowd, M. K., French, A. D. and Reilly, P. J. (1992) *Carbohydr. Res.* **233**, 15.

36 – Lippens, G, Wieruszeski, J.-M., Talaga, P., and Bohin, J.-P. (1996) *J. Biomol. NMR* **8**, 311-318.

37 – Poppe, L., and van Halbeek, H. (1991) *J. Magn. Res.* **93**, 214-217

38 – Uhrin, D., Mele, A., Boyd, J., Wormald, M. R. and Dwek, R. A. (1992) *J. Magn. Res.* **97**, 411-418

39 – Montelione, G. T., Winkleer, M. E., Rauenbuehler, P., and Wagner, G. (1989) *J. Magn. Res.* **82**, 198-204

40 – Wider, G., and Wüthrich, K.(1993) *J. Magn. Res. Series B* **102**, 239-241

41 – Harbison, G. S. (1993) *J. Am. Chem. Soc.* **115**, 3026-3027

42 – Lippens, G., Wieruszeski, J.-M., Talaga, P., Bohin, J.-P., and Desvaux H. (1998) *J. Am. Chem. Soc.* **118**, 7227-7228.

43 – Otting, G. and Wüthrich, K. (1990) *Q. Rev. Biophys.* **23**, 39-96

44 – Fesik, S. W., Gampe, R. T., Eaton, H. L., Gemmecker, G., Olejniczak, E. T., Neri, P., Holzman, T. F., Egan, D. A., Edalji, R., Simmer, R., Helfrich, R., Hochlowski, J. and Jackson, M. (1991) *Biochemistry* **30**, 6574-6583

45 – Weber, C., Wider, G., Von Freyberg, B., Traber, R., Braun, W., Widmer, H., and Wüthrich, K. (1991) *Biochemistry* **30**, 6563-6574

46 – Lee, W., Harvey, T.S., Yin, Y., Yau, P. Litchfield, D., and Arrowsmith (1994) *Nat. Struct. Biol* **1**, 877-890.

47 – Zhang, H., Zhao, D., Revington, M., Lee, W., Jia, X., Arrowsmith, C., and Jardetzky, O. (1994) *J. Mol. Biol.* **238**,592-614.

48 – Tvaroska, I., Hricovini, M., and Petrakova, E. (1989) *Carbohydr. Res.* **189**, 359-362

49 – Horvath, D., and Lippens, G. (unpublished).

50 – Horvath, D., Lippens, G., and Van Belle, D. (1996) *J. Chem. Phys.* **105**, 4197-4210.

51 – Horvath, D.,Van Belle, D., Lippens, G., and. Wodak, S. J. (1996) *J. Chem. Phys.* **104**, 17.

52 – Palmer III, A. G., Williams, J., McDermott, A. (1996) *J. Phys. Rev.* **100**, 13293-13310,

53 – Lipari, G., Szabo, A. (1982) *J. Am. Chem. Soc.* **104**, 4546-4559.

54 – Clore, G. M., Driscoll, P. C., Wingfield, P. T., Gronenborn, A. M. (1990) *Biochemistry* **29**, 7387-7401.

55 – Brüschweiler, R., Liao, X., Wright, P. (1995) *Science* **268**, 886-889.

56 – Yamazaki, T., Muhandiram, R., Kay, L. E. (1994) *J. Am. Chem. Soc.* **116**, 8266-8278.

57 – Carr, H. Y., Purcell, E. M. (1954) *Phys. Rev.* **94**, 630-638.

58 – Meiboom, S., and Gill, D. (1958) *Rev. Sci. Instrum.* **29**, 688-691.

59 – Jones, J. A., Hodgkinson, P., Barker, A. L. , Hore, P. J. (1996) *J. Magn. Res. B* **113**, 25-34.

60 – Kaplan, J. I. (1995) in " Encyclopedia in Nuclear Magnetic Resonance" , Ed. Grant and Harris, vol. 2, pp. 1252.

61 – Lippens, G., Wieruszeski, J.-M., Horvath,, D., Talaga, P., and Bohin, J.-P. (1998) *J. Am. Chem. Soc.* **120**, 170-177.

62 – Kessler, H., Griesinger, C., Lautz, J., Müller, A., van Gunsteren, W. F., Berendsen, H. J. C. (1988) *J. Am. Chem. Soc.* **110**, 3393-3396.

63 – Jarvis, M. C. (1994) *Carbohydr. Res.* **259**, 311.

64 – Lowe, I.J. (1959) Phys. Rev. Lett. **2**, 285.

65 – Andrew, E. R., Bradbury, A., and Eades, R. G. (1958) Nature **182**, 1659

66 – Anderson, R. C., Jarema, M. A, Shapiro, M. J., Stokes, J. P., and Ziliox, M. (1995) J. Org. Chem. **60**, 2650.

67 – Pop, I., Dhalluin, C., Depréz, B., Melnyk, P., Lippens, G., and Tartar, A. (1996) Tetrahedron **52**, 12209 - 12222.

68 – Valles, J. M., Lin, K., Denegre, J. M., and Mowry, K. L. (1997) Biophys. J. **73**, 1130-1133.

NMR STUDIES OF THE 269 RESIDUE SERINE PROTEASE PB92 FROM *BACILLUS ALCALOPHILUS*

YASMIN KARIMI-NEJAD[1], FRANS A.A. MULDER[1], JOHN R. MARTIN[1], AXEL T. BRÜNGER[2], DICK SCHIPPER[3] AND ROLF BOELENS[1]

[1] *Bijvoet Center for Biomolecular Research, Utrecht University, Padualaan 8, 3584CH Utrecht, The Netherlands*

[2] *Dept. Molecular Biophysics and Biochemistry, Yale University, New Haven, CT 06520, USA*

[3] *Gist-Brocades, CT&S/ARS, PO Box 1, 2600MA Delft, The Netherlands*

Abstract

Serine protease PB92 is an enzyme of the subtilase family which is used commercially as a protein degrading component of washing powders. The 269 residue protease PB92 (molecular weight 27 kDa) is one of the largest monomeric proteins studied by NMR in detail. The assignment of almost all ^{1}H, ^{13}C and ^{15}N resonances has been described previously [Fogh, R.H. *et al* (1995), J. Biomol. NMR 5, 259-270]. A solution structure structure of PB92 using NOE distances extracted from several ^{13}C and ^{15}N 3D NOE spectra recorded at 600 MHz and 750 MHz has been determined recently [Martin, J.R. *et al* (1997), Structure 5, 521-532]. The tertiary structure and location of secondary structure elements of PB92 appear to be very similar to the one found in the crystal structure of PB92 [Van der Laan J.M. *et al* (1992), Prot. Engng 5, 405-411]. The system has been used as a test-case for method development and a well working protocol using Torsion Angle Dynamics for structure determination of large proteins has been developed, which is robust and simple to use as compared to distance geometry/simulated annealing methods. Heteronuclear relaxation data are consistent with our current structural and biochemical knowledge of PB92. Long-lived water molecules have been localized using heteronuclear 2D NMR experiments [Knauf, M. *et al.* (1996), Eur. J. Biochem. 238, 423-434]. Twenty-two unambiguous protein-water contacts were observed in isotope-filtered NOE experiments, indicating the presence of bound water molecules in PB92. The hydration sites identified by NMR were compared to the results of crystallography.

1. Introduction

NMR structural investigations of proteins of increasingly larger size have become feasible due to the recent progress in NMR methodology [1]. Advanced measurement techniques in combination with sophisticated labeling strategies have made possible studies of systems as large as 64 kDa, as the recently reported resonance assignments for a $^2H/^{13}C/^{15}N$-labeled Trp repressor-operator complex demonstrate [2]. However, full structure determinations for systems larger than 25 kDa are currently far from being routine. Until recently, all examples in this range were either oligomers, for which the spectra are simplified due to symmetry, or molecular complexes, the components of which can be analysed separately by isotope-filtering techniques. While there are several examples of large protein complex structures solved by NMR [3-5], to our knowledge, only two NMR structure determinations of monomeric systems larger than 25 kDa molecular weight have been reported thus far: that of the N-terminal domain of enzyme I from E. coli, comprising 259 residues [6], and that of the 269-residue protein serine protease PB92 from *Bacillus alcalophilus* from our laboratory [7]. In this overview we will present various aspects of our NMR studies on this protease.

High-alkaline serine protease PB92 (Maxacal™) is a subtilisin-like serine protease belonging to the subtilase family of proteases [8], which are of interest both as well-studied examples of enzyme catalysis [9] and as molecules of considerable industrial importance, being protein-degrading components of washing powders [10]. Subtilases are currently being used as model systems for protein-engineering studies (see [11] and references cited therein) to improve stability and performance of the mutant enzymes added to modern washing powders.

The crystal structure of serine protease PB92 has been determined at 1.75 Å resolution [12] and has a high degree of structural homolgy with other members of the subtilase family such as subtilisin BPN' [13], subtilisin Carlsberg [14], thermitase [15] and proteinase K [16]. Crystal structures of the closely related subtilisins 309 and BL from *B. Lentus* are also known at high resolution [17,18]. The substrate binding site of subtilisins shows considerable structural variability among these structures, a feature that has been attributed to flexibility [19,20]. However, it is known that this region is often distorted in crystal structures of subtilisins because of crystal contacts [21,22]. The presence of these distortions renders a functional interpretation of the structures in this key region difficult. Thus, apart from the challenge to study a protein of the size of PB92, and to validate NMR methods for studying large proteins, knowledge of the solution structure of a subtilase could provide greater insight into the structure-function relationships for this class of enzymes. Relating the structure with aspects on dynamics and stability in solution under different conditions might therefore be of immediate practical interest for developing new variants of these industrial enzymes.

2. Complete ^{1}H, ^{13}C and ^{15}N NMR assignments

Using 3D triple-resonance methods almost all ^{1}H, ^{13}C and ^{15}N resonances of PB92 have been assigned. As described in detail elsewhere [23], for the backbone atoms this was accomplished by means of 3D HNCA, HNCO, HN(CO)CA and HN(CA)CO spectra for defining sequential linkages, supplied with data from 3D TOCSY-(^{15}N,^{1}H)-HSQC and HCACO spectra for defining lead spins systems. Several of these assigments were checked using 3D HN(CA)HA and 3D CBCA(CO)NNH experiments. The sidechain assignments were mainly based on 3D HCCH-TOCSY and -COSY experiments, as presented previously [24]. The aromatic resonances were identified in 2D CT-HSQC experiments and connected with each other and the rest of the residue using 3D NOESY-(^{13}C,^{1}H)-HSQC experiments. In this way, all backbone carbonyl and CH$_n$ carbons, all amide (NH and NH$_2$) nitrogens, and 99.2% of the amide and CH$_n$ protons were assigned. Stereospecific assignments were made for 14 out of the 25 valine residues and 14 out of the 19 leucine residues using a 10% ^{13}C-labelled protein by the method of Neri et al. [25]. Fig. 1 shows the dispersion of the N,NH pairs in a 2D (^{15}N,^{1}H)-HSQC spectrum. The high resolution in this spectrum could be obtained by recording at an elevated temperature (42° C), since the autolytic activity of the enzyme was efficiently inhibited by diisopropylfluorophophonate. The high stability of the enzyme combined with the nice spectral properties, which led to a virtually complete assignment of all atoms in PB92, gave clearly good prospects for further detailed NMR studies.

3. Determination of the solution structure

There are several difficulties in determining solution structures of large proteins. The large number of NOE interactions, which has to be analyzed, seems a trivial bookkeeping problem, but since only a few automatic analysis tools exist, it can certainly not be underestimated. Multidimensional spectroscopy and high fields are essential for resolving the overlap and the ambiguities that would otherwise exist in 2D NOE spectra. The NOE data used for the structure determination of PB92 were obtained from three different 3D spectra, all with a high resolution in the f$_1$-domain and each with a recording time of approximately five days: a 3D-NOESY-(^{15}N,^{1}H)-HSQC spectrum for NOEs to the amide protons and two 3D-NOESY-(^{13}C,^{1}H)-HSQC spectra, one optimized for NOEs to aliphatic and the other for NOEs to aromatic nuclei. Even with these data the analysis of the NOE data remained complex and many of the (more than 8000) cross-peaks could not be unambiguously assigned by automated procedures. For PB92, some ambiguities were resolved by manual inspection of the 3D spectra (mainly relating to NOEs with the amide and methyl protons) and by resort to a 3D (^{13}C, ^{13}C, ^{1}H)-NOESY-HSQC spectrum. Additional information used came from H$_\alpha$, C$_\alpha$, C$_\beta$ and CO chemical shifts, evaluated using the chemical shift index (CSI) program [26] in concurrence with analysis of medium range NOEs. These data were converted into dihedral restraints for backbone torsion angles in the early calculations. In addition, hydrogen bond restraints were

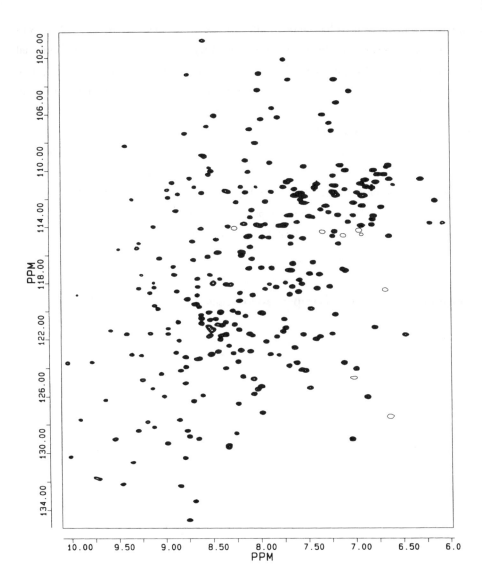

Figure 1. 2D (^{15}N,^{1}H) HSQC spectrum of serine protease PB92. Positive levels are drawn with a full set of contours; negative levels (showing folded-in peaks from Arg Nε and Lys Cζ) with one contour only.

assigned between slowly exchanging amide protons (described previously [23]) and carbonyl acceptors, whereever these could be defined by NOE data. This gave a minimal set of distance and angular restraints, which could be used to generate an ensemble of low resolution structures of PB92. This initial ensemble guided the assignment of many otherwise ambiguous NOE cross-peaks. With these additional restraints a second improved set of structures could then be made. For PB92 this so-called „bootstrap" approach [27,28] converged in eight cycles. During the iteration procedure the number of NOEs gradually increased from 1490 to 3372, while the backbone and all-heavy-atom r.m.s.d. *vs* the average (calculated over all 20 structures and residues 1-269) decreased from 1.58 to 1.16 Å and 2.03 to 1.45 Å, respectively. A more complete description of the procedure used for the structure determination of PB92 was given in [7].

In the final run (with the values for the initial calculation following directly in parentheses) a total of 3502 (1836) experimentally derived constraints were used giving an average of 13 (6.5) constraints per residue consisting of 3372 (1490) interproton distances, 130 (130) hydrogen bond constraints (representing 65 hydrogen bonds) and 0 (216) dihedral angle constraints (representing ϕ and ψ for 108 residues). The interproton distance constraints were obtained from 1027 (476) long range (|i-j| > 4), 341 (97) medium range ($2 \leq |i-j| \leq 4$), 753 (509) sequential and 1251 (408) intraresidue NOEs. The distribution of NOEs in the input data is shown in Figure 2*a*. In the final run a simulated annealing refinement protocol was performed on the family of 20 structures obtained from the previous iteration. A family of 18 structures was then selected from the output of the refinement.

4. The solution structure of serine protease PB92

The solution structure of serine protease PB92, presented in Figure 3, is represented by a family of 18 conformers which overlay onto the average structure with backbone and all-heavy-atom r.m.s.d. values of 0.88 and 1.22 Å, respectively. For the well-defined regions (cf. Figure 2*b*) these 18 structures overlay to the average with a backbone r.m.s.d. of 0.88 Å and an all-heavy-atom r.m.s.d. of 1.21 Å. Analysis of the ϕ and ψ angles shows that 98.2% of these angles fall within the allowed regions of the

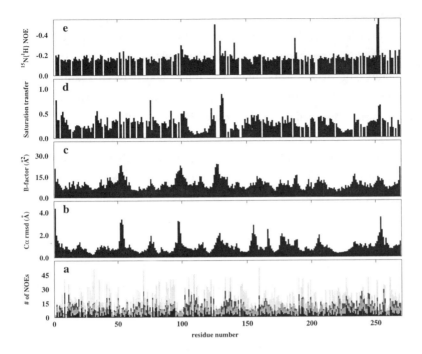

Figure 2 Structural data for the 18 NMR-derived serine protease PB92 structures (panels *a, b, d &e*) and corresponding B-factors (panel c) for the X-ray crystal structure of PB92 [12] plotted as a function of residue number. a, NOE distribution in which the intra-residue, sequential, medium range and long range constraints are indicated as black, grey, black and light-gray, respectively. b, C_α r.m.s.d. versus the mean solution structure. c, B-factors of the C_α atoms for the X-ray structure. d, Saturation transfer between water and amide protons. e, 15N{1H} NOEs.

Ramachadran map. A *cis* peptide bond is present between Y161 and P162, like in the crystal structures of all other subtilases.

Secondary structure analysis reveals nine β-strands and seven α-helices. These secondary structural elements can be divided into an internal core of a parallel β-sheet comprising seven strands (e1-e7) and two buried helices (hC and hF), which is

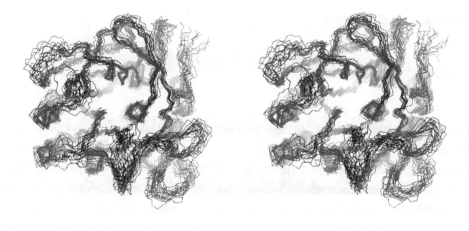

Figure 3. Stereoview of the family of 18 conformers of serine protease PB92, superimposed onto their average. The ensemble was visualized with the program MOLMOL [47].

encapsulated by five amphipathic helices (hB, hD, hE, hG and hI) and two anti-parallel β-strands (e8-e9). The less well-defined parts of the molecule coincide with loop regions containing residues with a relatively low number of constraints (Figure 2a), for which in most cases the presence of flexibility could be established from heteronuclear $^{15}N\{^{1}H\}$ NOE values as well as from amide proton exchange rates (see below).

The well-defined internal core of the solution structure of PB92 demonstrates close correspondence with the crystal structure [12], superimposing with an r.m.s.d. of 0.70 Å. A genuine difference in secondary structure between the solution and crystal structure of PB92 is found for the segment comprising residues 254-257, which forms an α-helix (hH) in the crystal but which is part of a highly mobile and disordered loop in the solution structure. In general, however, the results from the crystallographic studies and the current NMR studies are consistent and there is good agreement with the amount of disorder present in both the crystal and solution structure.

5. Mobility and disorder

5.1 HETERONUCLEAR RELAXATION

In order to determine whether the less well-defined segments of the molecule correlate with internal mobility or are due to lack of experimental constraints, heteronuclear relaxation experiments were performed: the $^{15}N\{^1H\}$ NOE values are plotted in Figure 2*e*. These relaxation measurements indicate that the majority of the backbone of serine protease PB92 is remarkably rigid with the presence of only a limited number of residues involved in localised mobility. This observation shows good correspondence with a study of the related protease subtilisin 309 from *Bacillus lentus* [29]. Most of the residues exhibiting strong heteronuclear NOEs (indicating mobility) are part of surface-exposed loops. The mobility observed within the substrate binding site (G100, S101, S126, and, more peripherally, S130) is consistent with the presence of conformational heterogeneity in the solution structure. A similar situation is found for A188 and G189, located in a loop region close to the second (low-affinity) calcium binding site.

For some regions of the protein, among which regions involved in calcium binding or close to the Ca^{2+} binding sites, the considerable disorder present in the solution structure was not mirrored by strong $^{15}N\{^1H\}$ NOEs. It is possible that these segments are dynamic on a time scale not sampled by heteronuclear NOE measurements, especially since all of them are located in potentially mobile, surface-exposed loops, also exhibiting a low density of long-range distance constraints.

5.2 HYDROGEN EXCHANGE

Chemical exchange between backbone amide hydrogens and the solvent was probed using saturation transfer experiments and from the observability of their amide protons in a freshly prepared D_2O sample [12]. Fast hydrogen exchange was identified for residues situated at the beginning or end of secondary structure elements, in loop regions (among which the substrate binding sites) and near the two Ca^{2+} binding sites. A comparison with the ^{15}N heteronuclear NOE data reveals that many of these residues are also involved in dynamics on the ps/ns timescale. In general, all slowly exchanging residues participate in secondary structure and turn elements, implying that they are involved in hydrogen bonding networks. Additional residues that displayed slow exchange, are close to the rigid S_1 pocket or to the high affinity Ca^{2+} binding site.

5.3 COMPARISON BETWEEN R.M.S.D. VALUES AND B-FACTORS

The r.m.s.d. values within the ensemble of conformers and the B-factors [16] of the crystal structure (with respect to the $C\alpha$ atoms, in both cases) are plotted as a function of residue number in Figure 2, panels b and c, respectively. In general, these two profiles show remarkably good correspondence. Those regions of the molecule that present increased B-factors in the crystal structure also exhibit higher r.m.s.d. values in the family of solution structures; this relationship also holds true for the low B-factor/r.m.s.d. segments. The plot of the B-factors presents three major peaks which correspond to loops at the surface of the molecule, and slightly higher values for the two termini. The three loops comprise residues 50-59, 98-100 and 126-130, the latter two of which are involved in substrate binding (see section on substrate binding site below). The observation of high B-factors for these segments in the crystal structure is mirrored by the presence of high r.m.s.d. values in solution.

6. Torsion angle dynamics of PB92

A drawback of the iterative "bootstrap" approach in determining structures is that it might artificially induce a self-consistency of the restraint set, enforced by the data collection procedure, eventually leading to an equally artificial precision of the resulting protein structure. Obviously, any progress in the computational area that improves the convergence of structure calculations is highly desirable especially for this type of NMR structure calculations - not only because the requirement for less iterative steps might speed up the procedure, but even more so, because it will eventually minimize the potential bias associated with this approach.

Molecular dynamics in torsion angle space, a recently introduced new method for protein structure determination [30], has been shown to improve the convergence of NMR structure calculations of proteins and DNA [31]. It was applied to NMR data sets of small to medium-size systems, and it was concluded that the method has better sampling properties as well as a higher success rate compared to conventional simulated annealing in Cartesian space [31]. These advantages should become even more significant with increasing protein size. As a test we applied the torsion angle dynamics (TAD) protocol to the NMR structure determination of PB92, the solution structure of which was described above and was based on a combination of metric matrix distance geometry and simulated annealing in Cartesian space known as the DG/SA hybrid method [32-34]. In the described iterative calculations, eventually a number of 3372 NOE distance restraints were determined. However, in the study presented here, we have used a significantly smaller set of restraints, namely only those that could be derived unambiguously from the spectra itself, i. e. without a structure-aided assignment of NOE intensities. For comparison, we applied the DG/SA protocol to the same input data set. This system, composed of 3721 atoms, is described by 11163 degrees of freedom in Cartesian and 1099 degrees of freedom in torsion angle space, respectively. Various measures of structure quality as well as an assessment of the sampling and convergence properties of the two methods have been applied to test the calculation protocols. The recently introduced torsion angle dynamics algorithm as well as the established DG/SA hybrid method are able to correctly determine the solution structure of the 269-residue protein serine protease PB92 from a relatively sparse input restraint set.

Both calculational methods, however, can be improved substantially in their performance on large molecules by introducing appropriate modifications that adapt the protocols to the size of the studied protein. From our comparative analysis it is evident that molecular dynamics in torsion angle space is the method of choice for NMR structure determination of larger biomolecules. The more favourable ratio of parameters to observables intrinsic to the TAD algorithm is reflected by its better performance as compared to the DG/SA protocol.

The RMSD profiles of the structure calculations of PB92 (cf. Fig. 4) demonstrate that the TAD calculations sample more conformational substates of the protein than the DG/SA method. At the same time, torsion angle dynamics has a better convergence behaviour, producing structures that exhibit higher coordinate precision and lower conformational energies. When torsion angle restraints are available, the protocol also

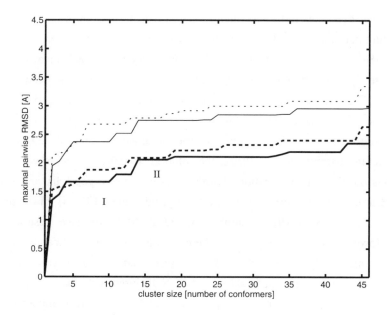

Figure 4 Energy-ranked RMSD profiles of structure ensembles of PB92, computed with calculational protocols optimized for large proteins, including constraints for the backbone angle φ. The RMSD profiles for the TAD ensembles are drawn in solid thick (backbone RMSD) and thin lines (all heavy atom RMSD). The corresponding profiles for the DG/SA ensemble are rendered in dashed tick (backbone RMSD) and dashed thin lines (all-heavy atom RMSD).

scores significantly better regarding measures of structure quality independent of the structure generation process. However, the improved quality of the TAD structures comes along with somewhat a higher computational cost in terms of the CPU time needed for the generation of one conformer. It should be noted, however, that up to now the TAD method has not been optimized yet in terms of CPU requirements. Furthermore, it can be expected that the iterative refinement of NMR structures using the bootstrap approach will require less iteration steps with the TAD method due to its superior convergence behaviour, a feature that will eventually speed up the structure determination process. A more detailed account on the application of torsion angle dynamics to serine protease PB92 can be found in [35].

7. NMR Studies of Bound Water

Most crystal structures of subtilisins contain a large number of internal water molecules, the function of which is not yet understood. Since NMR spectroscopy is an attractive tool for the investigation of protein hydration [36,37], we investigated the presence of bound waters in subtilisin PB92 by triple-resonance 2D NMR experiments. We also analysed high-resolution subtilisin X-ray structures with respect to the conservation of structural hydration sites. The results of this study were compared to those of our NMR investigation.

Ten crystal structures belonging to nine different subtilisins varying in sequence indentity between 58% and 99%, have been analysed. These included serine protease PB92 [12], the subilisins 309 [17], BL [18], Carlsberg [38], BPN' (D.T. Gallagher, J.D. Oliver, R. Bott, C. Betzel, and G.L.Gilliland, private communication) and four variants of subtilisin 309 (R. Bott, private communication). We found that of the 31 internal water molecules present in the PB92 X-ray structure, 16 are completely and 5 partially conserved among the subtilisins examined.

In order to localize bound water molecules in PB92 in solution, we used a combination of isotope-filtered ($^{15}N,^{1}H$)-correlated 2D NMR experiments termed HAWAII (Hydration And WAter-protein exchange In Isotopically labelled proteins), for which the pulse sequences are given Figure 5. The pulse sequences are basically two-dimensional $^{15}N,^{13}C$-filtered ^{15}N-NOESY-HSQC experiments based on the MEXICO experiment devised to measure amide proton exchange rates in double-labelled proteins [39]. A double-tuned isotope filter [40] at the beginning of the sequence eliminates all signals of the $^{15}N,^{13}C$-labelled protein, ensuring that the observed magnetization originates from the solvent. In one experiment both protein-water NOEs and protein-water exchange are observed. In a second, otherwise completely identical experiment, cross-relaxation is suppressed by a low power pulse train [41,42], and only pure exchange with the solvent is observed. Such a combination of two experiments has been used previously to study the hydration of *Desulfovibrio vulgaris* flavodoxin [43]. These experiments enabled us to discriminate between protein-water NOEs and protein-water exchange in PB92. Exchange-relayed NOEs to labile protons in the protein were identified from an analysis of the structure. Twenty-two protein-water NOEs were identified unambiguously from the spectra. These NOEs correspond to a minimum of 12 hydration sites, all of

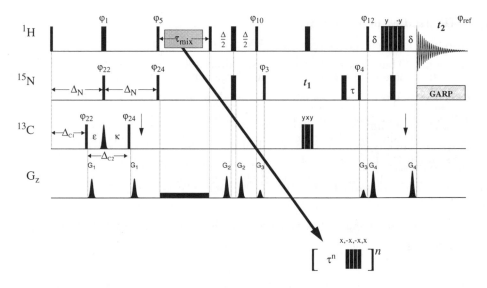

Figure 5 Pulse scheme for the HAWAII experiments. Narrow and wide bars denote hard 90° and 180° square pulses, respectively. The shaped 180° ^{13}C pulse had a duration of 360 µs and a hyperbolic secant envelope with a squareness level $\mu = 8$, resulting in an inversion bandwidth of 25 kHz [44,45]. The pulse was applied with an RF field strength of 20 kHz. The RF field strength for the low-power pulse train in the second experiment was 12 kHz. All z-gradient pulses had a half-sine-bell shape and a duration of 400 µs, with the exception of the square-shaped z-gradient which was applied for the whole duration of the mixing time. Gradient strengths in G cm^{-1} were: G1 = 6, G2 = 10.8, G3 = 1.8, G4 = 42, and 0.6 G cm^{-1} for the square-shaped z-gradient applied during the mixing time. The carbon carrier is switched from 75 ppm (i.e. a position halfway between the methyl and the aromatic resonances) for the isotope filter part to a position half-way between the C$_\alpha$ and the carbonyl region (130 ppm) for ^{13}C decoupling in F_1, as indicated by the small arrows. In the NOE experiment, the proton carrier is placed at the position of the H$_2$O resonance during the whole sequence, while for the pure-exchange experiment, the proton carrier is switched from the position of the H$_2$O resonance to a position between the middle of the amide region and the H$_2$O resonance for the duration of the mixing time, and then for the rest of the sequence set back to its old position. Phase cycling is as follows: $\phi_1 = y$; $\phi_5 = \phi_{12} = (x,x,-x,-x)$; $\phi_{10} = y$; $\phi_{22} = 8(x), 8(-x)$; $\phi_{24} = 16(x), 16(-x)$; $\phi_3 = x,-x$; $\phi_4 = 4(x), 4(-x)$; rec. = 2(x,-x), 2(-x, x). Quad detection in the ^{15}N dimension was achieved by applying States-TPPI to the phase ϕ_3 and the receiver phase. Delay durations are $\Delta_N = 5.4$ ms, $\Delta_{C1} = 3.20$ ms, $\Delta_{C2} = 3.94$ ms, $\Delta = 5.0$ ms, $\tau = t_1(0) + 4\tau_{90}$ (^{13}C). τ_n was set to twice the length of the low-power 90°1H pulse. The delay δ was adjusted such that the total refocusing time was 5.0 ms, and the composite WATERGATE pulse was a 3-9-19 binomial pulse [46].

which are also present in the X-ray structure. Moreover 8 additional buried X-ray waters may also have long residence times (i.e. > 300 ps) in solution according to these data. The location of the 20 buried X-ray waters corresponding to our NOE data is shown in Figure 6.

The bound waters we have localized by NMR include *i) a network of hydrogen-bonded water molecules*, starting at residue Trp 6, *ii) four waters in/near the catalytic site*, one of which is highly coordinated by five hydrogen bonds to the catalytic residue Asp 32, its neighbours, and Ser 125 in the substrate recognition site, *iii) three water molecules located in/near the substrate binding site, iv) two water molecules near the beginning of the C-terminal helix, v) two water molecules* located between strands e1 and e2 of PB92, *vi) two water molecules in the vicinity of the weak second Ca^{2+} binding site* of PB92.

One conserved structural water present in the active site of all initially examined crystal structures is either absent in solution or has a very short residence time: a possibly catalytic water molecule located in the vicinity of the nucleophilic serine 221. Most likely, the absence of this water molecule is due to the bulky DIFP inhibitor covalently linked to this serine side chain in our NMR sample.

8. Conclusions

The solution structure of serine protease PB92 from *Bacillus alcalophilus* presented here demonstrates the feasibility of NMR structural studies on proteins approaching 30 kD. The structure closely resembles the crystal structure of PB92, only differing for the segment comprising residues 254-257, which forms an α-helix in the crystal but which is part of a highly mobile and disordered loop in the solution structure.

The solution structure of the serine protease PB92 presents a remarkably well-defined global fold which is rigid with the exception of only a restricted number of sites. The high rigidity of the molecule may have evolved as a protective measure against autolysis. From the few residues in the protein exhibiting significant internal mobility a large fraction is involved in substrate binding: the presence of flexibility within the binding site supports the proposed induced fit mechanism of substrate binding.

Figure 6 Backbone representation of the X-ray structure of PB92 [12]. Internal water molecules in agreement with NOE data are shown as balls. Dark shaded balls indicate the bound waters minimally needed to explain the NOE data, balls in shaded in light gray indicate X-ray waters which may also have long residence times according to the NOE data. The structure was drawn with the program MOLMOL [47].

Both torsion angle dynamics as well as the established DG/SA hybrid method (after adapting the calculation protocols to the size of the proteins) are able to correctly determine the solution structure of the 269-residue protein PB92 from a relatively sparse input restraint set. It is evident however, that in the future molecular dynamics in torsion angle space will be the method of choice for NMR structure determination of larger biomolecules. The more favourable ratio of parameters to observables intrinsic to the TAD algorithm is reflected by its better performance as compared to the DG/SA protocol.

Most of the long-lived waters in PB92, that were observed by NMR, are conserved among the subtilisins we investigated, implying that they are of structural and/or functional importance for the enzyme. This certainly holds for the waters observed in the active center of the enzyme. From the X-ray structure it is known that they are involved in hydrogen bonds to the catalytic residues and their neighbours, thus maintaining the correct conformation of these residues with respect to catalysis. The waters found in the substrate binding site might modulate the binding specificity of PB92, enabling the enzyme to adapt to different substrates. An intruiging feature is the water chain in the vicinity of the N-terminus of the protein, knowon to be conserved among all subtilisin structures available so far. The long residence times of the water molecules involved in this network, together with their high degree of conservation, strongly suggests a possible structural and functional role for these waters, although it is not clear yet what this role might be.

9. Acknowledgements

We are grateful for enlightening discussions with Robert Kaptein and we thank Rick Bott and Wolfgang Aehle for their encouragement during the project and their critical reading of the manuscript. The 750 MHz NMR spectra were recorded at the SON NMR Large Scale Facility (Utrecht, The Netherlands) which is supported by the Large Scale Facility programme of the European Union.

10. References

1. Bax, A. and Grzesiek, S. (1993). Methodological advances in Protein NMR. *Acc. Chem. Res.* **26**, 131-138.

2. Shan, X., Gardner, K. H., Muhandiram, D. R., Rao, N. S., Arrowsmith, C. H. and Kay, L. E. (1996). Assignment of ^{15}N, ^{13}C, and H_N Resonances in an $^{15}N,^{13}C,^{2}H$ Labeled 64 kDa Trp Repressor-Operator Complex Using Triple-Resonance NMR Spectroscopy and ^2H-Decoupling, *J. Am. Chem. Soc.* **118**, 6570-6579.

3. Zhang, H., Zhao, D., Revington, M., Lee, W., Jia, X., Arrowsmith, C. and Jardetzky, O. (1994). The solution structures of the trp repressor-operator DNA complex. *J. Mol. Biol.* **238**, 592-614.

4. Matsuo, H., Li, H., McGuire, A. M., Fletcher, C. M., Gingras A.-C., Sonenberg, N. and Wagner, G. (1997). Structure of translation factor eIF4E bound to m^7GDP and interaction with 4E-binding protein. *Nature Struct. Biol.* **4**, 717-724.

5. Berg, B. van den, Tessari, M., Boelens, R., Dijkman, R., Kaptein, R., de Haas, G. H. and Verheij, H. M., (1995). Solution structure of porcine pancreatic phospholipase A2 complexed with micelles and a competitive inhibitor. *J. Biomol. NMR* **5**, 110-121.

6. Garrett, D. S., Seok, Y.-J., Liao, D.-I., Peterkofsky, A., Gronenborn, A. M. and Clore, G. M. (1997). Solution structure of the 30 kDa N-terminal domain of enzyme I of the Escherichia coli phosphoenolpyruvate:sugar phosphotransferase system by multidimensional NMR. *Biochemistry* **36**, 2517-2530.

7. Martin, J. R., Mulder, F. A. A., Karimi-Nejad, Y., v.d. Zwan, J., Mariani, M., Schipper, D. and Boelens, R. (1997). The Solution Structure of Serine Protease PB92 from *Bacillus alcalophilus*. *Structure* **5**, 521-532.

8. Siezen, R.J., de Vos, W.M., Leunissen, J.A.M. and Dijkstra, B.W. (1991). Homology modelling and protein engineering strategy of subtilases, the family of subtilisin-like serine proteases. *Protein Engng.* **4**, 719-737.

9. Kraut, J. (1977). Serine proteases: structure and mechanism of catalysis. *Ann. Rev. Biochem.* **46**, 331-358.

10. Shaw, W.V. (1987). Protein engineering: the design synthesis and characterisation of factitious proteins. *Biochem. J.* **246**, 1-17.

11. Teplyakov, A.V., van der Laan, J.M., Lammers, A.A., Kelders, H., Kalk, K.H., Misset, O., Mulleners, L.J.S.M. and Dijkstra, B.W. (1992). Protein Engineering of the high-alkaline serine protease PB92 from *Bacillus alcalophilus*: functional and structural consequences of mutations at the S4 substrate binding pocket. *Protein. Eng.* **5**, 413-420.

12. Laan, J.M. van der, Teplyakov, A.V., Kelders, H., Kalk, K.H., Misset, O., Mulleners, L.J.S.M. and Dijkstra, B.W. (1992). Crystal structure of the high-alkaline serine protease PB92 from *Bacillus alcalophilus*. *Protein Engng.* **5** 405-411.

13. Bott, R., Ultsch, M., Kossiakoff, A., Graycar, T., Katz, B. and Powers, S. (1988). The three-dimensional structure of *Bacillus amyloliquefaciens* subtilisin at 1.8 Å and an analysis of the structural consequences of peroxide inactivation. *J. Biol. Chem.* **263**, 7895-7906.

14. Bode, W., Papamokos, E. and Musil, D. (1987). The high-resolution X-ray crystal structure of the complex formed between subtilisin Carlsberg and eglin c, an elastase inhibitor from the leech *Hirudo medicinalis*. *Eur. J. Biochem.* **166**, 673-692.

15. Gros, P., Betzel, Ch., Dauter, Z., Wilson. K.S. and Hol, W.G.J. (1989). Molecular dynamics refinement of a thermitase-eglin c complex at 1.98 Å resolution and comparison of two crystal forms that differ in calcium content. *J. Mol. Biol.* **210**, 347-367.

16. Betzel, Ch., Pal, G.P. and Saenger, W. (1988). Synchrotron X-ray data collection and restrained least-squares refinement of the crystal structure of proteinase K at 1.5 Å resolution. *Acta Cryst.* B**44**, 163-172.

17. Betzel, Ch., Klupsch, S., Papendorf, G., Hastrup,S., Branner, S. and Wilson, K.S. (1992). Crystal structure of the alkaline proteinase savinase™ from *Bacillus lentus* at 1.4 Å resolution. *J. Mol. Biol.* **223**, 427-445.

18. Goddette, D.W., Paech, C., Yang, S.S., Mielenz, J.R., Bystroff, C., Wilke, M.E. and Fletterick. R.J. (1992). The crystal structure of the *Bacillus lentus* alkaline protease, subtilisin BL, at 1.4 Å resolution. *J. Mol. Biol.* **228**, 580-595.

19. Lange, G., Betzel, Ch., Branner, S. and Wilson, K.S. (1994). Crystallographic studies of savinase, a subtilisin-like proteinase, at pH 10.5 *Eur. J. Biochem.* **224**, 507-518.

20. McPhalen, C.A. and James, M.N.G. (1988). Structural comparison of two proteinase-protein inhibitor complexes: eglin c-subtilisin Carlsberg and CI2-subtilisin Novo. *Biochemistry* **27**, 6582-6598.

21. Sobek, H., Hecht, H.-J., Aehle, W. and Schomburg, D. (1992). X-ray structure determination and comparison of two crystal forms of a variant (Asn115Arg) of the alkaline protease from *Bacillus alcalophilus* refined at 1.85 Å resolution. *J. Mol. Biol.* **228**, 108-117.

22. Dauberman, J.L, Ganshaw, G., Simpson, C., Graycar, T.P., McGinnis, S. and Bott, R. (1994). Packing selection of *Bacillus lentus* subtilisin and a site-specific variant. *Acta Cryst D* **50**, 650-656.

23. Fogh, R.H., Schipper, D., Boelens, R. and Kaptein, R. (1994). ^{1}H, ^{13}C and ^{15}N NMR backbone assignments of the 269-residue serine protease PB92 from *Bacillus alcalophilus*. *J. Biomol. NMR* **4**, 123-128.

24. Fogh, R.H., Schipper, D., Boelens, R. and Kaptein, R. (1995). Complete ^{1}H, ^{13}C and ^{15}N NMR assignments and secondary structure of the 269-residue serine protease PB92 from *Bacillus alcalophilus*. *J. Biomol. NMR* **5**, 259-270.

25. Neri, D., Szyperski, T., Otting, G., Senn, H. and Wüthrich, K. (1989). Stereospecific nuclear magnetic resonance assignments of the methyl groups of valine and leucine in the DNA-binding domain of the 434 repressor by biosynthetically directed fractional ^{13}C labelling. *Biochemistry* **28**, 7510-7516.

26. Wishart, D.S. and Sykes, B.D. (1994). The ^{13}C chemical-shift-index: a simple method for the identification of protein secondary structure using ^{13}C chemical-shift data. *J. Biomol. NMR* **4**, 171-180.

27. Güntert, P., Berndt, K. D. and Wüthrich, K. (1993). The program ASNO for computer-supported collection of NOE upper distance constraints as input for protein structurere determination. *J. Biomol. NMR* **3**, 601-606.

28. Meadows, R. P., Olejniczak, E. T. and Fesik, S. W. (1994). A computer-based protocol for semiautomated assignments and 3D structure determination of proteins. *J. Biomol. NMR* **4**, 79-96.

29. Remerowski, M.L., Pepermans, H.A.M., Hilbers, C.W. and van de Ven, F.J.M. (1996). Backbone dynamics of the 269-residue protease savinase determined from ^{15}N-NMR relaxation measurements. *Eur. J. Biochem.* **235**, 629-640.

30. Rice, L. M. and Brünger, A. T. (1994). Torsion angle dynamics: reduced variable conformational sampling enhances crystallographic structure refinement. *Proteins* **19**, 277-290.

31. Stein, E. G., Rice, L. M. and Brünger, A. T. (1997). Torsion-Angle Molecular Dynamics as a New Efficient Tool for NMR Structure Calculation. *J. Magn. Reson.* **124**, 154-164.

32. Nilges, M., Clore, G.M. and Gronenborn, A.M. (1988). Determination of three-dimensional structures of proteins from interproton distance data by hybrid distance-dynamical simulated annealing calculations. *FEBS Lett.* **229**, 317-324.

33. Nilges, M., Kuszewski, J. and Brünger, A. T. (1991). In: Computational Aspects of the Study of Biological Macromolecules by Nuclear Magnetic Resonance Spectroscopy, ed. J. C. Hoch, F. M. Poulsen and C. Redfield. Plenum Press, New York, pp 451-455.

34. Kuszewski, J., Nilges, M. and Brünger, A. T. (1992). Sampling and efficiency of metric matrix distance geometry: a novel partial metrization algorithm. *J. Biomol. NMR* **2**, 33-56.

35. Karimi-Nejad, Y., Warren, G.T., Schipper, D., Brünger, A.T. and Rolf Boelens, R. (1998). NMR Structure Calculation Methods for Large Proteins: Application of Torsion Angle Dynamics and Distance Geometry/Simulated Annealing to the 269-Residue Protein Serine Protease PB92. *Mol. Phys.*, in the press.

36. Otting, G., Liepinsh,E. and Wüthrich, K. (1991). Protein hydration in aqueous solution.*Science* **254**, 974-980.

37. Otting, G. (1997), NMR studies of water bound to biological molecules. *Prog. NMR Spectrosc.* **31**, 259-185 .

38. Steinmetz, A.C., Demuth, H.U. and Ringe, D.(1994), Inactivation of subtilisin Carlsberg by N-((tert-butoxycarbonyl)alanylprolylphenylalanyl)-O-benzoylhydroxyl- amine:formation of a covalent enzyme-inhibitor linkage in the form of a carbamatederivative. *Biochemistry* **33**, 10535-10544.

39. Gemmecker, G., Jahnke, W. and Kessler, H.(1993), Measurement of fast protein exchange rates in isotopically labeled compounds. *J. Am. Chem. Soc.* **115**, 11620-11621.

40. Gemmecker, G., Olejniczak, E.T. and Fesik, S.W.(1992), An improved method for selectively observing protons attached to ^{12}C in the presence of ^{1}H-^{13}C spin pairs. *J. Magn. Reson.* **96**, 199-204.

41. Fejzo, J., Westler, W. M., Macura, S. and Markley, J.(1990), Elimination of cross-relaxation effects from two-dimensional chemical exchange spectra of macromolecules. *J. Am. Chem. Soc.* **112**, 2574-2577.

42. Fejzo, J., Westler, W. M., Macura, S. and Markley, J.(1991), Strategies for eliminating unwanted cross-relaxation and coherence-transfer effects from two-dimensional chemical exchange spectra. *J. Magn. Reson.* **92**, 20-29.

43. Knauf, M.A., Löhr, F., Blümel, M., Mayhew, S.G. and Rüterjans, H.(1996), NMR investigation of the solution conformation of oxidized flavodoxin from *Desulfovibrio vulgaris*. Determination of the tertiary structure and detection of protein-bound water molecules. *Eur. J. Biochem.* **238**, 423-434.

44. Silver, M. S., Joseph, R. I. and Hoult, D. I.(1984), Highly selective π/2 and π pulse generation. *J. Magn. Reson.* **59**, 347-353.

45. Folmer, R. H. A., Hilbers, C.W., Konings, R. N. H. and Hallenga, K.(1995), A ^{13}C double-filtered NOESY with strongly reduced artefacts and improved sensitivity. *J. Biomol. NMR* **5**, 427-432.

46. Sklenar, V., Piotto, M., Leppik, R. and Saudek, V.(1993), Gradient-tailored water suppression for ^{1}H-^{15}N HSQC experiments optimized to retain full sensitivity. *J. Magn. Reson.* **102**, 241-245.

47. Koradi, R., Billeter, M. and Wüthrich, K. (1996). MOLMOL, a program for display and analysis of macromolecular structures. *J. Mol. Graphics* **14**, 51-55.

INTERMOLECULAR INTERACTIONS OF PROTEINS INVOLVED IN THE CONTROL OF GENE EXPRESSION

G. WAGNER[1], H. MATSUO[1], H. LI[1], C.M. FLETCHER[1], A.M. MCGUIRE[1], A.-C. GINGRAS[2], N. SONENBERG[2]
[1]Department of Biological Chemistry and Molecular Pharmacology, Harvard Medical School, 240 Longwood Avenue, Boston, Massachusetts 02115, USA
[2]Department of Biochemistry, McGill University, 3655 Drummond Street, Montreal, Quebec H3G 1Y6, Canada

1. Introduction

Gene expression in eukaryotic cells is controlled at many levels. Usually gene expression is activated by external signals, such as hormones or growth factors that bind to cell surface receptors and stimulate signaling cascades. The action of kinases and phosphatases may lead to transport of transcription factors to the nucleus and activation of transcription. Translation of mRNA is also highly regulated [1]. In eukaryotic cells, mRNA is usually modified at the 5' end with m^7Gppp, the so called cap. The mRNA is capped in the nucleus by a capping enzyme, or a group of enzymes, that have triphosphatase, guanylate kinase and methylase functions. The cap seems to be important for pre-mRNA splicing, nuclear export and translation initiation [2]. A dimeric cap-binding protein associates with the mRNA for transport through the nuclear pores. In the cytoplasm, the eukaryotic translation initiation factor eIF4E associates with the cap. eIF4E binds in the N-terminal portion of the large scaffold protein eIF4G (1000 to 1400 residues), which binds the RNA helicase eIF4A at a different site. The helicase is thought to unwind secondary structure in the 5' untranslated region (5'UTR) of the mRNA. Thus, the primary function of eIF4E appears to be to direct the helicase towards the 5'UTR via the interaction with eIF4G, so the secondary structure can be unwound and ribosomes can bind.

Translation initiation is highly regulated. eIF4E is one of the least abundant of the translation initiation factors. This downregulates translation. Furthermore, formation of the complex between eIF4E and the scaffold protein eIF4G is inhibited by the regulatory 4E-binding protein (4E-BP). This occupies the eIF4G-binding site on eIF4E [3]. This inhibitory complex dissociates only when 4E-BP becomes phosphorylated as a consequence of a signaling event.

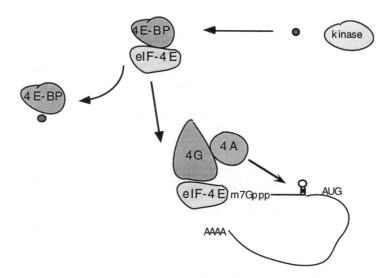

Fig 1: Regulation of translation initiation by 4E-BP. The cap-binding protein eIF4E is thought to direct the RNA-helicase eIF4A (4A) towards the 5' untranslated region of the mRNA. This is achieved by coupling eIF4E with 4A via the scaffold protein eIF4G (4G). This is downregulated by the binding of the regulatory protein 4E-BP. The latter complex can be dissociated by phosphorylation of 4E-BP and translation initiation can occur.

We asked whether we can solve by NMR the solution structure of eIF4E, in order to understand the details of cap binding. Since the protein is very unstable we had to find ways to stabilize it. This was achieved by addition of mild detergents (CHAPS) and the structure of the eIF4E complex was solved in a CHAPS micelle [4]. The location of the protein in the micelle was characterized. Furthermore, we have studied the regulatory protein 4E-BP and its interaction with eIF4E [5].

2. Preparation of eIF4E - Use of CHAPS to Prevent Aggregation and Precipitation

Initially, both yeast and mouse eIF4E were expressed to perform NMR studies. Both the yeast and the mouse protein are very unstable and have limited solubility of less than 0.5 mM concentrations. To increase the solubility and stability, we have examined addition of non-denaturing detergents or amino acids. Best results are obtained for the yeast protein with addition of CHAPS [4]. The concentration of eIF4E can then be increased to ca. 1 mM and the protein is stable for several months. The small changes in chemical shifts observed upon addition of CHAPS stabilized at a concentration of 25 to 30 mM CHAPS, the concentration used for our experiments. CHAPS has a critical micelle concentration of around 6 mM. Thus, it appears that at 25 to 30 mM CHAPS, there is at least one micelle available per protein, and at a protein concentration of 1 mM, we estimate that there are approximately 25 to 30 molecules of CHAPS in one micelle embedding one protein. With the molecular weights of CHAPS and yeast eIF4E as 612 and 24,300, respectively, the total molecular weight of the protein/micelle complex was estimated around 40 to 43 kDa. Dynamic light scattering experiments yield an approximate molecular weight of 50 kDa, however, the experiments suffer from polydispersity which is indicated by large baseline parameters. No satisfactory conditions were found for mouse eIF4E, and all structural efforts were pursued with the yeast protein.

3. NMR Structure Determination of Yeast eIF4E

The yeast protein was expressed and purified as described in [4] using an m^{7}GDP-agarose affinity resin [6]. Complexes of the protein with cap analogs were prepared by eluting with m^{7}Gpp, m^{7}Gppp, or m^{7}GpppA, respectively. All structural studies were carried out on cap-analog complexes as the apoprotein could not be prepared in a way suitable for NMR studies. For resonance assignments of the micelle-bound yeast eIF4E, the protein was labeled with ^{15}N, ^{13}C and ^{2}H. Triple resonance experiments optimized for deuterated proteins [4] were employed and almost complete backbone assignments were obtained. Assignments were facilitated with ^{15}N labeling of Lys, Val, Phe, Ala, Asp, Gly, and Leu residues. Side chain assignments and NOESY experiments were obtained from 75% or 50% deuterated protein samples. NOE constraints were measured from 3D ^{15}N dispersed NOESY experiments of fully or 75% deuterated protein.

The structure of yeast eIF4E consists of an eight-stranded curved antiparallel β-sheet that is covered on the convex hydrophobic side by three long helices, a2, a4 and a5 (Figs. 2,3,4). These three helices are roughly anti-parallel and are slightly tilted relative to the β-strands. Three short helices, a1, a3 and a6 are inserted in loops connecting b1 and b2, b3 and b4, and b7 and b8. These short helices are oriented perpendicular to the plane of the sheet, and are primarily responsible for the concave shape of this face of the protein. Two tryptophans (W58 and W104), located at the N-terminal (distal) ends of the two helices a1 and a3 (Figs. 2,3), sandwich the methylated base of the cap (Fig. 4).

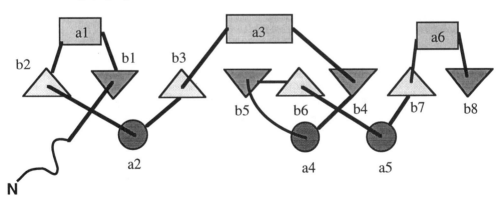

Fig. 2: Topology of the eIF4E fold. Triangles pointing up or down represent β-strands coming out or going into the plane of the figure.

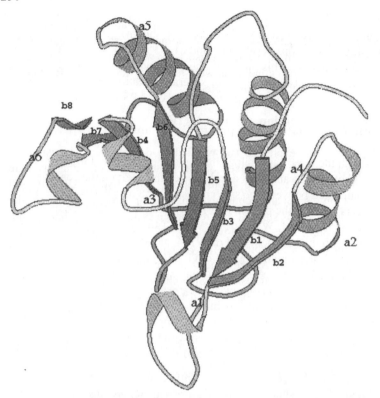

Fig. 3 . Ribbon diagram of the eIF4E structure generated with the program Molscript [7].

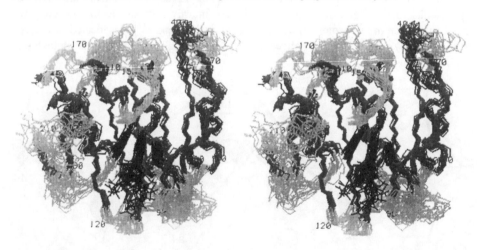

Fig. 4. Stereodiagram of an ensemble of structures of yeast eIF4E. Backbone elements with regular secondary structure are shown in dark gray, loops of non-regular secondary structure are in light gray, the bound m^{7}Gpp and the two tryptophan side chains complexing the m^{7}G group (W104 left, W58 right) are in black.

The N-terminal 35 residues are unstructured as indicated by the absence of long-range NOEs and by ^{15}N relaxation studies (data not shown). The relaxation data further indicate mobility of the loops between b1 and a1, b3 and a3, b4 and a4, a5 and b7, and a6 and b8. Interestingly, this includes the two loops preceding the crucial tryptophans in positions 58 and 104. Analysis of several relaxation experiments yields an overall correlation time of 16 ns for the protein in the CHAPS micelle.

4. Interaction with Cap Analogs m^{7}Gpp, m^{7}Gppp and m^{7}GpppA

Complexes with the three cap analogs m^{7}Gpp and m^{7}Gppp and m^{7}GpppA were studied. The m^{7}G of the cap analog is deeply buried in a cleft on the concave side of the protein. It is complexed between the side chains of W58 and W104 (Figs. 4,5). The methyl group points towards the protein interior where it forms NOE contacts with the side chains of W58, W104, W166, H94, V153, T48, and L62. The O^6 atom forms a hydrogen bond with the backbone NH of W104, and the side chain of E105 forms a hydrogen bond with the amino group of m^{7}G. E105 is conserved in all known eIF4E homologues and mutation of this residue to alanine in human eIF4E abrogates function. The imino group of the m^{7}G is likewise ideally situated to form a hydrogen bond with the side chain of E105. The side chain of nearby E103, which is not conserved in mammalian eIF4Es, appears to form a hydrogen bond to the N$^\varepsilon$H of W58. Furthermore, the sugar moiety of the m^{7}G is oriented with the H1' directed towards W58 and the H2' and H3' towards W104. The H4' and H5' resonances have been assigned, but show no NOEs to the polypeptide. There are numerous NOEs between

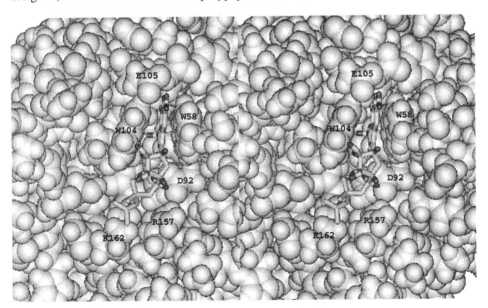

Fig. 5: Stereodiagram of the cap binding site. The protein surface is shown in a space-filling representation, the bound m^{7}Gpp is given in a wire representation. Several residues that are crucial for the interaction are labeled.

the side chains forming the cap-binding pocket. This narrow pocket is highly hydrophobic as its sides are formed by the side chains of W58 and W104 while T48, L62, V153 and W166 line the bottom. H94, located at the bottom of the pocket, is the only non-hydrophobic residue. This residue is not conserved as it is a serine in most other eIF4Es. Comparing spectra of eIF4E complexed with m^{7}Gpp, m^{7}Gppp and m^{7}GpppA indicate that the terminal phosphate group in m^{7}Gppp is close to K162 and R157, and the second base in m^{7}GpppA is close to the loop between helix a6 and strand b8, which is in the upper left hand corner of Fig. 5.

5. Specificity of the m^{7}GpppX Interaction

Methylation of the guanosine nucleotides enhances their affinity for eIF4E by 4 to 6 fold compared to non-methylated analogs [8]. This is attributed to an enhanced π-π-stacking interaction of the methylated base with the tryptophans: The methylation leads to electron deficient π-orbitals that interact favorably with the electron-rich π-orbitals of the tryptophan indols [9]. The stacking of the m^{7}G with the side chains of W104 and W58 is very similar to that observed in crystal structures of cap analogs with tryptophan derivatives [9]. Examination of the structure reveals that the hydrophobic interactions of the N7-methyl within the binding pocket (with W166, V153, T48, L62) are important in stabilizing cap binding. Likewise, the hydrogen bond between the side chain of the highly conserved E105 and the N2 amino group stabilizes binding, consistent with the observation that elimination of the N2 amino group, as in m^7-inosine, or methylation of N2 decreases the affinity [8, 10].

However, there are several observations regarding the ionization state of the m^{7}G that cannot yet be fully rationalized. While the N7-C8-N9 group of the methylated base is hypothesized to have a delocalized positive charge, there is no negative charge nearby to neutralize this. Our current NMR data is unable to elucidate the ionization state of the bound m^{7}G. Both the H8 and N1H (imino proton) are absent from the spectra acquired on the free and bound m^{7}G. The possibility of chemical or conformational exchange prevents us from concluding the absence of these protons. However, it is unusual not to observe hydrogen bonded imino protons at 5°C, providing evidence of either a weak or no hydrogen bond. We are currently conducting further experiments to determine the ionization state of m^{7}G in the complex with eIF4E.

6. Regulation of Translation Initiation via the 4E-BP interaction

In mammalian cells, translation initiation is regulated by the interaction of eIF4E with 4E-binding proteins (4E-BP1 and 4E-BP2) [11]. A 4E-binding protein, p20, has also been found in yeast, and it has been shown to repress cap-dependent translation initiation [12]. We have expressed 4E-BP1, 4E-BP2 and p20. The latter could not be prepared in a form suitable for NMR experiments since the GST-fusion protein could not be cleaved. Thus, we titrated yeast eIF4E with 4E-BP2 and followed the ^{1}H-^{15}N correlated spectrum of eIF4E. Addition of 4E-BP2 causes the disappearance of ca. 20 cross peaks in a ^{1}H-^{15}N correlated spectrum of yeast eIF4E and the appearance of a similar number of new peaks. Smaller changes are observed for many other cross peaks which remain close to their original position. The spectra indicate slow exchange between free and complexed eIF4E, typical of tight binding ($K_d < 10$ μM). Resonances with the largest changes include residues 32 - 50 and 62 - 79. Most of these residues cluster around one edge of the β-sheet (b2 and b1), the helix a2, and the four adjacent loops connecting b2 and a2, a4 and b5, b6 and a5, and the conserved loop region N-terminal to b1. These results have established the 4E-BP binding site on eIF4E and further structural studies on the complex are in progress. The amino acid sequence of the 4E-BP binding site is highly conserved between yeast and human/mouse. Thus, it is likely that the 4E-BP binding site in mammalian eIF4E resides at the analogous position. 4E-BP-binding only occurs in the absence of CHAPS. Addition of CHAPS reversed the spectral changes indicating that CHAPS dissociates the eIF4E/4E-BP complex. The 4E-BP binding site overlaps with the region of the protein surface immersed in the CHAPS micelle (see below).

All spectral properties of free 4E-BP1 indicate that this 118-residue protein is unfolded[5, 13]: The resonances show little spectral dispersion, there are only intraresidue and sequential but no long-range NOEs, and the CD spectrum is typical for a random coil structure. Using isotope labeling and triple resonance experiments, the NMR spectrum of 4E-BP1 in its free state was almost completely assigned with the exception of a stretch of 10 residues that includes the eIF4E-binding site [13]. Thus, it is possible that this small region may contain partially ordered states in multiple conformations that are in intermediate exchange and are thus not observable. The heteronuclear ^{1}H-^{15}N NOEs of all HSQC cross peaks that can be analyzed and are not exchange-broadened are typical for an unfolded state [13]. Upon addition of mouse eIF4E [5], or yeast eIF4E (unpublished result), about 10 cross peaks in the ^{1}H-^{15}N HSQC spectrum of 4E-BP1 shift significantly, while the bulk of the cross peaks remains unchanged. This indicates that only a small portion of 4E-BP1 becomes ordered upon interaction with eIF4E while the rest of the protein remains flexible.

7. Topology of the CHAPS Interaction

Since the eIF4E protein was placed in a CHAPS micelle, we asked whether we could locate the region of the eIF4E surface that is immersed in the micelle. This was achieved with a CHAPS titration experiment. The residues affected by CHAPS binding are located in a contiguous surface area consisting of helices a2, a4 and a5, β strands b1 and b2, and the two loops connecting a2 to b3 and a4 to b5. This indicates that the three long helices and adjacent regions are immersed in the CHAPS micelle while the cap-binding site and the three short helices are outside. This is consistent with the finding that cap-binding is not affected by addition of CHAPS. Furthermore, we have recorded 3D NOESY spectra of the m^{7}Gpp/eIF4E complex with and without CHAPS. The NOEs to W104 are essentially unchanged, while those to W58 are conserved, but have significantly lower intensities. This indicates that the curvature of the CHAPS micelle tends to open up the cap-binding site and that the cap is associated more tightly with W104 than W58. This is also consistent with the observation that the W104F mutation reduced cap binding affinity by more than 95% while the W58F mutation reduced affinity by only 50% [14]. The 4E-BP-binding site is partially covered by the CHAPS micelle, explaining why addition of CHAPS causes dissociation of 4E-BP [4].

8. Conclusion

We have solved the solution structure of yeast eIF-4E with NMR spectroscopy. For solubility and stability reasons, the protein was immersed in a CHAPS micelle leading to an overall molecular weight of 40 - 50 kDa. The interaction with cap analogs was characterized. The cap is sandwiched between two tryptophan side chains. The binding site for the regulatory 4E-binding protein (4E-BP) is distinct from the cap-binding site and exhibits a shallow groove. The 4E-BP by itself is unstructured. Upon binding eIF4E only a small portion of the 4E-BP becomes ordered. Insertion of eIF4E in the CHAPS micelle dissociates the complex with 4E-BP.

Acknowledgments

This work was supported by the National Institute of Health (grant CA68262).

References

1. Merrick, W.C. and Hershey, J.W.B. (1996) The pathway and mechanism of eukaryotic protein synthesis. in Hershey, J.W.B., Mathews, M.B. & Sonenberg, N. (eds.) *Translational Control*, Cold Spring Harbor Laboratory Press, Plainview, New York, pp. 31-69).
2. Varani, G. (1997) A cap for all occasions. *Structure* **5**, 855-858.
3. Haghighat, A., Mader, S., Pause, A. and Sonenberg, N. (1995) Repression of cap-dependent translation by 4E-binding protein 1: competition with p220 for binding to eukaryotic initiation factor 4E. *EMBO J.* **14**, 5701-5709.
4. Matsuo, H., Li, H., McGuire, A.M., Fletcher, C.M., Gingras, A.-C., Sonenberg, N. and Wagner, G. (1997) Structure of Translation Factor eIF4E Bound to m7GDP and Interaction with 4E-Binding Protein. *Nature Structural Biology* **4**, 717-724.
5. Fletcher, C.M., McGuire, A.M., Li, H., Gingras, A.-C., Matsuo, H., Sonenberg, N. and Wagner, G. (1998) 4E binding proteins interact with the translation factor eIF4E without folded structure. *Biochemistry* **37**, 9-15.
6. Edery, I., Altmann, M. and Sonenberg, N. (1988) High-level synthesis in Escherichia coli of functional cap-binding eukaryotic initiation factor eIF-4E and affinity purification using a simplified cap-analog resin. *Gene* **74**, 517-525.
7. Kraulis, P.J. (1991) Molscript - a program to produce both detailed and schematic plots of protein structures. *J. Appl. Crystall.* **24**, 946-950.
8. Ueda, H., Maruyama, H., Doi, M., Inoue, M., Ishida, T., Morioka, H., Tanaka, T., Nishikawa, S. and Uesugi, S. (1991) Expression of a synthetic gene for human cap binding protein (human IF-4E) in Escherichia coli and fluorescence studies on interaction with mRNA cap structure analogues. *J. Biochem (Tokyo)* **109**, 882-889.
9. Ishida, T., Kamiichi, K., Kuwahara, A., Doi, M. and Inoue, M. (1986) Stacking and hydrogen bonding interactions between phenylalanine and guanine nucleotide: crystal structure of L-phenylalanine-7-methylguanosine-5'-monophosphate complex. *Biochem Biophys Res Commun* **136**, 294-299.
10. Carberry, S.E., Darzynkiewicz, E., Stepinski, J., Tahara, S.M., Rhoads, R.E. and Goss, D.J. (1990) A spectroscopic study of the binding of N-7-substituted cap analogues to human protein synthesis initiation factor 4E. *Biochemistry* **29**, 3337-3341.
11. Pause, A., Belsham, G.J., Gingras, A.C., Donze, O., Lin, T.A., Lawrence, J.C., Jr. and Sonenberg, N. (1994) Insulin-dependent stimulation of protein synthesis by phosphorylation of a regulator of 5'-cap function. *Nature* **371**, 762-767.
12. Altmann, M., Schmitz, N., Berset, C. and Trachsel, H. (1997) A novel inhibitor of cap-dependent translation initiation in yeast: p20 competes with eIF4G for binding to eIF4E. *EMBO J.* **16**, 1114-1121.
13. Fletcher, C.M. and Wagner, G. (1998) The interaction of eIF4E with 4E-BP1 ia an induced fit to a completely disordered protein. *Protein Science* **7**, 1639-1642.
14. Altmann, M., Edery, I., Trachsel, H. and Sonenberg, N. (1988) Site-directed mutagenesis of the tryptophan residues in yeast eukaryotic initiation factor 4E. Effects on cap binding activity. *J Biol Chem* **263**, 17229-17232.

NMR-BASED MODELING OF PROTEIN-PROTEIN COMPLEXES AND INTERACTION OF PEPTIDES AND PROTEINS WITH ANISOTROPIC SOLVENTS

H. KESSLER, G. GEMMECKER
Technische Universität München, Institut für Organische Chemie und Biochemie Lichtenbergstr. 4, D - 85747 Garching, Germany

Intermolecular interactions are the main cause of conformational changes in flexible molecules. Hence, the detailed study of the structure and dynamics of flexible molecules yields experimental evidence for the role of the environment in molecular structures. Weak intermolecular complexes with the solvent or between molecules in solution can be investigated by NMR spectroscopy.

1. Influence of Isotropic Environment on the Molecular Conformation

The conformation in the crystal can depend on the crystallization conditions and different conformations are often found in different crystal structures. But sometimes even several different conformations of a flexible molecule are found in a single unit cell. When these crystals are dissolved a rapidly exchanging ensemble of conformations is observed which may, but need not, contain the conformations observed in the crystal. Sometimes drastical conformational differences are found between crystal and solution.

More than two decades ago we have demonstrated for the first time in a clear way that the conformation in solution can be different from the one in the crystal by a combination of X-ray analysis, solid state NMR and dynamic NMR spectroscopy [1].

We have chosen two cases where a barrier between the crystal conformation and that in the solvent exists, which is high enough that (at least at low temperatures) the interconversion is slow on the NMR shift time scale.

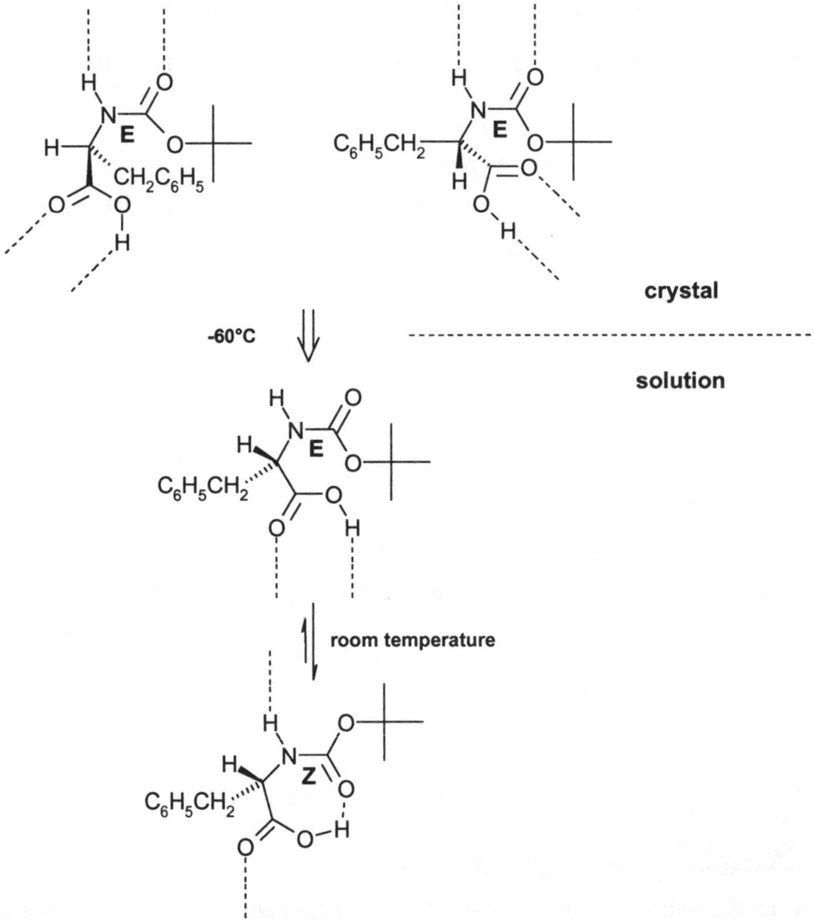

Fig. 1 Conformational change of Boc-Phe-OH upon transition from the crystalline to the solution state at -60°C and subsequent warming up.

The simple Boc-protected amino acid phenylalanine (Boc-Phe-OH) provides such an example (Fig. 1). In the crystal two conformations exist which mainly differ in the ϕ-angle (between N and C^α). A double set of signals is also found in solid state NMR.

Dissolving the crystal at -60°C yields a single signal set (rotation about φ is fast). The E conformation at the urethane bond of the crystal structure is still retained. Warming up the solution yields another signal set resulting from the Z conformation, which is the dominant rotamer in solution.

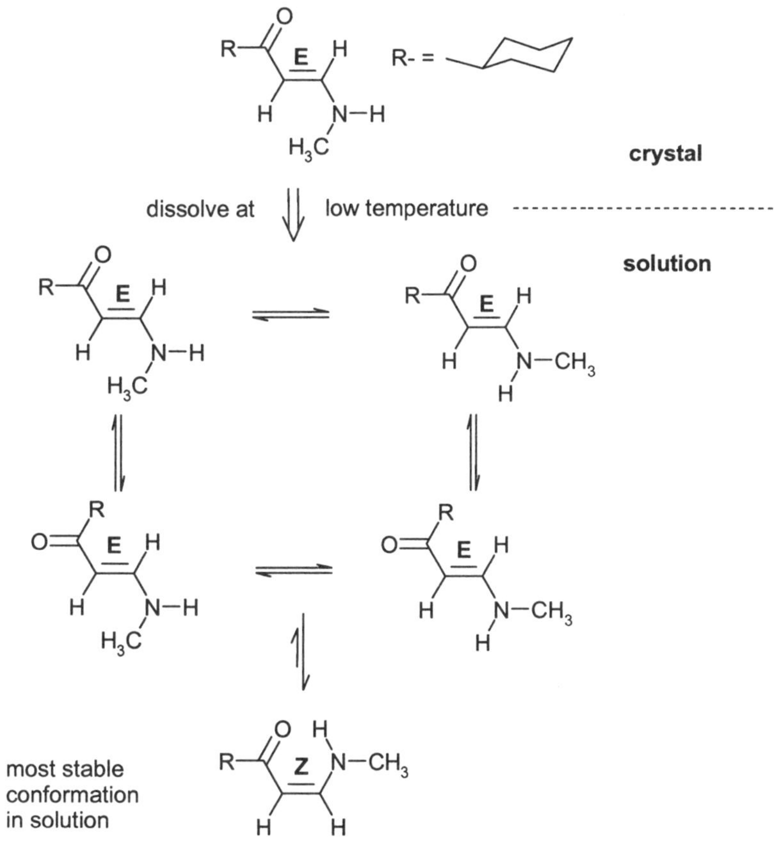

Fig. 2 *Different conformations of a vinilogous amide in crystal and in solution.*

The other example was a vinylogous amide, found in a single conformation in the crystal (Fig. 2). However, in solution at low temperature four distinct rotamers were observed, which interconverted rapidly at room temperature. All these solution conformations finally end up in a single conformation by changing the configuration at the double bond from E to Z.

Of course also the nature of the solvent influences the conformational equilibrium. A representative example is cyclosporin A, which exhibits in nonpolar solvents ($CDCl_3$, C_6D_6, THF) a single dominant conformation very similar to the conformation in the crystal [2,3]. In more polar solvents (MeOH, DMSO) [4, 5] or bound to cyclophilin, its receptor protein [6, 7], the conformation is drastically changed. Such changes, even if there are only small differences, are clearly detectable by NMR spectroscopy. E.g., in the cyclic octapeptide hymenistatin, when going from $CDCl_3$ to DMSO a twist of one amide bond can be established and rationalized by the different solvation of this amide bond in the molecular dynamics calculation in the two different solvents [8]. The lesson learned by these studies is that restrained or free molecular dynamics should always be performed with explicit solvents [9]. It was also demonstrated that *in vacuo* calculations exhibit distinct artifacts which disappear when explicit solvents are used in the calculations [10, 11].

The application of distance dependent dielectricity constants is at least a step into the right direction for small molecules, which are the most sensitive ones for effects of the surrounding, due to the relatively large surface area (in relation to the number of atoms).

2. Behaviour of Flexible Molecules in Lipid/Water Interphases

The importance of the interaction of peptides with cellular interphases is often discussed and still an open question. Schwyzer proposed a partial intrusion into the lipid part of the membrane before the reaction with surface recpetors occurs [12]. The function of the membrane is considered as pre-forming a conformation which can interact better or more specific with the receptor and as a medium for two-dimensional diffusion to facilitate the meeting of the molecules.

Hence we were further interested in studying the behavior of flexible molecules at the interphase between water and a lipophilic solvent by molecular dynamics calculations as a mimic for the insertion of peptides into biological membranes. As high-resolution NMR spectroscopy needs isotropically fast tumbling molecules, a mimic for the membrane has to be used for experimental studies. Wüthrich et al. have shown a long time ago that micellar solution can be used for this purpose and that such conditions can induce structure in oligopeptides which are random coil in pure water [13]. MD calculations with real membranes require extremely long calculation times, so we developed a simpler two-phase box with CCl_4 and water [14]. It can be used for reasonably fast calculations as a more realistic mimic for the membrane than an isotropic solvent would be [15]. The induction of a distinct conformation in a bradykinin analogue in the anisotropic environment could be demonstrated via NMR measurements in SDS micelles and MD calculations in the two-phase box [16].

In a search for a case where a single prefered conformation in an isotropic solvent is changed into a single but different conformation in micellar solution, several cyclic penta- and hexapeptides have been synthesized in which two aminohexadecanoic acid (amino acid with a long lipophilic side chain) were introduced. One of the peptides, cyclo(–D-Asp–Asp–Gly–Ahd–Ahd–Gly–) with Ahd = L-α-aminohexadecanoic acid, adopts in isotropic $CDCl_3$ solution a conformation clearly different from that in SDS micelles / H_2O, as shown by NMR data [17]. In Fig. 3 the movement of the amphiphilic peptide from the lipophilic CCl_4 phase into the CCl_4 / H_2O interphase is demonstrated, as calculated in a 250 μs MD simulation in a CCl_4 / H_2O two-phase system.

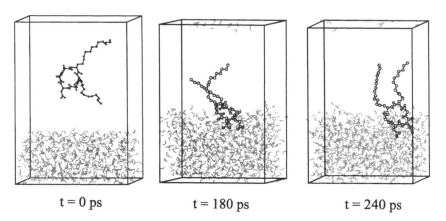

Fig. 3 Interdigitation, orientation and conformational rearrangement of cyclo(–D-Asp–Asp–Gly–Ahd–Ahd–Gly–) during the free MD simulation in the H_2O / CCl_4 two-phase system (H_2O molecules are shaded in grey, the CCl_4 molecules are omitted for clarity).

3. Interactions between proteins

Intermolecular interactions are the basis of all important functions of biomolecules (proteins, nucleic acids, sugars and small molecules) such as substrate-receptor, enzym inhibitor, protein-protein, protein-nucleic acid, and between nucleic acids. In the last part of this article we will concentrate on some studies of short living protein-protein complexes which are very big for NMR spectroscopy. The proteins are part of the phosphotransferase system (PTS proteins), which is one of the active sugar transport mechanisms into bacteria through the bacterial membrane (permeases) [18]. During this import the sugar (glucose, mannose, mannitol or others) is phosphorylated. The phosphate is delivered from phosphoenolpyruate (PEP) in several steps via different enzymes (PEP → enzyme I → HPr → IIA → IIB → sugar, Fig. 4), the phosphorylated forms of which are rather shortlived in solution. Also, the protein-protein complexes occuring during phosphate transfer are relatively weak and the big supramolecular assembly is in rapid equilibrium. We were interested in studying the structure of these complexes for the glucose and mannose transporters to understand how the process works.

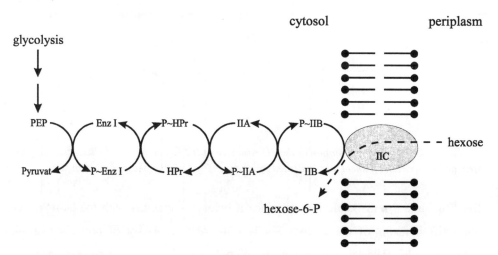

Fig. 4 Phosphate transfer scheme for the bacterial phosphotransferase system (PTS).

There were already several structural studies by X-ray analysis or NMR spectroscopy for the cystosolic proteins, especially HPr from different bacteria and the glucose specific IIA^{Glc} protein. Erni et al. from the university of Bern demonstrated that the IIB^{Glc} domain of the glucose transporter can be cleaved off from the large transmembrane IICB^{Glc} protein, yielding a stable, soluble and fully functional protein suitable for NMR studies [19]. Our high-resolution NMR structure of IIB^{Glc} exhibited a new protein fold, consisting of a four stranded antiparallel sheet which is covered from one side by three helices [20]. The Cys^{35} residue, which is transiently phosphorylated, is located on the surface in a β-turn between strands 1 and 2 (Fig. 5).

Fig. 5 Cartoon representation of the 3D structure of IIB^{Glc}. The three α-helices and the four-stranded antiparallel β-sheet are indicated by ribbons; Cys35 is the phosphorylation site.

IIB^{Glc} is quite a remarkable protein, since it belongs - together with the homologous other IIB domains of the glucose family - to the few classes of proteins that are transiently phosphorylated at a Cys residue. However, it shares no sequence nor structural homology with the other large group of Cys-phosphorlated proteins, the protein phosphotyrosin phosphatases.

The phosphorylated enzyme IIB^{Glc} hydrolyses fast and therefore cannot be studied by NMR in a straightforward way. To keep the enzyme phosphorylated during the measurement we treated the isotopically labelled IIB^{Glc} (1.5 mM solution) with a 20 fold excess of phosphoenolpyruvate (PEP) together with catalytic amounts of the intermediate phosphocarrier proteins enzyme I, HPr and IIA in non-labelled form [21].

A comparison of the amide chemical shifts and NOE patterns in the phosphorylated and the unphosphorylated form of IIB^{Glc} allowed to map the changes caused by the introduction of a thiophosphate at Cys35. It became clear that phosphorylation did not affect the overall structure of the protein, but induced only local conformational changes, esp. a reduction of flexibility in the turn region between β1 and β2 due to additional hydrogen bonds to the phosphate group [22].

Next we investigated the interactions between both PTS proteins IIA^{Glc} and IIB^{Glc} with NMR-spectroscopic methods. The chemical shift changes observed for one ^{15}N-labeled protein upon addition of its unlabeled complex partner allowed to map the mutual contact sites. For both proteins, the affected surface area corresponds to a contagious region roughly centered about the phosphorylation sites His90 in IIA^{Glc} and Cys35 in IIB^{Glc} (Figs. 6 & 7). However, for the larger rest of the amide signals no signal shifts could be observed upon complex formation, so that a global structural rearrangement of the two domains can be excluded.

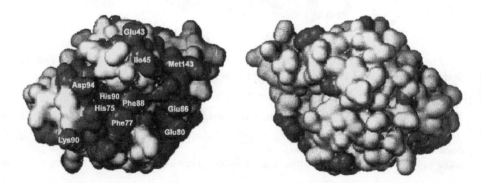

Fig. 6 Contact site (dark) of IIA^{Glc} for the complexation with IIB^{Glc}, derived from amide signal shifts upon complex formation (left); the part opposite to the phosphorylation site His90 remains unaffected (right).

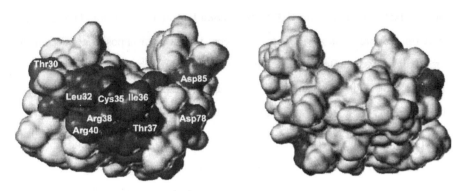

Fig. 7 Contact site (dark) of IIB^{Glc} for the complexation with IIA^{Glc}, derived from amide signal shifts upon complex formation (phosphorylation site Cys35 shown in black).

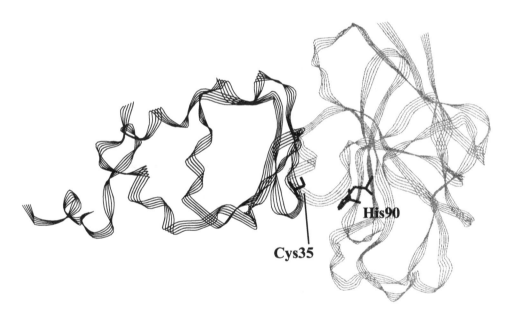

Fig. 8 Model of the complex between the two PTS domains IIA^{Glc} (grey) and IIB^{Glc} (black), derived from the NMR data on contact sites (cf. Figs. 6 & 7). For clarity, only the protein backbones are shown in ribbon mode, with the exception of the two phosphorylation sites His90 and Cys35.

Both contact sites exhibit an essentially complementary structure: His90 in IIA^{Glc} is located in a depression of the protein surface, while Cys35 is positioned on a protrusion of the IIB^{Glc} surface. Both contact sites consist mainly of nonpolar residues, so that the complex formation will strongly rely on hydrophobic interactions. However, the contact sites also comprise a few polar amino acids, which are highly conserved among homologous PTS proteins and are supposed to be essential for the specifity of the intermolecular interaction.

It is known that the phosphate transfer from IIA^{Glc} to IIB^{Glc} proceeds via an intermediate with a bridging phosphate group, thus His90 and Cys35 have to come into close contact in the complex. Taking this additional information into account, a model for the IIA^{Glc} - IIB^{Glc} complex can be derived from the known contact sites, in which both proteins make contact through their β-sheets oriented parallel to each other (Fig. 8).

References

1. Kessler, H., Zimmermann, G., Förster, H., Engel, J., Oepen, G., and Sheldrick, W.S. (1981) Does a molecule have the same conformation in the crystalline state and in solution? Comparison of NMR results for the solid state and solutions with those of the X-ray structural determination, *Angew. Chem., Int. Ed.* **20**, 1053-1055

2. Loosli, H.R., Kessler, H., Oschkinat, H., Weber, H.P., Petcher, T.J., and Widmer, A. (1985) Peptide Conformations 31. The conformation of cyclosporin A in the crystal and in solution, *Helv. Chim. Acta* **68**, 682-704.

3. Kessler, H., Köck, M., Wein, T., and Gehrke, M. (1990) Reinvestigation of the conformation of cyclosporin A in chloroform, *Helv. Chim. Acta* **73**, 1818-1832.

4. Ko, S.Y. and Dalvit, C. (1992) Conformation of cyclosporin A in polar solvents. *Int. J. Pept. Protein Res.* **40**, 380-382.

5. Kessler, H., Loosli, H.R., and Oschkinat, H. (1985) Peptide Conformations 30. Assignment of the ^{1}H-, ^{13}C-, and ^{15}N-NMR spectra of cyclosporin A in CDCl$_3$ and C$_6$D$_6$ by a combination of homo- and heteronuclear two-dimensional techniques, *Helv. Chim. Acta* **68**, 661-681.

Kessler, H., Gehrke, M., Lautz, J., Köck, M., Seebach, D., and Thaler, A. (1990) Complexation and medium effects on the conformation of cyclosporin A studied by NMR spectroscopy and MD calculations, *Biochem. Pharmacol.* **40**, 169-173.

6. Fesik, S.W., Gampe, R.T. Jr., Eaton, H.L., Gemmecker, G., Olejniczak, E.T., Neri, P., Holzman, T.F., Egan, D.A., Edalji, R., Simmer, R., Helfrich, T., Hochlowski, J., and Jackson, M. (1991) NMR studies of [U-^{13}C]cyclosporin A bound to cyclophilin: bound conformation and portions of cyclosporin involved in binding, *Biochemistry* **30**, 6574-6583.

7. Weber, C., Wider, G., von Freyberg, B., Traber, R., Braun, W., Widmer, H., and Wüthrich, K. (1991) The NMR structure of cyclosporin A bound to cyclophilin in aqueous solution, *Biochemistry* **30**, 6563-6574.

8. Konat, R.K., Mierke, D.F., Kessler, H., Kutscher, B., Bernd, M., and Voegeli, R. (1993) Synthesis and solvent effects on the conformation of hymenistatin 1, *Helv. Chim. Acta* **76**, 1649-1666.

9. van Gunsteren, W.F. and Berendsen, H.J.C. (1990) Computer simulations of molecular dynamics: Methodology, applications and perspectives in chemistry. *Angew. Chem. Int. Ed. Engl.* **29**, 992-1023.

10. Kurz, M., Mierke, D.F., and Kessler, H. (1992) Calculation of molecular dynamics for peptides in dimethyl sulfoxide: Elimination of vacuum effects, *Angew. Chem. Int. Ed.* **31**, 210-212.

11. Kessler, H., Mronga, S., Müller, G., Moroder, L., and Huber, R. (1991) Conformational analysis of a IgG1 hinge peptide derivative in solution determined by NMR spectroscopy and refined by restrained molecular dynamics simulations, *Biopolymers* **31**, 1189-1204.

12. Schwyzer, R. (1995) 100 Years Lock-and-Key Concept: Are Peptide Keys Shaped and Guided to Their Receptors by the target Cell Membrane? *Biopolymers (Peptide Science)* **37**, 5-16.

13. Braun, W., Wider, G., Lee, K. H., and Wüthrich, K. (1983) Conformation of glucagon in a lipid-water interphase by 1H nuclear magnetic resonance, *J. Mol. Biol.* **169** (4), 921-948.

14. Guba. W. and Kessler, H. (1994) A novel computational mimetic of biological membranes in molecular dynamics simulation, *J. Phys. Chem.* **98**, 23-27.

15. Moroder, L., Romano, R., Guba, W., Mierke, D.F., Kessler, H., Delporte, C., Winand, J., and Christophe, J. (1993) New evidence for a membrane-bound pathway in hormone receptor binding, *Biochemistry* **32**, 13551-13559.

16. Guba, W., Haeßner, R., Breipohl, G., Henke, S., Knolle, J., Santagada, V., and Kessler H. (1994) A novel combined approach of NMR and molecular dynamics with a biphasic membrane mimetic: Conformation and orientation of the bradykinin antagonist Hoe 140, *J. Am. Chem. Soc.* **116**, 7532-7540.

17. Koppitz, M., Mathä, B., and Kessler, H. (1998) Structure investigation of amphiphilic cyclopeptides in isotropic and anisotropic environments - A model study simulating peptide-membrane interactions, *J. Am. Chem. Soc.* submitted.

18. Postma, P. W., Lengeler, J. W., and Jacobson, G. R. (1993) Phosphoenolpyruvate:carbohydrate phosphotransferase systems of bacteria, *Microbiol. Rev.* **57** (3), 543-594.

19. Buhr, A., Flükiger, K., and Erni, B. (1994) The glucose transporter of *Escherichia coli*. Overexpression, purification, and characterization of functional domains. *J. Biol. Chem.* **269** (38), 23437-23443.

20. Eberstadt, M., Golic Grdadolnik, S., Gemmecker, G. Kessler, H., Buhr, A., and Erni, B. (1996) Solution structure of the IIB domain of the glucose transporter of *Escherichia coli*, *Biochemistry* **35**, 11286-11292.

21. Pelton, J.G., Torchia, D.A., Meadow, N.D. and Roseman, S. (1992) Structural Comparison of Phosphorylated and Unphosphorylated Forms of IIIGlc, a Signal-Transducing Protein from *Escherichia coli*, Using Three-Dimensional NMR Techniques, *Biochemistry* **31**, 5215-5224.

22. Gemmecker, G., Eberstadt, M., Buhr, A., Lanz, R., Golic Grdadolnik, S., Kessler, H. and Erni, B.(1997) Glucose Transporter of Escherichia coli: NMR Characterization of the Phosphocysteine Form of the IIB^{Glc} Domain and Its Binding Interface with the IIA^{Glc} Subunit. *Biochemistry* **36** (24), 7408-7417.

PROTEIN SURFACE RECOGNITION

X. SALVATELLA[1], T. HAACK[1], M. GAIRÍ[2], J. DE MENDOZA[3], M.W. PECZUH[4], A.D. HAMILTON[4] AND E. GIRALT[1]

[1]*Departament de Química Orgànica*
Universitat de Barcelona
Martí i Franqués 1-11, 08028 Barcelona
Spain
[2]*Unitat de Ressonància Magnètica Nuclear*
Serveis Científico-Tècnics
Universitat de Barcelona
Martí i Franqués 1-11, 08028 Barcelona
Spain
[3]*Departamento de Química Orgánica*
Universidad Autónoma de Madrid
Cantoblanco, 28049 Madrid
Spain
[4]*Department of Chemistry*
Yale University
350 Edward Street
New Haven,
Connecticut 06520
USA

1. Introduction

Molecular recognition is at the core of the design of any supramolecular system. Proteins, because of their biological relevance and structural complexity, provide us with exquisite examples of molecular recognition in a natural setting. Most of the effort reported in the literature towards the design of artificial compounds able to bind proteins with high affinity and selectivity has been targeted to rather hydrophobic areas: the design of enzyme inhibitors could be the paradigm of such studies, the inhibitor being designed to occupy a groove on the protein surface where the active center of the enzyme is located [1].

Molecular recognition of the "external" protein surface constitutes a much more challenging target. Difficulties arise because of the high degree of solvation and the hydrophilic character of protein surfaces, where intermolecular interactions such as hydrogen bonds and electrostatic forces play a key role. The ability of the ligand to make hydrogen bonds or to establish electrostatic interactions with the protein surface will be seriously affected by the presence of solvating water. This is due firstly to the excellent properties of water as both hydrogen bond donor and acceptor and secondly to the rather high local dielectric constant of the surface (compared to a hydrophobic pocket such as an active site). In spite of these difficulties the design of compounds able to recognize protein surfaces is important considering that the biological role of most proteins depends on their interaction with other molecules, and in particular with other proteins. Molecules with protein surface recognition ability have the potential to interfere with such control processes and constitute a promising new area in drug design.

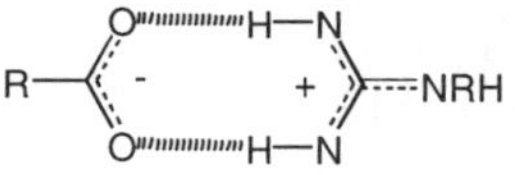

Figure 1: Interaction between carboxylate and guanidinium

The interaction of guanidinium ions with oxoanions such as sulphate, phosphate and carboxylate is well known[2]. As shown in figure 1 the two driving forces for the oxoanion-guanidinium interaction are the electrostatic attraction and the simultaneous formation of two hydrogen bonds. We have recently described in detail the interaction between a bisguanidinium compound equivalent to. compound **2** in figure 2 and sulphate anion[3]. The NMR studies have shown that recognition of sulphate by the bisguanidinium compound takes place through dimerization, i.e. every sulphate ion is interacting with two molecules of the bisguanidinium compound simultaneously. More importantly, circular dicroism has shown that the recognition process is concomitant with a high increase in the helicity of the bisaguanidinium compound. By means of molecular mechanics we have suggested a model where two molecules of **2** form a double-helix which wraps around four sulphate anions, maximizing the anion-guanidinium interaction. This double-helix has a pitch of 22 A and is left or right-handed depending on whether the configuration of the tetraguanidinium compound is all-*R* or all-*S*.

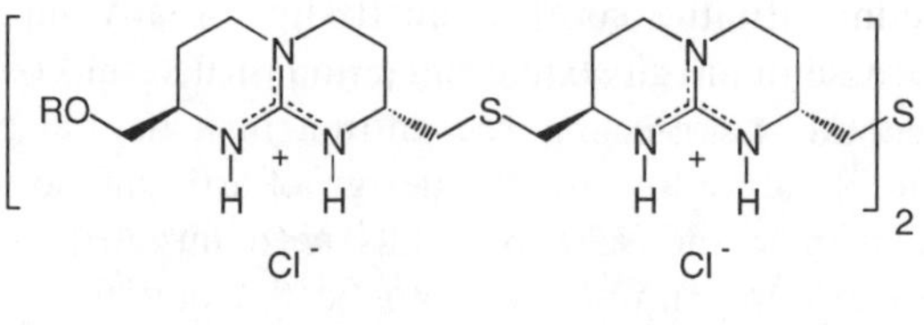

compound **1** R = Si^{t}BuPh$_2$
compound **2** R = H

Figure 2: Oligoguanidinium compounds

2. Molecular Recognition of Anionic Helical Superhelices

The polycationic character of compounds **1** and **2** together with their tendency to adopt helical structures suggested that these compounds could be used as receptors for the recognition of peptides or proteins with anionic helical structures.

Although the most common helical structure in protein chemistry is the α-helix there is a total lack of complementarity between an anionic α-helix and the double-helix defined by the sulphate salt of tetraguanidinium compound **2**. Indeed the pitch of an α-helix is 5.4 Å, far smaller than the pitch of the double-helix, 22 Å.

Since the periodicity of an α-helix is 3.6 residues per turn, an i/i+3/i+6/i+9 arrangement of residues does not form an array parallel to the axis of the helix but a left-handed superhelix with a pitch of 27 Å, 5 times longer than the pitch of an α-helix. An arrangement of i/i+4/i+8/i+12 defines a very similar superhelix, in terms of translation distance, which is right-handed and has a pitch of 57 Å. Such superhelices are shown in figure 3, where (Ala)$_{17}$ has adopted an α-helical secondary structure and the methyl groups of alanine at i/i+3... and i/i+4... respectively have been rendered as space-filling spheres.

Figure 3: Superhelices defined by an i/i+3/i+6/i+9 and i/i+4/i+8/i+12 arrangement of residues.

The introduction of negatively charged residues (aspartic or glutamic acid) at the positions which define these superhelices yields structures which can be complementary to the helices defined by the oligoguanidinium compounds **1** and **2**. Indeed a comparison between the translation distances and the rotation angles of the two structures shows a good match. The validity of such a hypothesis has been recently confirmed through the synthesis and study of a series of polyanionic peptides with the sequences shown in scheme 1[4]. NMR studies have shown the formation of a complex between peptide **3** and compound **2** in which the peptide adopts a clearly helical structure. Circular dicroism has allowed us to measure the 1:1 stochiometry of the complex, to estimate the affinity constant (1.6×10^5 M^{-1} for **1-3** and 3.4×10^5 M^{-1} for **2-3**) and to observe a decrease in affinity when the tetraanionic peptide **3** is replaced by the tri- and dianionic peptides **4** and **5**.

(3) Ac-A-A-A-<u>D</u>-Q-L-<u>D</u>-A-L-<u>D</u>-A-Q-<u>D</u>-A-A-Y-NH$_2$
(4) Ac-A-A-A-<u>D</u>-Q-L-<u>D</u>-A-L-<u>D</u>-A-Q-N-A-A-Y-NH$_2$
(5) Ac-A-A-A-<u>D</u>-Q-L-<u>D</u>-A-L-N-A-Q-N-A-A-Y-NH$_2$
(6) Ac-F-R-<u>E</u>-L-N-<u>E</u>-A-L-<u>E</u>-L-K-<u>D</u>-A-Q-A-NH$_2$

Scheme 1: Sequences of the polyanionic peptides.

TABLE 1. Protein yielded by the sequence search.

Name of protein	Species	pdb code	size	anionic path
adenovirus type II hexon	human	0ad2	967	EEEDEDEEEE
c-AMP dependent protein kinase	bovine	1apk	379	EYFEKLEKEE
coagulation factor IX	human	1cfh	47	EAREVFEDTE
human class histocompatibility antigen A2	human	1has	770	ERIEKVEHSD
P53 (oligomerization domain)	human	1sal	168	ELNEALELKD
DNA polymerase III β subunit	E.coli	2pol	732	EQEEAEEILD
pyruvate decarboxylase	brewer's yeast	1pvd	1074	EVKEAVESAD
aldose reductase	pig	1dla	1256	DESDFVETWE
elongation factor Tu	Thermus aquaticus	1eft	405	DMVDDPELLD
glycogen phosphorylase	rabbit	1abb	3312	DMEELEEIEE
hemocyanin	lobster	1lla	628	EAGDFNDFIE
protein R2 of ribonucleotide reductase	E.coli	1mrr	750	DDPEMAEIAE
tropomyosin	rabbit	2tma	568	DEQEKLELAE
annexine IV	bovine	1ann	173	DAQDLYEAGE
plasminogen	human	1cea	144	DYCDILECEE
lactoferrin	human	1lcf	691	DLSDEAERDE
lysyl t-RNA synthetase	E.coli	1lyl	1515	DYHDLIELTE
procarboxypeptidase B	pig	1pba	81	DENDISELHE
3-α-hydroxysteroid dehydrogenase	rat	1ral	308	DICDTWEAME
deoxyuridine 5'-triphosphate nucleotidohydrolase	E.coli	1dup	152	EDFDATDRGE
myosin regulatory domain	bay scallop	1scm	354	DVFELFDFWD
endonuclease III	E.coli	2abk	211	DLCEYKEKVD
creatine amidinohydrolase	Pseudomonas putida	1chm	802	DTFEDVELME
ferredoxin II	Desulfovibrio gigas	1fxd	58	DVFEMNEXGD
methylamine dehydrogenase	Paracoccus denitrificans	2bbk	960	ERTDYVEVFD
phosphoglucomutase	rabbit	3pmg	1122	EKADNFEYHD
HIV type I virus matrix protein	HIV virus type 1	2hmx	133	DTKEALDKIE
repressor of primer of DNA replication	E.coli	1rpr	126	DADEQADICE
aconitase	pig	5can	754	EPFDKWDXKD
beta B2 crystallin	bovine	2bb2	118	EKGDYKDSGD
scytalone dehydratase	Magnaporthe grisea	1std	172	EWADSYDSKD
GreA transcript cleavage factor	E.coli	1grj	158	EAREHGDLKE

3. Occurrence of Helical Anionic Paths in Native Protein Surfaces

The fact that a synthetic molecule such as **2** was able to selectively recognize an anionic superhelical path in a purpose-built peptide prompted us to use it as a first approach to the recognition of protein surfaces. The suitability of **2** as a protein surface receptor critically depended on the occurrence of such superhelical anionic paths in the surface of native protein structures. A computer-driven search was carried out in the Brookhaven Protein Data Bank in order to find out which proteins showed such an arrangement of acidic residues on their surface.

The search was carried out using QUEST3D, the software from the Cambridge Structural Database which allows the search of specific amino acidic sequences within the approximately 8000 structures stored in the Brookhaven Protein Data Bank. As already described both an i/i+3/i+6/i+9 and an i/i+4/i+8/i+12 arrangement of acidic residues define the superhelix which can be recognized by **2**. We have however only performed the following search: D/E-X-X-D/E-X-X-D/E-X-X-D/E, where D and E are the one-letter codes for aspartic and glutamic acid and X represents any of the 20 natural amino acids. Such a search yielded 31 Brookhaven Protein Data Bank entries shown in table 1, which we studied according to the most relevant criteria for recognition by **2**: the helicity of the 10 residue stretch and the accessibility of the oxygen atoms of their oxoanionic residues.

3.1 HELICITY SCORE AND ACCESSIBILITY

Two main features of the recognition process are key to the possible interaction between a protein surface and **2**. As the CD and NMR experiments of the complex between **2** and the model peptides have shown, the interaction is concomitant with an induction of helicity on the peptide and a relative loss of conformational freedom of the side chains of the acidic residues. We have therefore sorted the structures of proteins which show a helical anionic path sequence and good accessibility of the oxygen atoms of the aspartic and glutamic side chains.

A preliminary visual study of the protein structures helped us discard about 60 % of the hits generated by the search: only those which showed a pseudo-helical and partially accessible anionic path were retained. These 10 proteins, listed in table 2, were considered good candidates for recognition and were therefore studied using a more systematic approach.

P53 (oligomerisation domain)	Procarboxypeptidase B
pyruvate decarboxylase	3-α-hydroxysteroid dehydrogenase
protein R2 of ribonucleotide reductase	myosin regulatory domain
annexine IV	repressor of primer of DNA replication
lysyl t-RNA synthetase	scytalone dehydratase

Table 2: Candidate proteins after visual inspection

The helicity of a given residue i can be assessed by measuring the deviation of its Ψ and Φ dihedral angles from the Ψ_h and Φ_h angles of a residue in an ideal α helix. In order to evaluate the global helical content of the anionic paths of the proteins we have designed a parameter we name helicity score which is defined by the following equation:

$$\text{helicity score} = 10 - \frac{\sqrt{\sum_{i=1}^{10} f(i)(\Psi_i - \Psi_h)^2} + \sqrt{\sum_{i=1}^{10} f(i)(\Phi_i - \Phi_h)^2}}{20}$$

where $f(i) = 0.35 + 0.2i - 0.01538 i^2$

and $\Psi_h = -47°$ and $\Phi_h = -57°$ are the dihedral angles in a canonical α-helix and Ψ_i and Φ_i are the actual dihedral angles for a residue at position i in the sequence

The parabolic weighting function has been used to account for the major importance of the central residues in the context of cooperativity and its parameters have been selected using a heuristic approach. The helicity score parameter has been designed in order to give values from 0 to 10 and has provided us with an assessment tool which is far more reliable than simple visual inspection of the surface of the protein. The values we have obtained for the 10 candidates range from 5.8 for scytalone dehydratase to 9.5 for the tetramerisation domain of P53.

In order to estimate the accessibility of each amino acidic path to the tetraguanidinium ligand **2**, we have used a variation of the method described by Connolly to calculate solvent-accessible surfaces of proteins and nucleic acids[5]. The Connolly model defines the accessible surface as the points of contact between the Van der Waals surface of the macromolecule and a solvent-like probe-sphere which is rolled over it. As shown in figure 2 compound **2** can be considered to be a tetramer formed by four monomeric and rigid subunits linked by a more flexible -CH$_2$-S-CH$_2$- spacer. In analogy with the Connolly model, we can consider a probe sphere of radius 4.7 A to have an accessibility behavior similar to that of compound **2**. By means of a molecular modeling package we have measured the area of the oxygen atoms of the acidic side-chains accessible to such sphere. This has enabled us to compare the accessibility of the 10 candidates obtained after visual inspection. We have also studied the accessibility at three (1.4, 2.5 and 3.6 A) other radii in order to gain information on the size dependency of the accessibility and check the validity of our model.

3.2 P53

P53 is a protein of enormous importance in the regulation of the cell cycle and a major target of cancer research. It is a three domain protein and displays an anionic path in its tetramerization domain (residues 343 to 352). Being a tetrameric protein, four copies of the anionic path are available for complexation and its helicity and accessibility features are reasonably good. Figure 4 shows the tetrameric structure of the oligomerization domain as obtained by NMR and the accessibility profile of its acidic residues[*] [6].

Figure 4: Structure and accessibility estimates of the tetramerisation domain of P53

3.3 PYRUVATE DECARBOXYLASE

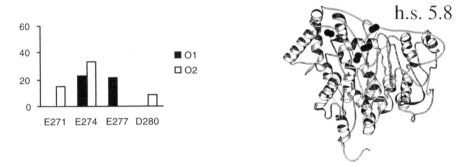

Figure 5: Structure and accessibility estimates of pyruvate decarboxylase

Pyruvate decarboxylase catalyzes the decarboxylation of pyruvate to acetaldehyde and is a homodimer with each monomer being formed by three domains, (α, β and γ). Its structure has been solved by X-ray diffraction with the exception of residues 106 to 133 and 292 to 30, which could not be defined, and displays an anionic

---

[*] The accessibility profile shows the accessible area in square angstroms for the two oxygen atoms of each acidic residue. O1 and O2 are the PDB definitions for the two carboxylate oxigens. The helicity score is shown by the structure of the protein.

path located between residues 271 and 280, far away from both the dimer interface and the active site of the enzyme[7]. Both its helicity and accessibility, as shown in next figure, are far from adequate: its complexation by **2** therefore seems unlikely.

3.4 ANNEXIN IV

Annexin IV is a placental anticoagulant protein which strongly binds calcium and phospholipids[8]. It has a monomeric structure which is formed by four homologous repeats with a consensus sequence. The anionic path of annexin IV is very helical but its aspartic acid residues at positions 173 and 176 are completely inaccessible to the probe sphere and hence the tetraguanidinium receptor.

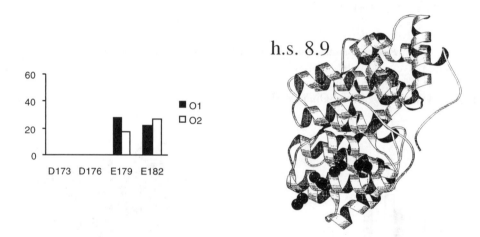

Figure 6: Structure and accessibility estimates of annexin IV

3.5 LYSYL t-RNA SYNTHETASE

Lysyl t-RNA synthetase is a heat-inducible enzyme which catalyses the reaction of L-Lysine and t-RNA in *E.coli* and is thought to act as a modulator of the heat-shock and stress responses[9]. The homodimeric structure of this protein shows an anionic path from residues 285 to 294: its helicity score and accessibility are however not optimal and in principle preclude any binding to molecule **2**.

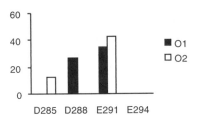
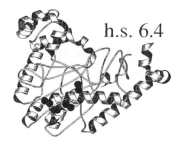

Figure 7: Structure and accessibility estimates of lysyl t-RNA synthetase

3.6 3-α-HYDROXYSTEROID DEHYDROGENASE

3-α-Hydroxysteroid dehydrogenase is a key enzyme in the biosynthesis of steroid hormones and is thought to be responsible for the carcinogenicity of some aromatic polycyclic hydrocarbons through its conversion to reactive quinones[10]. This protein is monomeric and has an anionic path between residues 143 and 152. Its helicity and accessibility features are however inadequate for complexation.

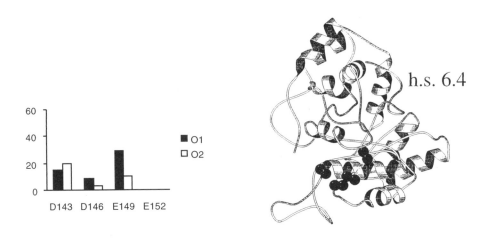

Figure 8: Structure and accessibility estimates of 3-α-hydroxysteroid dehydrogenase.

3.7 MYOSIN REGULATORY DOMAIN

Calcium regulation of the skeletal muscle in molluscs takes place through myosin. In these species the heavy chain of myosin forms a globular complex with two other chains, called the essential and the regulatory light chains, where the calcium regulation and the ATP-ase activity centers are located[11]. The anionic path of this protein, placed between residues 13 and 22 of the essential light chain, is close to both the interfacial and the calcium regulation center and is indeed thought to be involved in the interaction with calcium. The parametrization of the protein did however yield poor results as shown in figure 9.

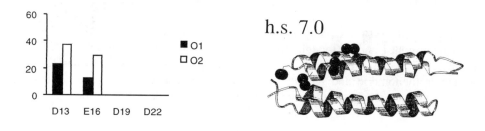

Figure 9: Structure and accessibility estimates of myosin essential light chain.

3.8 REPRESSOR OF PRIMER OF DNA REPLICATION

The repressor of primer of DNA replication of *E.coli* is responsible for regulation of the replication of plasmidic DNA by modulating the initiation of the transcription of the primer RNA precursor[12]. This protein is an antiparallel dimer and shows an anionic path between residues 30 and 39. Its helicity score and accessibility features are clearly unsatisfactory for recognition.

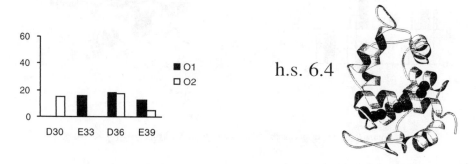

Figure 10: Structure and accessibility estimates of the repressor of primer of DNA replication.

3.9 PROTEIN R2 OF RIBONUCLEOTIDE REDUCTASE

Ribonucleotide reductase in an essential enzyme in the biosynthesis of DNA: it catalyzes the first step in the DNA replication pathway; it is a tetrameric protein of the $\alpha_2\beta_2$ type, the β_2 sub-unit sometimes being called protein R2 of ribonucleotide reductase. The redox activity of this protein takes place through several redox centers which are however far away from the anionic path, which is located between residues aspartic acid 257 and glutamic acid 266[13]. The helicity and accessibility features of this protein are constant along the sequence but too low to attempt its complexation by **2**.

Figure 11: Structure and accessibility estimates of protein R2 of ribonucleotide reductase.

3.10 PROCARBOXYPEPTIDASE B

Procarboxypeptidases are the inactive precursors of the proteolytic enzymes carboxypeptidase A and B. Activation takes place through the loss of a 100 residue tail which has an anionic sequence between aspartic acid 19 and glutamic acid 28[14].

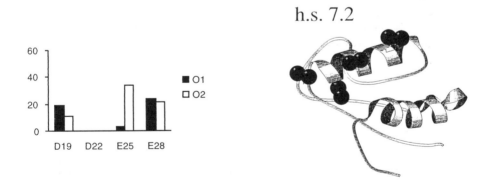

Figure 12: Structure and accessibility estimates of procarboxypeptidase B.

In spite of a high helicity score the accessibility profile does not meet the requirements for complexation.

3.11 SCYTALONE DEHYDRATASE

Scytalone dehydratase is an enzyme which catalyzes an important step in the biosynthetic pathway of melanin in *Magnaporthe grisea*, a pest which is responsible for the low yield of rice plantations. This enzyme is a homotrimer which displays an anionic path per monomer, between aspartic acids 25 and 28 which is located far from both the interface between the monomers and the active site[15]. Moreover, the position of aspartic acid causes a serious distortion of the helix which yields a surface which very unlikely will be complexed by receptor **2**.

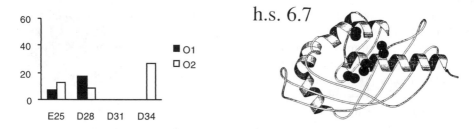

Figure 13: Structure and accessibility estimates of scytalone dehydratase.

3.12 COMPARATIVE ANALYSIS

The parametrizations we have carried out, which are summarized in table 3, have yielded a general result: the accessibility and helicity of the anionic paths are too low to expect any binding to our tetraguanidinic receptor. There is however a significant exception to this general statement: the anionic path of the tetramerisation domain of P53. Its exceptional accessibility can in principle be related to the preponderance of the more accessible glutamic acid residues in its sequence, but a calculation of the expected total accessible area based on the average glutamic and aspartic contributions over the different proteins yields an area which represents only 55% of the measured one. P53 is therefore the protein we have chosen to assess the surface-binding properties of receptor **2**.

4. Future Approaches

As shown in the previous section P53 fulfills the two requirements that we expect to be important for complexation by **2**. Its helicity score is so high that we can consider its anionic path to be almost perfectly helical. Moreover, although the aspartic acid side-chain at position 352 is partially buried, the accessibility of the oxygen atoms is also high, when compared to the rest of candidate proteins.

Mutations of the tumor suppressor protein P53 are found in about 55% of lung cancers and 50% of colon and rectal cancers; indeed the only role of P53 seems to be the protection against tumors. Levels of P53 in the nucleus of the cell are increased whenever its DNA is damaged: the action of P53 will then depend on the extent of such damage: if the DNA can be repaired P53 will act as a "cell cycle brake" to prevent the spread of the mutation until the DNA has been deleted. However, if the DNA damage is very serious P53 will cause apoptosis, i.e. programmed cell death[16].

P53 is a three-domain tetrameric protein. The three domains are the DNA-binding, the regulatory and the tetramerization domain. The anionic path of P53 is located in the tetramerization domain and spans from residue 343 to 352. The main role of this domain seems to be stabilization of the tetrameric structure of the protein, which is particularly prone to denaturation. The tetramerization domain of P53 is in fact a dimer of dimers. Two monomers form a dimer mainly by the formation of a β-sheet whereas the interactions which stabilize the tetramer take place between the two α-helices of the dimers[17].

Although most mutations leading to cancer are found in the DNA-binding and regulatory domains a modulation of the tetramerization equilibrium by interaction of a receptor with the tetramerization domain appears to be an interesting goal both from the methodological and therapeutic points of view[18].

Our first approach to attempting the complexation of P53 by **2** was carried out by the synthesis compound **6** in scheme 1. This peptide is a fragment of the tetramerization domain of P53 (residues 341 to 355) which shows the E E E D arrangement of acidic residues. The interaction between **2** and **6** was studied following a strategy parallel to the one followed with peptide **3**.

Although peptide **6** adopts a partially helical secondary structure in solution we have been unable to notice any helicity increase by addition of the tetraguanidinic receptor **2**. CD experiments, however, show that some interaction is taking place between both molecules. The binding constant is nevertheless far lower than the one obtained for compound **3**. Such preliminary results suggest that the interaction between peptide **6** and compound **2**, if any, does not induce any helicity in the peptide.

The main two differences between peptide **3** and **6** in the context of the interaction with **2** are the following: the presence in peptide **6** of two positively charged residues (arginine at position 2 and lysine at position 10) and the preponderance of glutamic acid residues in peptide **6**. The presence of two positively charged residues in the peptide may, by changing the global charge of the peptide from -4 to -2, make the initial stages of complexation difficult. The added conformational freedom due to the presence of an extra CH_2 in the side chain of the acidic residues could make the important changes of conformation detected in other peptides unnecessary for complexation. The relative importance of these two differences is being assessed by the synthesis, structural and interaction studies of several mutants of peptide **3** and **6**. We are also presently carrying interaction experiments with the whole tetramerisation domain of P53.

5. Conclusions

The search and subsequent detailed analysis of the protein surfaces we have carried out has provided us with a very interesting candidate for protein surface recognition. P53 is an excellent target for such studies because of its first glance accessible anionic path and the potential applications of a molecule able to interact with its tetramerization domain.

Although the results we have obtained so far are not showing any interaction between our model peptide **6** and receptor **2** these studies and work which is currently under progress will help improving the design of guanidinium-based protein surface interacting molecules.

6. Acknowledgments

This work was supported by CICYT(PB95-1131), Generalitat de Catalunya (Grup Consolidat (1995GR494) and Centre de Referència en Biotecnologia) and Comisión de Intercambio entre España y Estados Unidos.

7. References

1. Silverman, R.B. (1988) *Mechanism-Based Enzyme Inactivation: Chemistry and Enzymology*, CRC Press, Boca Raton, FL
2. Echavarren, A., Galán, A., Lehn, J.-M. and de Mendoza, J. (1989) Chiral Recognition of Aromatic Carboxylate Anions by an Optically Active Abiotic Receptor Containing a Rigid Guanidinium Binding Subunit, *J. Am.Chem .Soc.* **111**, 4994
3. Sánchez-Quesada, J., Seel, C., Prados, P., de Mendoza, J., Dalcol, I. and Giralt, E. (1996) Anion Helicates: Double Strand Helical Self-Assembly of Chiral Bicyclic Guanidinium Dimers and Tetramers around Sulfate Templates, *J.Am.Chem.Soc.* **118**, 277
4. Peczuh, M. W., Hamilton, A.D., Sánchez-Quesada, J., de Mendoza , J., Haack, T. and Giralt, E. (1997) Recognition and Stabilisation of a α-Helical Peptide by a Synthetic Receptor, *J.Am.Chem.Soc* **119**, 9327
5. Connolly, M.L. (1983) Solvent Accessible Surfaces of Proteins and Nucleic Acids, *Science* **221**, 709
6. Lee, W., Harvey, T.S., Yin, Y., Yau, P., Litchfield, D. and Arrowsmith, C.H. (1994) Solution Structure of the tetrameric minimum transforming domain of p53, *Nature Struct. Biol* **1**, 877
7. Dyda, F., Furey, W., Swaminathan, S., Sax, M., Farrenkopf, B. and Jordan, K. (1993) Catalytic Centers in the Thiamin Diphosphate Dependent Enzyme Pyruvate Decarboxylase at 2.8 A resolution, B*iochemistry* **32**, 6165
8. Sutton, R.B. and Sprang, S.R., PDB entry pdb1ann.ent (unpublished data)
9. Onesti, S., Theoclitou, M.-E., Wittung, E.P.L., Miller, A.D., Plateau, P., Blanquet, S. and Brick, P. (1994) Crystallisation and Preliminary Diffraction Studies of *E. Coli* Lysyl t-RNA Synthetase, *J.Mol.Biol* **243**, 123
10. Bennett, M.J., Schlegel, B.D., Jez, J.M., Penning, T.M. and Lewis, M. (1996) Structure of 3-α-Hydroxysteroid/Dihydrodiol Dehydrogenase Complexed with $NADP^+$, B*iochemistry* **35**, 10702
11. Xie, X., Harrison, D.H., Schlichting, I., Sweet, R.M:, Kalabokis, V.N., Szent-Györgyi, A.G. and Cohen, C. (1994) Structure of the Regulatory Domain of Scallop Myosin at 2.8 A Resolution, *Nature* **368**, 306
12. Eberle, W., Klaus, W., Cesareni, G., Sander, C. and Roesch, P. (1990) Proton Nuclear Magnetic Resonance Assignments and Secondary Structure Determination of the *E.Coli* ROP Protein, *biochemistry* **29**, 7402
13. Nordlund, P., Sjöberg, B.M., and Eklund, H. (1990) Three-dimensional structure of the free radical protein of ribonucleotide reductase, *Nature* **345**, 593
14. Vendrell, J., Wider, G., Avilés, F.X. and Wüthrich, K. (1990) Sequence-specific ^{1}H NMR assignments and determination of the secondary structure for the activation domain isolate from pancreatic procarboxypeptidase B, *Biochemistry* **29**, 7515
15. Lundqvist, T., Rice, J., Hodge, C.N., Basarab, G.S., Pierce, J. and Lindqvist, Y. (1994) Crystal structure of Scytalone dehydratase – a disease determinant of the rice pathogen *Magnaporthe Grisea*, *Structure* **2**, 937
16. Hale, A.J., Smith, C.A., Sutherland, L.C., Stoneman, V.E.A., Longthorne, V.L:, Culhane, A.C. and Williams, G. (1996) Apoptosis: molecular regulation of cell death, *Eur.J.Biochem* **236**, 1
17. Clubb, R.T., Omichinski, J.G., Sakaguchi, K., Apella, E., Gronenborn, A.M. and Clore, G.M. (1995) Backbone Dynamics of the oligomerisation domain of p53 determined from ^{15}N NMR relaxation measurements, *Protein Sci.* **4**, 855

NMR STUDIES OF PROTEIN-LIGAND AND PROTEIN-PROTEIN INTERACTIONS INVOLVING PROTEINS OF THERAPEUTIC INTEREST

J. FEENEY
National Institute for Medical Research, Mill Hill, London, NW7 1AA

1. Introduction

Currently there is great interest in trying to understand the molecular recognition processes involved in protein-ligand and protein-protein interactions. Such interactions are of crucial importance in many areas of biology including enzyme catalysis and regulation, the control of gene expression and in drug-receptor interactions. In all cases the specificity of the binding is central to the biological function. Much of our recent work has been aimed at trying to understand the molecular basis for binding specificity in drug-receptor complexes where the findings can have practical significance by providing the basis for rational structure-based drug design. The NMR method is well-suited to studies of such systems in solution and we have been applying NMR and other spectroscopic techniques, together with molecular modelling and biochemical approaches, to characterise the structures and ligand interactions in several complexes. Our specific aims are firstly to determine the solution structures of specific proteins and their complexes and to detect and characterise any mixtures of conformations and, secondly, to investigate the specificity of binding by identifying and characterising individual interactions between the ligand and protein and measuring the rates of dynamic processes within the complexes.

Some of the complexes studied are sufficiently small (less than 35 kDa) to allow detailed structural work to be carried out in solution. For example, we have determined the structures of several complexes of *Lacotobacillus casei* dihydrofolate reductase (162 residues) with antifolate drugs and used NMR measurements to characterise specific interactions, conformational equilibria and dynamic processes within the complexes. However, many other proteins of therapeutic interest form complexes that are too large for complete structural determination by NMR. In such cases an alternative approach is to examine smaller domains of the proteins which have retained their structural and functional properties. In some cases, studies of complexes formed by functional domains of large proteins can also provide useful information and we have used this approach to examine interactions involving matrix metalloproteinases and their inhibitors (for example tissue inhibitors of metalloproteinases, TIMPS) We have carried out a structural determinations on a trucated form of one of these, Δ-TIMP-2, and defined its interaction surface in the complex formed with a 19 kDa catalytic domain from stromelysin. There are many other proteins for which receptors have not yet been isolated. In these cases structural information can be obtained for the known partner and used, in combination with data from mutagenesis and functional studies, to probe its binding sites for a putative receptor. Such approaches have been used in our structural studies of trefoil proteins such as PSP (pancreatic spasmolytic polypeptide) and pNR-2/pS2 (a breast cancer associated trefoil protein)

2. Dihydrofolate Reductase
Collaborators:- B. Birdsall, V.I. Polshakov, A.R. Gargaro, W.D. Morgan, P. M. Nieto, G. Martorell, A.Soteriou, M.D. Carr, T.A.Frenkiel, M. J.Gradwell, H.T.A. Cheung, J.E. McCormick, G.C.K. Roberts and J. Feeney.

Dihydrofolate reductase (DHFR) is an essential enzyme in all cells and is responsible for catalysing the NADPH-linked reduction of dihydrofolate to tetrahydrofolate [1]. The enzyme is of fundamental pharmacological importance being the "target" for an important group of drugs including antineoplastic (methotrexate), antibacterial (trimethoprim) and antimalarial (pyrimethamine) agents. These drugs act by inhibiting the enzyme in malignant or parasitic cells. An improved understanding of the specificity of ligand binding could assist in the design of improved antifolate drugs. The small size of the protein makes it amenable to detailed structural study [2-20]. Many X-ray [2-9] and NMR [10-15] structural studies have been reported on complexes formed with substrates and inhibitors and several reviews containing structural information have been published [1,16-20]. Early NMR studies provided a great deal of information about specific interactions, multiple conformations and dynamic processes in various complexes of dihydrofolate reductase in solution [17-20]. However, it is only relatively recently that it has become possible to use NMR data to obtain complete three-dimensional structures of complexes of the enzyme in solution and to start addressing detailed questions relating to the specificity of ligand binding.

Methotrexate (1)

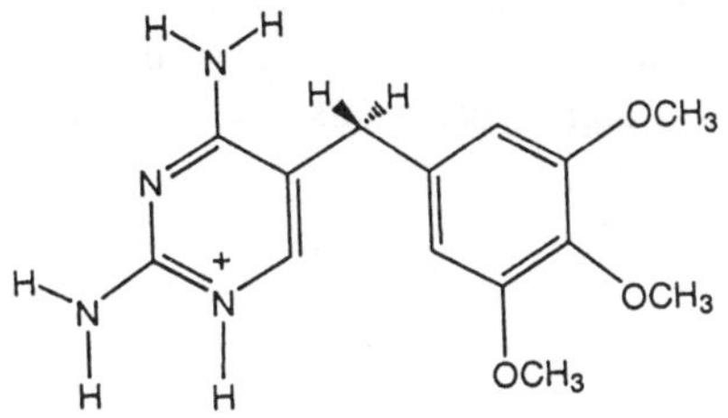

Trimethoprim (2)

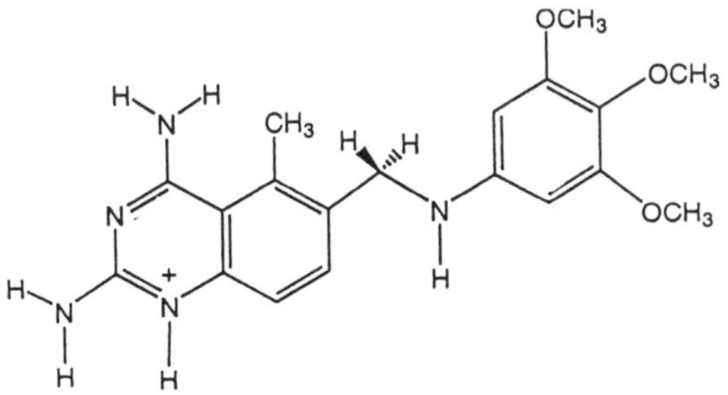

Trimetrexate (3)

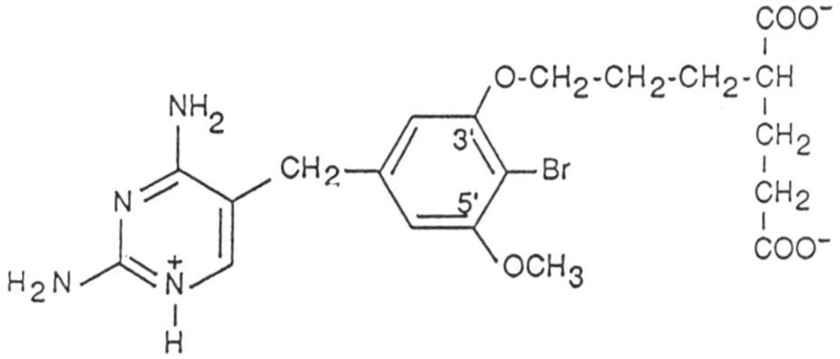

Brodimoprim-4,6-dicarboxylic acid (4)

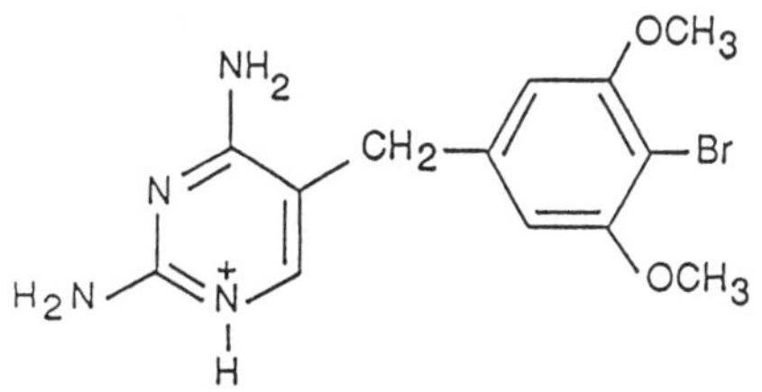

Brodimoprim (5)

284

We have examined complexes of substrates and inhibitors with the enzyme from *L casei* (Mr 18,300) and in several cases have determined the three-dimensional structures based on distance and torsion angle constraints determined from NMR data [10-13]. We have made extensive use of the program AngleSearch [21] which was developed to provide stereospecific assignments and torsion angle constraints from NOE and coupling constant related NMR data. Essentially complete NMR assignments for uniformly ^{15}N and ^{15}N/^{13}C labelled *L. casei* dihydrofolate reductase in complexes formed with the inhibitors methotrexate (**1**), trimethoprim (**2**), trimetrexate (**3**) and brodimoprim 4,6-dicarboxylate (**4**) have been obtained [10-13]. Several ^{13}C and ^{15}N analogues of methotrexate and trimethoprim, synthesised by our collaborators, were used in these studies to assist in obtaining isotope-edited or isotope-filtered spectra for defining protein-ligand NOEs. The structures of all of the complexes have been determined using distance and torsion angle constraints obtained from the NMR data in conjunction with either conventional simulated annealing molecular dynamics calculations of the full structure or by using various docking methodologies. Our highest quality structure to date is that of the binary complex of *L. casei* DHFR formed with the anticancer drug methotrexate [12]. The three-dimensional solution structure of this complex was determined using 2531 distance, 361 dihedral angle and 48 hydrogen bond restraints obtained from analysis of multidimensional NMR spectra. Simulated annealing calculations produced a family of 21 structures fully consistent with the constraints (see Figure 1) The structure has four α-helices and eight β-strands with two other regions, comprising residues 11-14 and 126-127, also interacting with each other in a β-sheet manner. High quality structures of the ligand in its binding site require a large number of protein-ligand distance constraints (see Figure 2) Earlier simulation studies [22] using NMR type distance constraints (with a strong, medium, weak classification) extracted from a high quality crystal structure of folate bound to human DHFR [4] had indicated

Figure 1. Stereoview of a superposition over the backbone atoms of residues 1-162 of the final 21 structures of the *L casei* reductase.methotrexate complex. Reproduced with permission from Gargaro *et al.*, [12].

that molecules of this type would require at least 50 ligand-protein distance constraints to produce acceptable structures. The methotrexate binding site in the solution structure of the methotrexate.DHFR complex is very well-defined and the structure around its glutamate moiety was improved by including restraints reflecting the previously determined specific interactions between the glutamate α-carboxylate with Arg 57 and the γ-carboxylate with His 28 [23,24]. The overall fold of the binary complex in solution is very similar to that observed in the X-ray studies of the ternary complex of *L. casei* DHFR formed with methotrexate and NADPH [2] (the structures of the binary and ternary complexes have an r.m.s. difference over the backbone atoms of 0.97 Å (see Figure 3)) In general terms, the NADPH binding site appears to be essentially pre-formed in the binary complex. This may contribute to the tighter binding of coenzyme in the presence of methotrexate. Thus no major conformational change takes place when

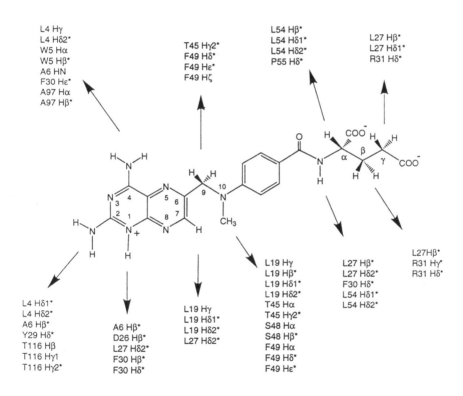

Figure 2. Protein-ligand intermolecular NOEs in the DHFR.methotrexate complex. Reproduced with permission from Gargaro *et al.*, [12].

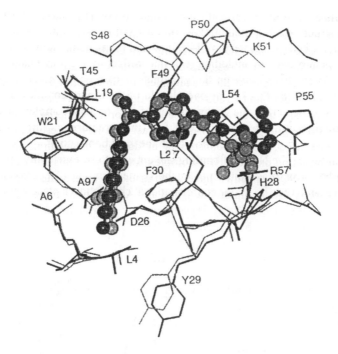

Figure 3. A comparison of the methotrexate binding site in the NMR determined solution structure (dark shading) in the *L.casei* DHFR.methotrexate structure with that in the crystal structure (light shading) of the ternary complex DHFR.methotrexate.NADPH (Bolin *et al.*, [2]).
Reproduced with permission from Gargaro *et al.*, [12].

NADPH binds to the binary complex. In the binary complex, the loop comprising residues 9-23 which forms part of the active site has been shown to be in the 'closed' conformation as defined by Sawaya and Kraut [3]. These workers considered the corresponding loops in crystal structures of various complexes of DHFRs from several organisms and classified the different structural possibilities for the loop as 'closed', 'open' or 'occluded'. They suggested that an 'occluded' loop is formed in complexes lacking the coenzyme. Our findings of a 'closed' conformation for the loop in the binary complex with methotrexate indicates that the absence of the NADPH does not necessarily result in the 'occluded' form of the loop as seen in crystal studies of complexes of other DHFRs examined in the absence of coenzyme.

In addition to the conventional structural determinations carried out using simulated annealing calculations in conjunction with distance and torsion angle constraints, other approaches have been explored including various methods for docking ligands into the protein [10,11]. Methods for validating the results of docking calculations have also been considered [22]. We have obtained the structure of the DHFR.trimetrexate complex by using both docking methods and by conventional simulated annealing approaches [26]: the agreement between the structures obtained from the two approaches provided further validation of the docking protocols used in these studies [27].

In an earlier collaboration with colleagues from Hoffman-la-Roche (I. Kompis), we examined the binding of some rationally designed trimethoprim analogues which have an extra side-chain containing two carboxylate groups aimed at interacting with two basic conserved residues on the protein that are not involved in trimethoprim binding [23,24]. One of these analogues, bodimoprim-4,6-dicarboxylate (Structure 4), was found to bind up to 1000-fold more tightly to DHFR than the parent molecule (Brodimoprim (5))while retaining its specificity for the bacterial enzyme and [24]. One of its carboxylate groups was shown to interact with His 28 perturbing its pK by one unit. Later, we used distance constraints in docking calculations to obtain detailed structural information on the DHFR complex formed with this analogue [11]. More recently, $^1H/^{15}N$ HSQC spectra have been used to investigate interactions between the second carboxylate group on the ligand and Arg 57 in the protein [11] (see Section 2.1) Similar studies using X-ray crystallography to monitor specific ligand interactions with Arg 57 have also been reported [28].

2.1. SPECIFIC INTERACTIONS IN DHFR-INHIBITOR COMPLEXES

In earlier studies, several specific contributions to the overall binding were identified by combining NMR and binding data for a series of complexes formed with structurally related ligands [29-31]. For example, the N1 position of trimethoprim was shown to be protonated in its DHFR complex and it was demonstrated that its interaction with the conserved Asp 26 residue increases the pK of the N1 protonation by more than 2 units [29,30]. Another study showed that the γ-carboxlate groups of folate or methotrexate interact with His 28 and increase its pK value by 1 unit [23]. More recently, studies of $^1H/^{15}N$ signals from arginine residues have proved to be particularly informative for monitoring protein ligand interactions [32,33]. For example, in $^1H/^{15}N$ HSQC spectra of the DHFR.methotrexate complex, it was possible to detect the interaction of the methotrexate α-carboxylate group with the guanidino group of Arg 57. In this case, hindered rotation about the N^ε-C^ζ and C^ζ-N^η bonds of the guanidino group of Arg 57 could be observed by detecting four separate signals for the NH^η nuclei in HSQC spectra. All four signals could be specifically assigned to their particular NH^η protons in Arg 57 (see Figure 4) The large downfield chemical shifts of two of these signals indicate that two of the NH^η protons ($NH^{\eta 12}$ and $NH^{\eta 22}$) are forming strong hydrogen bonds with the charged oxygen atoms of the α-carboxylate group of the glutamate moiety of methotrexate in a symmetrical end-on interaction [32] (see Figure 4) These effects involving Arg 57 were not seen for the complex formed with trimethoprim, which does not contain any carboxylate groups. The HSQC spectrum showing the four resolved NH^η signals for Arg 57 was recorded at 1^0C (see Figure 4) and increasing the temperature causes exchange line-broadening and coalescence of signals [33]. Rotation rates for the $N^\varepsilon C^\zeta$ and $C^\zeta N^\eta$ bonds can be calculated from lineshape analysis and, for the $N^\varepsilon C^\zeta$ bond, from zz-HSQC exchange experiments. The interactions between the methotrexate α-carboxylate group and the Arg 57 guanidino group decrease the rotation rates for the $N^\varepsilon C^\zeta$ bond by about a factor of 10 and those for the $C^\zeta N^\eta$ bonds by more than a factor of 100 with respect to their

288

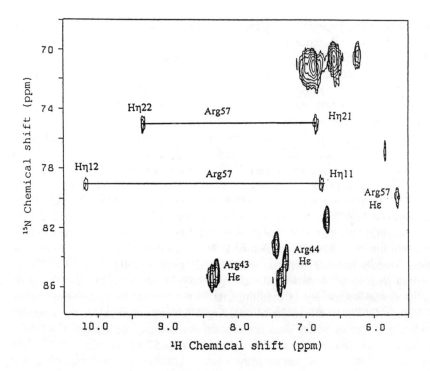

Figure 4. The ^{1}H-^{15}N HSQC spectrum of the Arg 57 NH$^\eta$ protons from the DHFR.Methotrexate complex. The chemical shifts indicate a symmetrical end-on interaction of the α-carboxylate oxygen atoms of methotrexate with the guanidino group of Arg 57
Reproduced with permission from Gargaro *et al.*, [32] and Nieto *et al.*, [33].

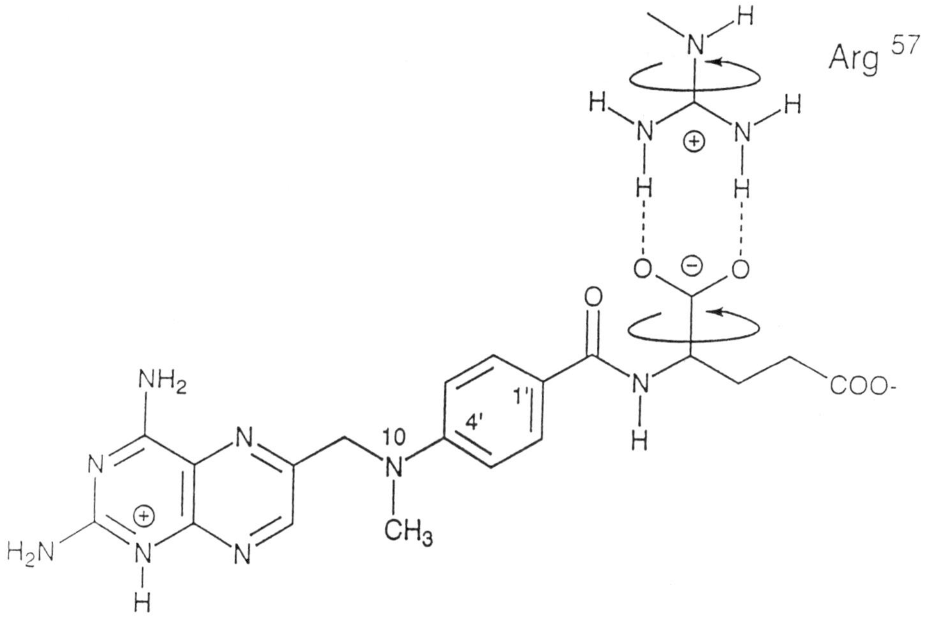

Figure 5. Correlated rotations about the $N^\varepsilon C^\zeta$ bond of the Arg 57 guanidino group and the $C'C^\alpha$ bond of the glutamate α-carboxylate group of methotrexate.
Reproduced with permission from Nieto et al., [33].

values in free arginine. Interestingly, the relative rates of rotation about these two bonds are reversed in the protein complexes compared with their values in free arginine. The simplest explanation for this behaviour is that there are correlated rotations about the $N^\varepsilon C^\zeta$ bond of the Arg 57 guanidino group and the $C'C^\alpha$ bond of the glutamate α-carboxylate group of methotrexate (See Figure 5) Such correlated rotations would provide a mechanism for rotation about the $N^\varepsilon C^\zeta$ bond without breaking the interactions of the NH^η protons with the carboxylate oxygen atoms which would then preferentially retard the rotations about the $C^\zeta N^\eta$ bonds [33].

In the DHFR complex formed with both methotrexate and NADPH present, Arg 43 shows a large downfield chemical shift for its NH^ϵ proton and also a retardation of its rate of exchange with water [32]. This pattern of deshielding contrasts with that seen for Arg 57 and is that expected for a side-on interaction of the guanidinium NH^ϵ proton with the charged oxygen atoms of the ribose 2'-phosphate group of NADPH. The 2'-phosphate group has been shown to bind in its dianionic form from NMR studies both in solution [34] and in the solid state [35].

Structure determinations of the DHFR complex with trimethoprim [10,36] indicated that Asp 26 is the only negatively charged residue within 5 Å of the protonated N1 position of trimethoprim. The Asp 26 side chain carboxylate oxygens were observed within 3 to 4 Å of the N1 proton in all the calculated structures but the calculations do not indicate the pattern of hydrogen bonding in the complex. We have used PM3 SCF calculations to further characterise the interactions between the protonated N1 position of trimethoprim and the carboxylate group of Asp 26 [36] and used the results to refine the structure of the DHFR.trimethoprim complex.

2.2. MULTIPLE CONFORMATIONS IN DHFR.FOLATE COMPLEXES

NMR has proved to be very useful for detecting multiple conformations in solutions of substrate and inhibitor complexes formed with DHFR. In many cases separate spectra have been detected for two or more co-existing conformations and NMR could also characterise the pH and temperature dependence of the equilibria. Examples include the binary complexes with folate [37,38] and pyrimethamine analogues [39] and ternary complexes with trimethoprim and $NADP^+$ [40] and with folate and $NADP^+$ [41-44]. In several cases the flexible ligands are occupying essentially the same binding site but having different conformations. The DHFR.folate.$NADP^+$ complex was shown to exist in three conformations (Form I (low pH form) and forms IIa and IIb (high pH forms)) in slow exchange on the NMR time scale [41,44]. More recently we have used ^{13}C labelled folates to demonstrate that multiple conformations exist also in the binary DHFR.folate complex [37,38]. Intermolecular NOEs, some obtained from ^{13}C isotope editing experiments, show that the pteridine ring in Form IIb has the orientation expected for the 'active' folate complex while Forms I and IIa have the pteridine ring turned over by 180°. Detailed ^{13}C studies of [2,4a,7,9-^{13}C]- and [4,6,8a-^{13}C]-folates bound to DHFR indicated that the active folate conformation (Form IIb) binds in the keto form while the methotrexate-like forms I and IIa bind in the enolic form with N1 protonated [44]. NMR studies of DHFR.folate.$NADP^+$ complexes using the Asp26Asn variant and Asp ^{13}C-γ-CO_2^- labelled DHFR indicate that while the Asp 26 residue is implicated in the pH dependence of the equilibrium it operates indirectly by influencing the pK of the protonation at the 4-position of bound folate [37]. The frequent occurrence of multiple conformations in protein ligand complexes in solution (seen in more than 25 DHFR complexes) is an important finding with implications not only for inhibitor design but also for understanding complex structure-function relationships. Each different conformation provides a new starting point for exploring drug design.

Complexes formed from antifolate drugs binding to dihydrofolate reductase provide excellent systems for carrying out detailed ligand binding studies. In addition to its intrinsic pharmacological interest, DHFR has proved to be an excellent test-bed for exploring ideas and methods of tackling the problem of binding specificity and some of the findings will have general applicability (for example, methodology for characterising specific protein ligand interactions involving charged residues, such as arginine, histidine and aspartic acid, and for detecting multiple conformations)

3. Tissue Inhibitor of Metalloproteinases (TIMP-2)
Collaborators:- R.A. Williamson, M.D.Carr, R.B. Freedman, F.W. Muskett, T.A. Frenkiel, A.J.P. Docherty, G. Murphy and J. Feeney.

The degradation the extracellular matrix in connective tissues is promoted by several metalloproteinases (collagenase, gelatinase, stromelysin) which are inhibited by a family of protein inhibitors called TIMPs (tissue inhibitor of metalloproteinases) Uncontrolled activity of such metalloproteinases leads to tissue destruction which can have important implications for a range of diseases including arthritis and cancer. Metalloproteinases play an important role in cancer metastasis by breaking down matrix proteins of blood vessels thus allowing penetration of invasive tumour cells through the basement membrane. The TIMPs act as natural suppressors of metastasis [45] and there is great interest in finding drug molecules which would act in the same way. Some inhibitors of metalloproteinases are currently being tested in clinical trials as agents to slow down the spread of cancer cells in metastasis. The availability of detailed structural information for the TIMPs and their complexes with proteinases could assist in the design of drugs with comparable inhibitory properties for use as such agents.

We have used NMR methods to obtain the 3-dimensional structure of a truncated form of one of the human TIMPs (TIMP-2) which retains most of its inhibitory activity [46]. This truncated protein, Δ-TIMP-2 (1-127 TIMP-2; Mr 14 kDa) was prepared initially by expressing the active recombinant protein in mammalian cells in culture [47]. It contains 3 disulphide bonds and is unglycosylated (unlike the related inhibitor TIMP-1) By using homonuclear 2D and 3D ^{1}H NMR experiments we obtained essentially all the signal assignments for Δ-TIMP-2 and extracted the NMR-based distance and torsion angle constraints required to determine the structure of this active domain (see Figure 6) [46]. The protein was found to contain a five-stranded antiparallel β-sheet rolled over on itself to form a closed β-barrel with two short helices packed close to one another on the same face of the barrel. This β-barrel topology is homologous with that of proteins of an oligosaccharide/oligonucleotide binding (OB) family. This structure was of modest resolution since it was determined using data obtained only from ^{1}H spectra but subsequently a higher resolution structure was obtained using ^{13}C/^{15}N labelled protein

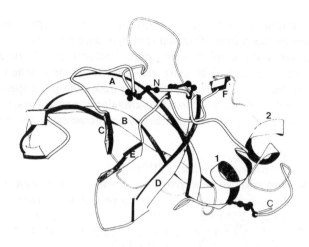

Figure 6. Ribbon diagram of Δ–TIMP
Reproduced with permission from Williamson *et al.*, [46].

expressed in *E. coli.* [48,49]. The structure revealed conserved surface regions which could be of potential functional importance. The structure is being used (i) to guide site-directed mutagenesis studies aimed at detecting the interaction sites on Δ-TIMP-2 which are important for its inhibitor action and (ii) to interpret the functional significance of available mutagenesis data. The binding face of the Δ-TIMP-2 molecule in its complex with a 19 kDa catalytic domain of stromelysin has been mapped by monitoring the $^1H/^{15}N$ chemical shift perturbations which accompany the complex formation [50]. Recently, X-ray data have been published for the complex of a related inhibitor TIMP-1 and the active domain of stromelysin. A comparison of this data with the structure of Δ-TIMP-2 in solution indicates that there is no major change in conformation of the inhibitor on binding. The data also show that the mapped binding surface of Δ-TIMP-2 is fully consistent with that determined in the X-ray structure study.

4. Trefoil Proteins
Collaborators:- M. A. Williams, V.I. Polshakov, M.D. Carr, F. E. May, B.R. Westley and J.Feeney.

Members of a family of "trefoil" domain proteins [51] have been implicated in several processes of medical importance such as in the repair of damaged endodermal tissue [52] possibly by controlling cell-migration [53-55]. Although the functions of the trefoil peptides are not yet known, early studies suggested they were involved in controlling intestinal smooth muscle contraction and gastric acid secretion [56]. Possible roles as growth factors have also been suggested based on their apparent growth stimulating effects in certain cell lines [57]. Understanding the biological function of such trefoil proteins, particularly their involvement in wound healing and in cancer, is the focus of major research efforts in several laboratories. Even in the absence of receptor molecules some useful information about the putative binding sites for the receptor can be deduced from structural studies of such proteins.

These proteins consist of one or more 'trefoil' or 'P' domains each comprising ~ 40 amino acids with well conserved sequence features. The most highly conserved features are six cysteine residues with essentially conserved spacings and an arginine residue located between the first two cysteines: each domain contains 3 disulphide bonds resulting in the three-loop structure characteristic of the trefoil motif.

4.1. PSP AND pNR-2/pS2

The first two trefoil proteins to be discovered were (i) pancreatic spasmolytic polypeptide (PSP) a 106 residue protein [51] composed of two highly homologous trefoil domains (with nearly 50% residue identity) and (ii) the 60 residue single domain protein pNR-2/pS2, a breast cancer associated protein [55]. We have used multidimensional NMR to determine the three-dimensional structures of both of these trefoil proteins [58-60]. Such information could help in elucidating the function of the proteins by identifying conserved structural elements which may be involved in mediating their biological effects. This structural information is also being used to guide mutagenesis studies aimed at mapping regions of the structure responsible for binding to putative cell-surface receptors and eliciting their biological response.

4.2. PSP (PANCREATIC SPASMOLYTIC PEPTIDE)

Carr and co-workers [58] have determined the solution structure of porcine pancreatic spasmolytic peptide, PSP, supplied by Dr L. Thim (Nova, Denmark). Its structure was concurrently determined using X-ray crystallography [61,62] and in one of these studies [61] the NMR structural information assisted in the analysis of the X-ray diffraction data. The two domains of PSP have a similar tertiary structure each containing a two stranded antiparallel β-sheet and a short α-helix. The solution structure was determined using only ^{1}H NMR (isotopically-labelled protein was not available) and its successful determination was a significant achievement for a protein of this size. Several conserved residues were found to be located on a single face of each domain forming a hydrophobic cleft with the potential for ligand/receptor binding [62]. The N- and C-termini of PSP are linked by an interdomain disulphide bond, which helps to fix the relative orientation of the two domains. The conserved hydrophobic residues on the surfaces of the two domains are found on opposite faces of the PSP structure which suggests that the molecule could be an adapter-type molecule used to bring together its binding ligands.

4.3. pNR-2/pS2 BREAST CANCER ASSOCIATED TREFOIL PEPTIDE

We recently used NMR to determine the solution structure of the human trefoil protein, pNR-2/pS2. This protein is expressed at low levels in normal breast epithelium and is also found at high levels in breast carcinoma, where it is a marker of hormonal responsiveness. It is also associated with a variety of other tumours. We initially used

NMR ^{1}H relaxation measurements to show that the protein is monomeric in solution [59]. This was later confirmed by using hydrodynamic measurements [63]. Subsequently, data from multidimensional NMR measurements on the unlabelled and ^{15}N-labelled recombinant pNR-2/pS2 Ser58 protein was used to determine its high resolution three-dimensional structure including the pattern of disulphide bonding (Figure 7) [60].

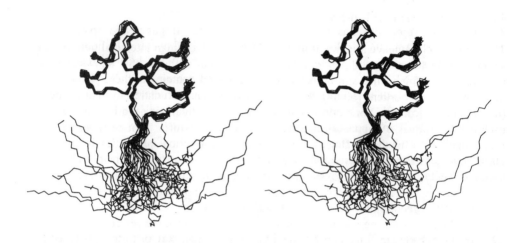

Figure 7. Stereoview of the superimposed backbone atoms of the trefoil protein pS2 for the final family of 17 structures. The superimpositions were made onto the backbone atoms of the core residues (residues 7-47). The conformation of the C-terminus is not well defined.
Reproduced with permission from Polshakov *et al.*, [60].

A large part of pNR-2/pS2 (residues 3-52) including the trefoil domain has an ordered structure in solution. The three disulphide bonds in the trefoil domain (7-47) form three sequential loops, which are packed together with the third loop sandwiched between the first and second. There is only a small amount of secondary structure, consisting of an α-helix (23-31) packed against a two stranded antiparallel β-sheet (33-35, 43-45) Three turns were characterised (residues (10-14), (19-22) and (38-42)). Outside the trefoil domain core, the N and C terminal strands are closely associated, forming an extended 'tail', which has some β-sheet type interactions between residues in strands 3-7 and 47-51, and residues in the strand become more disordered towards the termini: these tails are not present in PSP [60].

4.4. COMPARISON WITH THE PSP STRUCTURE

The pNR-2/pS2 structure has been compared with that of the two trefoil domains of PSP for which detailed structural information is also available [59-62]. A high degree of sequence conservation is seen between these mammalian trefoil peptides with 17 residues being conserved in all three of these domains, with 25 conserved between pNR-2/pS2 and domain-1 of PSP, 21 conserved between pNR-2/pS2 and domain-2 of PSP, and several other residues being conservatively mutated. Superposition of the structure of the trefoil domain of the pNR-2/pS2 onto the crystal structures of both domains of PSP [62] reveals very similar folds for all three domains. The antiparallel β-sheet and the α-helice observed for pNR-2/pS2 were also found in both domains of PSP. The overall conformation of the trefoil domain in pNR-2/pS2 was found to be more similar to that of domain-2 rather than to domain-1 of PSP.

4.5. CONSERVED FEATURES OF THE TREFOIL PEPTIDES

The amino acid sequence alignment of pNR-2/pS2, hITF (human intestinal trefoil factor, another single domain trefoil peptide), PSP and its human analogue hSP shows that there is a great deal of sequence similarity between the trefoil domains in these proteins [60]. Williams and co-workers [60,64] have examined the characteristic physical properties (hydrophobic, hydrogen bond donor and/or acceptor, positively or negatively charged) of each of the residues of the trefoil motif in all mammalian trefoil peptide sequences, and found that at 24 positions at least one characteristic physical properties is absolutely conserved. The conservation suggests that these residues probably have an essential structural and/or functional role. In particular, the degree of conservation of the surface features of the proteins is likely to be reflecting the importance of these features in ligand/receptor recognition processes. There is a contiguous hydrophobic patch of conserved residues on the surface of the protein comprising three residues (Pro 20, Pro 42, Trp 43) which are fully conserved in the known mammalian trefoil peptides, and another residue (Phe 19), which is conservatively replaced by Tyr in some domains (Figure 8) It is thought that this cluster of residues forms part of a ligand binding site [60,62]. Several of the surface residues around this patch have features which are conserved (i.e. hydrophobic residues are always present at the positions corresponding to Pro 11, Val 38 and Val 41 of pNR-2/pS2; a hydrogen bond donor is present at the position of Arg 12; hydrogen bond acceptors always replace Asn 16, Thr 23, Gln 26 and Asp 36). It has been suggested that this high degree of conservation of the physical properties of this region of the protein surface strongly suggests that the different human trefoil domains will bind similar target molecules in this region.

Gajhede and co-workers [62] have suggested that the cleft between the two loops of PSP containing the conserved hydrophobic patch, could provide a possible binding site for an oligosaccharide chain of the type found in mucin proteins. The cleft in pNR-2/pS2 is narrower than in PSP (6 Å for pNR-2/pS2, 10 Å for domain-1 and 8 Å for domain-2 of PSP). These differences could contribute to differences in binding specificities for these binding sites.

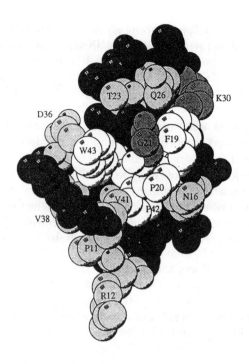

Figure 8. A CPK-type model of pNR-2/pS2 showing the surface residues on one face of the protein. The conserved residues Phe 19, Pro 20, Pro 42 and Trp 43 have been proposed as forming a ligand/receptor binding site are these residues are indicated in white. The physical characteristics of the flanking residues indicated in light grey are preserved in all mammalian trefoil peptides. Reproduced with permission from Polshakov *et al.*, [60].

4.6. IMPLICATIONS OF THE PNR-2/PS2 STRUCTURE FOR DIMER FORMATION

There is a seventh Cys residue at position 58 of pNR-2/pS2 which facilitates dimerisation of the native protein through formation of a disulphide bond. There is some functional evidence indicating that the dimeric form of the protein could be important. In our earlier studies we examined the variant of pNR-2/pS2 with Cys 58 replaced by serine and showed that it exists as a monomer in solution. Thus, there is no strongly interacting interface which would aid the dimerisation. The spasmolytic peptide PSP exists as a pair of covalently linked trefoil domains each having very similar structures to the trefoil domains of pNR-2/pS2 and this has encouraged speculation that a pNR-2/pS2 homodimer could adopt an overall structure similar to that of PSP [65]. Recently, Williams and co-workers examined the the solution structure of the pNR-2/pS2 dimer and found that the two trefoil domains have identical structures to the monomer: they are joined by a flexible linkage and there is no evidence for contact between the two domains [64]. Thus the spatial arrangement of its domains is different from those in PSP.

5. Conclusions

The NMR studies reported here indicate that detailed information about specific interactions, molecular conformations and multiple conformations, and dynamic processes can be obtained for protein-ligand complexes in solution. The methods would probably be appropriate for complexes of up to 35 kDa. For larger complexes monitoring the chemical shift changes which accompany complex formation can help in mapping the binding interface. In cases where only one partner in the interaction is known, structural studies can sometimes detect surface residues that are conserved across a family of related proteins. Such conserved surface regions can indicate likely ligand binding sites.

6. References

1. Blakley, R.L. (1985) Dihydrofolate Reductase, in Folates and Pterins (Blakley, R.L. and Benkovic, S.J., Eds.) Vol. 1, Chapt. 5, pp 191-253, J. Wiley, New York.
2. Bolin, J.T., Filman, D.J., Matthews, D.A., Hamlin, R.C. and Kraut, J. (1982) Crystal structures of *E.coli* and *L.casei* dihydrofolate reductase refined to 1.7 Å resolution. I. General features and binding of methotrexate. *J.Biol.Chem.* **257**, 13650-13662.
3. Sawaya, M.R. and Kraut, J. (1997) Loop and subdomain movements in the mechanism of *E. coli* dihydrofolate reductase: Crystallographic evidence. *Biochemistry* **36**, 586-603.
4. Oefner, C., D'arcy, A. and Winkler, F.K. (1988) Crystal structure of human dihydrofolate reductase complexed with folate. *Eur. J. Biochem.* **174**, 377-385.
5. Bystroff, C. and Kraut, J. (1991) Crystal structure of unliganded *Escherichia-coli* dihydrofolate reductase ligand induced conformational changes and cooperativity in binding. *Biochemistry* **30**, 2227-2239.
6. Bystroff, C., Oatley, S.J. and Kraut, J. (1990) Crystal structures of *Escherichia coli* dihydrofolate reductase - The $NADP^+$ holoenzyme and the folate- $NADP^+$ ternary complex - substrate binding and a model for the transition-state. *Biochemistry* **29**, 3263-3277.
7. Filman, D.J., Bolin, J.T., Matthews, D.A. and Kraut, J. (1982) Crystal structures of *Escherichia coli* and *Lactobacillus casei* dihydrofolate-reductase refined at 1.7 Å resolution. 2. Environment of bound NADPH and implications for catalysis. *J. Biol. Chem.* **257**, 13663-13672.
8. Matthews, D.A., Bolin, J.T., Burridge, J.M., Filman, D.J., Volz, K.W. and Kraut, J. (1985) Dihydrofolate reductase - the stereochemistry of inhibitor selectivity. *J. Biol. Chem.* **260**, 392-399.
9. Matthews, D.A., Bolin, J.T., Burridge, J.M., Filman, D.J., Volz, K.W., Kaufman, B.T., Beddell, C.R., Champness, J. N., Stammers, D.K., and Kraut, J. (1985a) Refined crystal structures of *Escherichia coli* and chicken liver dihydrofolate reductase containing bound trimethoprim. *J.Biol.Chem.* **260**, 381-391.
10. Martorell, G., Gradwell, M.J., Birdsall, B., Bauer, C.J., Frenkiel, T.A., Cheung, H.T.A., Polshakov, V.I., Kuyper, L. and Feeney, J. (1994) Solution structure of bound trimethoprim in its complex with *Lactobacillus casei* dihydrofolate reductase. *Biochemistry* **33**, 12416-12426.
11. Morgan, W.D., Birdsall, B., Polshakov, V.I., Sali, D., Kompis, I. and Feeney, J. (1995) Solution structure of a brodimoprim analogue in its complex with *Lactobacillus casei* dihydrofolate reductase. *Biochemistry* **34**, 11690-11702.
12. Gargaro, A.R., Soteriou, A., Frenkiel, T.A, Bauer, C.J., Birdsall, B., Polshakov, V.I., Barsukov, I.L., Roberts, G.C.K. and Feeney, J. (1998) The Solution Structure of the Complex of *Lactobacillus casei* Dihydrofolate Reductase with Methotrexate. *J. Mol. Biol.* **277**, 119-134.

13. Soteriou, A., Carr, M.D., Frenkiel, T.A., McCormick, J.E., Bauer, C.J., Sali, D., Birdsall, B. and Feeney, J. (1993) 3D ^{13}C/^{1}H NMR-based assignments for side-chain resonances of *Lactobacillus casei* dihydrofolate reductase. Evidence for similarities between the solution and crystal structures of the enzyme. *J. Biomolec. NMR* **3**, 535-546.
14. Falzone, C.J., Cavanaugh, J., Cowart, M., Palmer, A.G., III, Matthews, C.R., Benkovic, S.J. and Wright, P.E. (1994) ^{1}H, ^{15}N and ^{13}C resonance assignments, secondary structure, and the conformation of substrate in the binary folate complex of *Escherichia coli* dihydrofolate reductase. *J. Biomol. NMR*. **4**, 349-366.
15. Johnson J.M., Meiering E.M., Wright J.E., Pardo J., Rosowsky A., Wagner G. (1997) NMR solution structure of the anti tumour compound PT523 and NADPH in the ternary complex with human dihydrofolate reductase. *Biochemistry* **36**, 4399-4411.
16. Freisheim, J.H. and Matthews, D.A. (1984) In Sirotnak, F. M., Burchill, J. J., Ensminger, W. D. and Montgomery, J.A. (eds), Folate Antagonists as Therapeutic Agents. Vol. 1, pp 69-131, Academic Press Inc, New York.
17. Feeney, J. (1990) NMR studies of interactions of ligand with dihydrofolate reductase. *Biochem. Pharm.* **40**, 141-152.
18. Feeney, J. (1996) NMR studies of ligand binding to dihydrofolate reductase and their application in drug design. In: NMR in drug design. Edited by Craik, D. CRC Press.
19. Feeney, J. (1993) Isotope-aided NMR studies of protein-ligand interactions. In: Organic reactivity: physical and biological aspects. Edited by Golding, B.T. and Griffin, R.J., Royal Soc. Chem., pp. 161-184.
20. Feeney, J. and Birdsall, B. (1993) NMR studies of protein-ligand interactions. In: NMR of biological macromolecules: a practical approach. Edited by Roberts, G.C.K. Oxford University Press, pp. 183-215.
21. Polshakov, V.I., Frenkiel, T.A., Birdsall, B., Soteriou, A. and Feeney, J. (1995) Determination of stereospecific assignments, torsion-angle constraints, and rotamer populations in proteins using the program AngleSearch. *J. Magn. Reson. Series* X **108**, 31-43.
22. Gradwell, M.J. and Feeney, J. (1996) Validation of the use of intermolecular NOE constraints for obtaining docked structures of protein-ligand complexes. *J. Biomolec. NMR* **7**, 48-58.
23. Antonjuk, D.J., Birdsall, B., Burgen, A.S.V. Cheung, H.T.A., Clore, G.M., Feeney, J., Gronenborn, A., Roberts, G.C.K and Tran, W. (1984) A ^{1}H NMR study of the role of the glutamate moiety in the binding of methotrexate to dihydrofolate reductase. *Brit. J. Pharmacol.* **81**, 309-315.
24. Birdsall, B., Feeney, J., Pascual, C., Roberts, G.C.K., Kompis, I., Then, R.L., Muller, K. and Kroehn, A. (1984) A ^{1}H study of the interactions and conformations of rationally designed brodimoprim analogues in complexes with *Lactobacillus casei* dihydrofolate reductase. *J. Med. Chem.* **23**, 1672-1676.
25. Birdsall, B., Cassarotto, M.G., Cheung, H.T.A., Basran, J., Roberts, G.C.K. and Feeney, J. (1997) The influence of aspartate 26 on the tautomeric forms of folate bound to *Lactobacillus casei* dihydrofolate reductase. *FEBS Lett.* **402**, 157-161.
26. Brünger, A.T. (1992) X-PLOR 3.1. A system for X-ray crystallography and NMR. Yale University Press, New Haven, CT.
27. Polshakov, V.I., Birdsall, B. and Feeney, J. Unpublished results.
28. Kuyper, L.F., Roth, B., Baccanari, D.P., Ferone, R., Beddell, C.R., Champness, J.N., Stammers, D.K., Dann, J.G., Norrington, F.E.A., Baker, D.J. and Goodford, P.J. (1982) Receptor based design of dihydrofolate reductase inhibitors: Comparison of crystallographically determined enzyme binding with enzyme affinity in a series of carboxy-substituted trimethoprim analogues. *J. Med. Chem.* **25**, 1120-1122.
29. Roberts, G.C.K., Feeney, J., Burgen, A.S.V. and Daluge, S. (1981) The charge state of trimethoprim bound to *L. casei* dihydrofolate reductase. *FEBS Lett.* **131**, 85-88.
30. Bevan, A.W., Roberts, G.C.K., Feeney, J. and Kuyper, L. (1985) ^{1}H and ^{15}N NMR studies of protonation and hydrogen-bonding in the binding of trimethoprim to dihydrofolate reductase. *Eur. Biophys. J.* **11**, 211-218.
31. Cocco, L., Roth, B., Temple, C., Montgomery, J.A., London, R.E. and Blakley, R.L. (1983) Protonated state of methotrexate, trimethoprim, and pyrimethamine bound to dihydrofolate reductase. *Arch. Biochem. Biophys.* **226**, 567-577

32. Gargaro, A.R., Frenkiel, T.A., Nieto, P.M., Birdsall, B., Polshakov, V.I., Morgan, W.D. and Feeney, J. (1996) NMR detection of arginine-ligand interactions in complexes of Lactobacillus casei dihydrofolate reductase. *Euro. J. Biochem.* **238**, 435-439.
33. Nieto, P.M., Birdsall, B., Morgan, W.D., Frenkiel, T.A., Gargaro, A.R. and Feeney, J. (1997) Correlated bond rotations in interactions of arginine residues with ligand carboxylate groups in protein ligand complexes. *FEBS Lett.* **405**, 16-20.
34. Feeney, J., Birdsall, B., Roberts, G.C.K. and Burgen, A.S.V. (1975) ^{31}P NMR studies of NADP$^+$ and NADPH binding to L. casei dihydrofolate reductase. *Nature* **257**, 564-568.
35. Gerothanassis, I.P., Barrie, P.J., Birdsall, B. and Feeney, J. (1996) ^{31}P solid-state NMR measurements used to detect interactions between NADPH and water and to determine the ionization state of NADPH in a protein-ligand complex subjected to low-level hydration. *Euro. J. Biochem.* **235**, 262-266.
36. Polshakov, V.I., Birdsall, B., Gradwell, M.J. and Feeney, J. (1995) The use of PM3 SCF MO quantum mechanical calculations to refine NMR-determined structures of complexes of antifolate drugs with dihydrofolate reductase in solution. *J. Mol. Structure* **357**, 207-216.
37. Birdsall, B., Cassarotto, M.G., Cheung, H.T.A., Basran, J., Roberts, G.C.K. and Feeney, J. (1997) The influence of aspartate 26 on the tautomeric forms of folate bound to *Lactobacillus casei* dihydrofolate reductase. *FEBS Lett.* **402**, 157-161.
38. Birdsall, B., Feeney, J., Tendler, S.J.B., Hammond, S.J. and Roberts, G.C.K. (1989) Dihydrofolate reductase: multiple conformations and alternative modes of substrate binding. *Biochemistry* **28**, 2297-2305.
39. Birdsall, B., Tendler, S.J.B., Arnold, J.R.P., Feeney, J., Griffin, R.J., Carr, M.D., Thomas, J.A., Roberts, G.C.K. and Stevens, M.F.G. (1990) NMR studies of multiple conformations in complexes of *L. casei* dihydrofolate reductase with analogues of pyrimethamine. *Biochemistry* **29**, 9660-9667.
40. Gronenborn, A., Birdsall, B., Hyde, E.I., Roberts, G.C.K., Feeney, J. and Burgen, A.S.V. (1981) Direct observation by NMR of two coexisting conformations of an enzyme-ligand complex in solution. *Nature* **290**, 273-274
41. Birdsall, B., Gronenborn, A.M., Hyde, E.I., Clore, G.M., Roberts, G.C.K., Feeney, J. and Burgen, A.S.V. (1982) Hydrogen-1, carbon-13 and phosphorus-31 nuclear magnetic resonance studies of the dihydrofolate reductase- nicotinamide adenine dinucleotide phosphate- folate complex: characterization of three coexisting conformational states. *Biochemistry* **21**, 5831-5838.
42. Curtis, N., Moore, S., Birdsall, B., Bloxsidge, J., Gibson, C.L., Jones, J.R. and Feeney, J. (1994) ^{3}H-n.m.r. studies of multiple conformations and dynamic processes in complexes of folate and methotrexate with *Lactobacillus casei* dihydrofolate reductase. *Biochem. J.*, **303**, 401-405.
43. Cheung, H.T.A., Birdsall, B. and Feeney, J. (1992) ^{13}C NMR studies of complexes of *Escherichia coli* dihydrofolate reductase formed with methotrexate and folic acid. *FEBS Lett.*, **312**, 147-151.
44. Cheung, H.T. A., Birdsall, B., Frenkiel, T.A., Chau, D.D. and Feeney, J. (1993) ^{13}C NMR determination of the tautomeric and ionization states of folate in its complexes with *Lactobacillus casei* dihydrofolate reductase. *Biochemistry* **32**, 6846-6854.
45. Liotta, L.A. (1992), *Sci. American* **226**, 2, 34-41.
46. Williamson, R.A., Martorell, G., Carr, M.D., Murphy, G., Docherty, A.J.P., Freedman, R.B. and Feeney, J. (1994) Solution structure of the active domain of tissue inhibitor of metalloproteinases-2. A new member of the OB fold protein family. *Biochemistry* **33**, 11745-11759.
47. Docherty, A.J.P., O'Connell, J., Crabbe, T., Angal, S. and Murphy, G. (1992) The matrix metalloproteinases and their natural inhibitors: prospects for treating degenerative tissue diseases. *Trends Biotechnol.* **10**, 200-207.
48. Williamson, R.A.,Natalia, D., Gee, C.K. Murphy,G., Carr, M.D. and Freedman, R. B. (1966) Chemically and conformationally authentic active domain of human tissue inhibitor of metalloproteinases-2 refolded from bacterial inclusion bodies. *Eur. J. Biochem.* **241**, 476-483.
49. Muskett, F. W., Frenkiel, T.A., Feeney, J., Freedman, R.B., Carr, M.D. and Williamson, R.A. High resolution solution structure of the N-terminal domain of Tissue Inhibitor of Metalloproteinases-2 and Characterisation of its Interaction with matrix Metalloproteinase-3. Unpublished results.
50. Williamson, R.A., Carr, M.D., Frenkiel, T.A., Feeney, J. and Freedman, R.B. (1997) Mapping the binding site for metalloproteinase on the N-terminal domain of tissue inhibitor of metalloproteinases-2 by NMR chemical shift perturbation. *Biochemistry*. In press.

51. Thim, A. (1989) A new family of growth factor-like peptides. "Trefoil" disulphide loop structures as a common feature in breast cancer associated peptide (pS2), pancreatic spasmolytic polypeptide (PSP), and frog skin peptides (spasmolysins) *FEBS Lett.* **250**, 85-90.
52. Wright, N.A., Poulsom, R., Stamp, G.W., van Noorden, S., Sarraf, C., Elia, G., Ahnen, D., Jeffery, R., Longcroft, J., Pile, C., Rio, M. and Chambon, P. (1993) Trefoil peptide gene expression in gastrointestinal epithelial cells in inflammatory bowel disease. *Gastroenterology* **104**, 12-20.
53. Dignass, A., Lynch-Devaney, K., Kindon, H., Thim, L. and Podolsky, D.K. (1994) Trefoil peptides promote epithelial migration through a transforming growth factor β-independent pathway. *J. Clin. Invest.* **94**, 376-383.
54. Playford, R.J., Marchbank, T., Chinery, R., Evison, R., Pignatelli, M., Boulton, R.A., Thim, L. and Hanby A.M. (1995) Human spasmolytic polypeptide is a cytoprotective agent that stimulates cell migration, *Gastroenterology* **108**, 108-116.
55. Williams, R., Stamp. G.H.W., Gilbert, C., Pignatelli, M. and Lalani, E.-N. (1996) pS2 transfection of murine adenocarcinoma cell line 410.4 enhances dispersed growth pattern in a 3-D collagen gel. *J. Cell Sci.* **109**, 63-71.
56. Jørgensen, K. D., Diamant, B., Jørgensen, K. H. and Thim, L. (1982) Pancreatic spasmolytic polypeptide (PSP): III. pharmacology of a new porcine pancreatic polypeptide with spasmolytic and gastric acid secretion inhibitory effects. *Regul. Pept.* **3**, 231-243.
57. Hoosein, N.M., Thim, L., Jørgensen, K.H. and Brattain, M.G. (1989) Growth stimulatory effect of pancreatic spasmolytic polypeptide on cultured colon and breast tumor cells. *FEBS Lett.* **247**, 303-306.
58. Carr, M.D., Bauer, C.J., Gradwell, M.J. and Feeney, J, (1994) Solution structure of a trefoil-motif-containing cell growth factor, porcine spasmolytic protein. *Proc. Natl. Acad. Sci. USA*, **91**, 2206-2210.
59. Polshakov, V.I., Frenkiel, T.A., Westley, B., Chadwick, M., May, F., Carr, M.D. and Feeney, J. (1995) NMR-based structural studies of the pNR-2/pS2 single domain trefoil peptide. Similarities to porcine spasmolytic peptide and evidence for a monomeric structure. *Euro. J. Biochem.*, **233**, 847-855.
60. Polshakov, V.I., Williams, M., Gargaro, A., Frenkiel, T.A., Westley, B.R., Chadwick, M.P., May, F.E.B. and Feeney, J. (1997) High resolution solution structure of the human breast cancer oestrogen-inducible pNR-2/pS2: a single trefoil domain. *J. Mol. Biol.* **267**, 418-432.
61. De, A., Brown, D.G., Gorman, M.A., Carr, M.D., Sanderson, M.R. and Freemont, P.S, (1994) Crystal structure of a disulfide-linked "trefoil" motif found in a large family of putative growth factors. *Proc. Natl. Acad. Sci., USA*, **91**, 1084-1088.
62. Gajhede, M., Petersen, T.N., Henriksen, A., Petersen, J.F.W., Dauter, Z., Wilson, K.S. and Thim. L, (1993) Pancreatic spasmolytic polypeptide: first three-dimensional structure of a member of the mammalian trefoil family of peptides. *Structure* **1**, 253-262.
63. Lane, A.N. and Westley, B.R. Unpublished results.
64. Williams, M.A., Feeney, J., May, F.E.B. and Westley, B.R. Unpublished results
65. Chinery, R., Bates, P.A., De, A. and Freemont, P.S. (1995) Characterisation of the single copy trefoil peptides intestinal trefoil factor and pS2 and their ability to form covalent dimers. *FEBS Lett.* **357**, 50-54.

NMR DIFFUSION MEASUREMENTS IN CHEMICAL AND BIOLOGICAL SUPRAMOLECULAR SYSTEMS

YORAM COHEN*, ORNA MAYZEL, AYELET GAFNI, MOSHE GREENWALD, DANA WESSELY, LIMOR FRISH and YANIV ASSAF
School of Chemistry, Tel Aviv University, Ramat Aviv, Tel Aviv 69978, Israel

1. Introduction

Nature self-assembles molecules into large supramolecular aggregates which are capable of performing specific biological functions [1]. Therefore it is not surprising that supramolecular chemistry has become an important theme of chemistry in recent years [2]. One of the main problem of supramolecular sciences is the elucidation of the structure of the supramolecular aggregate that prevails in solution. Consequently, new analytical methods that may assist these structural determinations are required, especially those techniques that may provide some information regarding the mutual interactions between the different molecular components of the supramolecular system. In this context it seems to us that the diffusion coefficient, as measured by the pulsed gradient spin echo (PGSE) NMR technique [3], may provide such information.

2. Methodology

The PGSE NMR technique (Figure 1) is a simple, reliable, and flexible NMR method for measuring diffusion in solution and in biological tissue [4,5]. Indeed, in the last 30 years this technique has been used extensively to study the diffusion characteristics of many chemical and biological systems [4, 5], and more recently water diffusion was shown to be an important contrast mechanism in MR neuroimaging [6]. In addition, it has been demonstrated that the diffusion coefficients of a double quantum (DQ) transition can be determined using the pulse sequences shown in Figure 2 [7].

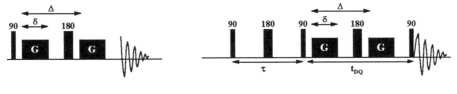

Figure 1   Figure 2

According to these sequences the ratio between the echo intensity in the absence (A_0) and in presence of pulsed gradient (A_g) is given by equation (1)

$$\ln(A_g/A_0) = -(\gamma g \delta p)^2 (\Delta - \delta/3) D \qquad (1)$$

Wherein γ is the gyromagnetic ratio (rad/gauss s), g is the pulsed gradient strength (gauss/cm), p is the coherence order (p = 1 and 2 for the SQ and DQ spectra, respectively), Δ and δ (s) are the time separation between the pulsed-gradients and the pulsed-gradient length, respectively and D is the diffusion coefficient ($cm^2 s^{-1}$). In this equation the expression ($\Delta-\delta/3$) is considered to be the diffusion time. In case of a single population and a totally isotropic and nonrestricted diffusion, the plot of $\ln(A_g/A_0)$ vs. g^2 should give a straight line which slope is proportional to the self-diffusion coefficient.

Host-guest complexes are usually composed of molecular species which differ in their size and consequently in their diffusion coefficients. In addition, supramolecular complexes are usually much larger from their building units and therefore changes in the diffusion coefficients are expected upon formation of the supramolecular system. From the difference between the diffusion coefficient of the guest in the free and the bound states the bound molar fraction can be calculated. From this molar fraction the association constant (K_a) can be calculated. These calculations assume that the free and the bound guest molecules are in fast exchange on the NMR time scale. If this is the case then the diffusion coefficients can be used in a similar way as chemical shifts in chemical shift titration experiments. The important advantage of diffusion measurements is that one can predict, *a priori*, what should be the diffusion coefficient of the small guest in the bound state. This should be, for the first approximation, the diffusion coefficient of the large host molecule. It should be noted however that the larger is the difference between the host and the guest the more accurate will be the previous statement and the more accurate will be also the translation of the changes in the diffusion coefficients into association constants.

3. NMR diffusion measurements in organic supramolecular systems

Because of the favorable NMR characteristics of the protons organic host-guest systems for which simultaneous determination of the diffusion coefficients is possible, are the most suitable for such measurements. Therefore the above approach was first used to determine the K_as of methylammonium chloride (**1**) to 18-crown-6 (**2**) and [2.2.2] cryptand (**3**) [7]. It was found that the association constants derived from the analysis of the diffusion coefficients were very similar to those obtained from other methods. It should be noted that the changes in the chemical shifts upon complex formation in these systems were rather small. In addition, we could easily determine the association constant between **1** and **3** both in D_2O and CD_3OD although partial protonation of **3** did occur under these conditions [8], a fact that complicated very much the extraction of these association constants from ^1H chemical shifts titration experiments.

Simultaneous determination of the diffusion coefficients in a high resolution ^1H NMR spectrum, provides also an elegant and quick mean of mapping mutual interactions between several molecular species and hence is very suitable for characterization of multicomponents aggregates and second sphere coordination complexes. For example, we could show that in the Cs$^+$ complex of the p-tert-butylcalix[4]arene (4/Cs$^+$) crystallized from CH$_3$CN, in which the coordination between the Cs$^+$ and the CH$_3$CN molecule is clearly observed in the solid state [9], there is no interaction whatsoever between the calix[4]arene moiety and the acetonitrile molecule of this cesium complex in CDCl$_3$ solution [10]. We have found that the diffusion coefficient of the calixarene moiety of the complex is $0.68 \pm 0.02 \times 10^{-5}$ cm^2 s^{-1} while that of the CH$_3$CN molecule originating from the complex was $1.67 \pm 0.03 \times 10^{-5}$ cm^2 s^{-1}. The diffusion coefficient of acetonitrile in CDCl$_3$ was found to be $1.68 \pm 0.02 \times 10^{-5}$ cm^2 s^{-1}, indicating no interaction between the complex and the acetonitrile molecule in the chloroform solution [10].

Cyclodextrins complexes in which the guest itself can be used as a host for an additional guest may have attractive applications. One type of such complexes are the complexes of γ-cyclodextrins (γ-CD, **5**) with small macrocycles [11]. These complexes which formation is difficult to follow by conventional spectroscopic methods (^{1}H NMR chemical shift, U.V-visible spectroscopy etc.) were studied by the PGSE diffusion technique [12]. It has been found that 12-crown-4 (**6**) and its tetraaza analogue (**7**) form complexes of moderate stability with γ-CD in D$_2$O (K$_a$s of 187 M^{-1} and 165 M^{-1}, respectively). In addition, it has been found that the hydrophobic interaction is not the main driving force for the formation of these complexes and indeed these complexes were found to form complexes of similar stability both in D$_2$O and in a 80:20 CD$_3$OD:D$_2$O solution. Even in pure DMSO$_{d6}$ K$_a$ for the **5/6** complex was found to be 111 M^{-1}. It has been also found that addition of salts to the macrocycle/γ-CD complexes caused a large decrease in the association constants of these complexes. Interestingly, this peculiar behavior is completely reversed when the macrocycle contains a single benzene ring [13]. In this case, the complex is much stronger in aqueous solution than in organic solvents as observed for most CDs complexes.

Diffusion measurements can also be used to study molecular weight changes and aggregation. Recently, we were interested in preparing double stranded helicates [14], from ligands that contain different coordination sites as in ligands **8** and **9**. The copper complex of **9** gave the expected ^{1}H NMR spectrum for [(**9**)$_2$Cu$_3$]$^{3+}$ and a FAB-MS peak at m/z of 1689 which corresponds to [(**9**)$_2$Cu$_3$(PF$_6$)$_2$]$^+$. In addition, [(**9**)$_2$Cu$_3$]$^{3+}$ has been characterized by x-ray crystallography [15].

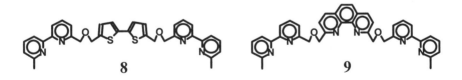

8 9

Contrary to that the copper complex of **8** gave a much more complex ^{1}H NMR spectrum but the corroboration that the formed copper complex is also a double stranded complex was obtained by studying the diffusion coefficients of **8**, **9**, [(**9**)$_2$Cu$_3$]$^{3+}$ and the copper complex of **8**. Under identical conditions the diffusion coefficients of **8**, **9**, [(**9**)$_2$Cu$_3$]$^{3+}$ and the copper complex of **8** were found to be $0.34 \pm 0.01 \times 10^{-5}$ cm^2 s^{-1}, $0.29 \pm 0.01 \times$

10^{-5} cm^2 s^{-1}, $0.22 \pm 0.01 \times 10^{-5}$ cm^2 s^{-1} and $0.18 \pm 0.01 \times 10^{-5}$ cm^2 s^{-1}, respectively. It is generally accepted that the diffusion coefficients ratio of two molecules is inversely proportional the cubic root of their molecular weights. Based on this relation the expected ratios of the diffusion coefficients for the formation of $[(8)_2Cu_3(PF_6)_2]^+$ $[(9)_2Cu_3(PF_6)_2]^+$ from **8** and **9** are 0.68 and 0.69, respectively. Interestingly the ratios found experimentally (0.65 and 0.62 for **8** and **9**, respectively) are both very similar to each other and very similar to the expected values thus corroborating the fact that a double strand complex was formed also in the case of **8**. Indeed, later a careful FAB-MS study of the copper complex of **8** revealed a peak at m/z of 1453 corresponding to $[(8)_2Cu_2(PF_6)]^+$.

This methodology was also used to evaluate the water hydration of 18-crown-6 (**10**) and its KI complex (**10**/KI) in CDCl$_3$ [16]. Interestingly, by following the changes in the diffusion coefficient and in the chemical shift of the water molecules upon changing the water/macrocycle or the water/complex ratios, it has been found that the average number of water molecules solvating **10**/KI is independent of this ratio but increases as the this ratio increases in the case of **10** [16]. It has been found that an average of ~0.3 water molecule is associated with each molecular complex, while the average of number of water molecule solvating **10** changes from ~0.5 to ~1.0 when the water/macrocycle ratio was changed from 0.8 to 3.8. Here the two methods provide complementary information. It should be noted however, that both methods cannot distinguish whether the obtained value of ~0.3 represents one water molecule that is attached to the complex for ~33% of its lifetime of whether there are 10 water molecules which each spend ~3.3% of their life time associated to the complex.

4. NMR diffusion measurements in biological tissue.

Hydration of biological molecules and the interactions of water molecules with different components in the living cells are the subject of many investigations. However, the study of water hydration in biological tissues or in aqueous solution is a much more difficult task than in non-aqueous solutions. In aqueous solution and in biological tissues most of the water peak originates from bulk water. Therefore one needs to filter out the majority of the bulk water peak in order to be able to characterize the water populations that form the hydration sphere or interact with the cellular components. This may be achieved, for example, by DQ filtered NMR spectroscopy of D$_2$O as it has been demonstrated that DQ spectrum of an isolated spin 1 nuclei may be observed if and only if the observed nucleus experiences a non-isotropic motion [17]. In addition, extraction of diffusion coefficients in biological systems is more difficult because of the intrinsic heterogeneity of the systems that may include different compartments, barriers, membranes etc. [18]. Therefore one generally speaks in terms of the apparent diffusion coefficient (ADC) in biological tissues [6]. Recently we have demonstrated that the combination of single-quantum (SQ) and DQ diffusion ^{2}H NMR spectroscopy enables to distinguish between free an bound water in biological tissues [19]. As the DQ signal originates from water molecules that spend, at least part of their life time, associated to much larger molecular entities one should expect the diffusion coefficient of the DQ signal to be significantly lower than that of the SQ signal which originate mainly from bulk water. Indeed this is the case as shown in Figure 3.[19] When the ^{2}H SQ and DQ diffusion were measured as a function of the diffusion time, (Δ-δ/3), we have found two water populations in each. Interestingly, it has been found that the ADC of the slow diffusing component of the SQ signal is very similar to the ADC of the fast diffusing component of the DQ signal. These results seem to corroborate the three components model (Figure 4). The ADC of the slow diffusing component of the DQ signal was found to be rather low (i.e. ~0.05 x 10^{-5} cm^2 s^{-1}).

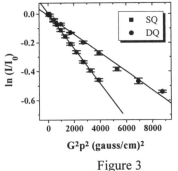

Figure 3

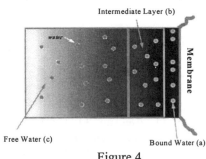

Figure 4

In summary diffusion coefficients as measured by the PGSE technique may assist the structural determination of supramolecular systems in solutions. It can be used to determine K_as, to probe multicomponents interactions, to study aggregation and molecular weight distribution, and to study water hydration in organic solvents, in aqueous solutions and in biological tissues. In aqueous solutions the combination of ^{2}H SQ and DQ diffusion offers a unique possibility to filter off the bulk water and to separate the water peak into the different populations that contribute to this peak. This increases our ability to study the interaction of water molecules with large molecular components. As the required hardware and software for such high resolution diffusion measurements are now available on conventional NMR spectrometers we believe that NMR diffusion measurements should be consider as an additional analytical technique for the characterization of supramolecular systems in solution.

5. References.

1. Varner, J. E. (Ed.) (1988) *"Self-Assembly Architecture"*, Alan R. Liss, New York.
2. a) Vogtle, F. (1991) *"Supramolecular Chemistry"*, John Wiley & Sons, Chichester. b) Lehn, J.-M. (1995) *"Supramolecular Chemistry, Concepts and Perspectives"*, VCH, Weinheim.
3. Stejskal, E. O. and Tanner, J. E. (1965) Spin diffusion measurements: spin echoes in the presence of a time-dependent field gradient, *J. Chem. Phys.* **42**, 288-292.
4. a) Stilbs, P. (1987) Fourier transform pulsed-gradient spin-echo studies of molecular diffusion, *Prog. NMR Spectrosc.* **19**, 1-45. b) Soderman, O. and Stilbs, P. (1994) NMR studies of complex surfactant systems, *Prog. NMR Spectrosc.* **26**, 445-482.

5. Lindblom, G. and Oradd, G. (1994) NMR studies of translational diffusion in lyotropic liquid crystals and in lipid membranes, *Prog. NMR Spectrosc.* **26**, 483-515.
6. a) Moseley, M. E., Cohen, Y., Minterovitch, J., Chileuitt, L., Shimizu, H., Kucharczyk, J., Wendland, M. F. and Weinstein, P. R. (1990) Early detection of regional cerebral ischemia in cats: Comparison of diffusion- and T2-weighted MRI and spectroscopy, *Magn Reson. Med.* **14**, 330-446. b) Le Bihan, D. (Ed.) (1995) "*Diffusion and Perfusion Magnetic Resonance Imaging*", Raven Press, New York.
7. Martin, J. F., Selwyn, L. S., Vold, R. R. and Vold, R. L. (1982) The determination of translational diffusion constants in liquid crystals from gradient double quantum spin echo decays, *J. Chem. Phys.* **76**, 2632-2634.
8. Mayzel, O. and Cohen, Y. (1994) Diffusion coefficients of macrocyclic complexes using the PGSE NMR technique: Determination of association constants, *Chem. Commun.* 1901-1902.
9. Harrowfield, J. M., Odgen, M. I., Richmond, W. R. and White, A. H. (1991) Calixarene-cupped cesium: a coordination conundrum ?, *J. Chem. Soc. Chem. Commun.*, 1159-1161.
10. Mayzel, O., Aleksiuk, O., Grynszpan, F., Biali, S. E. and Cohen, Y. (1995) NMR diffusion coefficients of *p-tert*-butylcalix[n]arene systems, *Chem. Commun.*, 1183-1184.
11. Vogtle F. and Muller, W. M. (1979) Complexes of γ-cyclodextrin with crown ethers, cryptands, coronates and crypatates, *Angew. Chem .Int. Ed. Engl.*, **18**, 623-624.
12. Gafni, A. and Cohen, Y. (1997) Complexes of macrocycles with γ-cyclodextrin as deduced from NMR diffusion measurements, *J. Org. Chem.*, **62**, 120-125.
13. Frish, L., Diodone, R., Plenio, H., Ueda, H. and Cohen, Y., unpublished results.
14. a) Lehn, J.-M., Rigault, A., Siegel, J., Harrowfield, J., Chevrier, B. and Moras, D. (1987) Spontaneous assembly of double-stranded helicates from oligobipyridine ligands and copper(I) cations: Structure of an inorganic helix, *Proc. Natl. Acad. Sci. USA*, 84, 2565-2569. b) Piguet, C., Bernardinelli, G. and Hopfgartner, G. (1997) Helicates as versatile supramolecular complexes, *Chem. Rev.*, **97**, 2005-2062.
15. Greenwald, M., Wessely, D., Goldberg, I. and Cohen, Y., submitted.
16. Mayzel, O., Gafni, A. and Cohen, Y. (1996) Water hydration of macrocyclic systems in organic solvents: an NMR diffusion and chemical shift study, *Chem. Commun.,* 911-912.
17. Sharf, Y., Eliav, U., Shinar, H. and Navon, G., (1995) Detection of anisotropy in cartilage using ^{2}H double-quantum-filtered NMR spectroscopy, *J. Magn. Reson.,* ***B*** **107**, 60-67.
18. Le Bihan, D. (1995) Molecular diffusion, tissue microdynamics and microstructure *NMR in Biomed.*, **8**, 375-385.
19. Assaf, Y.and Cohen, Y. (1996) Detection of different water populations in brain tissue using ^{2}H single- and double-quantum-filtered diffusion NMR spectroscopy, *J. Magn. Reson.* ***B*** **112**, 151-159.

DETERMINATION OF CALIXARENE CONFORMATIONS BY MEANS OF NMR TECHNIQUES

A. CASNATI,[a] J. de MENDOZA,[b] D.N. REINHOUDT[c], R. UNGARO.[a]
[a] *Dip.to di Chimica Organica e Industriale, Università degli Studi, Viale delle Scienze, I-43100 Parma, Italy;* [b] *Departamento de Química Orgánica, Universidad Autónoma de Madrid, Cantoblanco, E-28049 Madrid, Spain;* [c] *Laboratory of Organic Chemistry, Twente University, P.O. Box 217, 7500 AE Enschede, The Netherlands.*

Calixarenes have been widely used in the last ten years in Supramolecular Chemistry.[1] Their success is due to the fact that they are readily accessible in large quantities, can be easily functionalised both at the phenolic OH groups and at the aromatic nuclei, and, depending on the number of monomeric units in the macrocyclic structure, they present different size and conformational properties. The regio- and stereoselective control in the fuctionalisation of calix[4]- and calix[6]arenes,[1,2] provides the fundametal synthetic tool to shape these macrocycles and allows the obtainment of receptors able to selectively recognise cations, anions, and neutral molecules.[1] NMR techniques have proven to be of fundamental importance in the clarification of the conformational properties of calixarene-based ligands and their complexes. Calix[4]arenes, may exist in four limiting conformations namely the *cone, partial cone, 1,3-alternate* and *1,2-alternate*. When the steric bulkiness of the R groups is smaller or equal to ethyl, a conformational isomerism among the four conformations exists and, with the aid of dynamic ^{1}H NMR or two-dimensional EXSY experiments, it is possible to study the flexibility and the conformational preferences of these derivatives.[3] On the other hand, when the R groups is larger than ethyl, the macrocycle is fixed in one or more of the four structures of Figure 1.

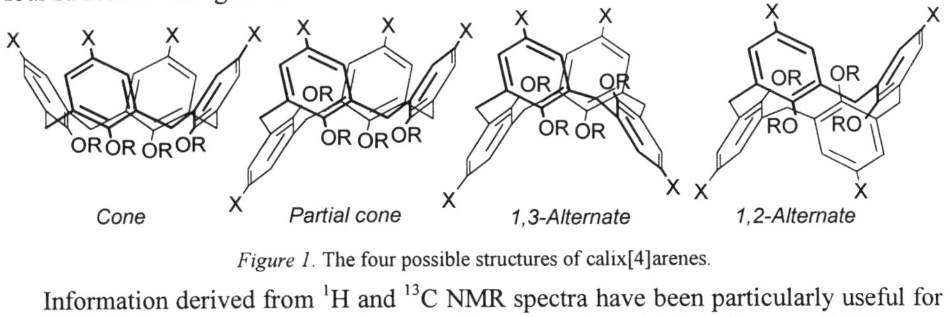

Figure 1. The four possible structures of calix[4]arenes.

Information derived from ^{1}H and ^{13}C NMR spectra have been particularly useful for understanding the conformational and binding properties of 1,3-dialkoxy-calix[4]arene-crowns-5[4] and -crowns-6[5], which were shown to be powerful and selective complexing agents for alkali metal ions.

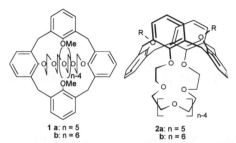

The dimethoxy derivatives (**1a** and **1b**) are conformationally mobile in solution as clearly evidenced from their ^{1}H NMR spectra at different temperature. Their conformational mobility is due to the small size of the methoxy groups which are not large enough to prevent isomerization and it is only slightly dependent on the length of the polyether ring.

Upon complexation of potassium or caesium ion respectively, compounds **1a** and **1b** rearrange their conformations mainly to the *1,3-alternate*.

In the case of compound **1b** in CD$_3$CN solution, for example, the *cone* conformations is more stable than the *partial cone* and *1,3-alternate* as clearly evidenced by the presence of only two doublets for the methylene bridge protons (ArCH$_2$Ar) in the ^{1}H NMR spectrum (Fig. 2a) and from a single resonance around δ = 31 ppm for the methylene bridge carbon (ArCH$_2$Ar) in the ^{13}C NMR spectrum. This is in accordance with the simple rules on the multiplicity and chemical shifts of the ^{1}H or ^{13}C nuclei of the methylene bridges,[6] exemplified by the typical patterns in the spectra represented in Figure 3 which easily allow to assign the conformation of calix[4]arenes.

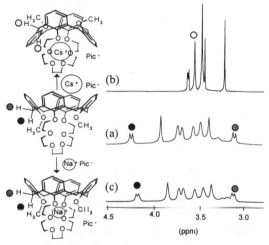

Figure 2: ^{1}H NMR spectrum in CD$_3$CN at room temperature of (a) ligand **1b**, (b) its caesium complex, (c) its sodium complex.

Upon complexation of caesium ion, ligand **1b** rearranges its structure to the *1,3-alternate* (Fig. 2b) as indicated by the singlet for the methylene bridge at 3.55 ppm in the ^{1}H and by a singlet around δ = 37 ppm in the ^{13}C NMR spectrum. Complexation of sodium ion fixes the ligand in the *cone* conformation (Fig. 2c). This observation prompted us to synthesise calix[4]arene-crown-5 (**2a**) and -crown-6 (**2b**) derivatives already preorganised in the *1,3-alternate* structure, which greatly improves their efficiency and selectivity in cation binding. Compounds **2a** and **2b** (R = *n*-Pr, *i*-Pr) are among the most efficient and selective ionophores for potassium and caesium, respectively, known so far. They have been used also as active part of ion selective electrodes (ISE's)[7a] or of chemically modified field effect transistors (CHEMFET's)[7b] for the detection of potassium and caesium ions or in supported liquid membranes (SLM's) for separation of alkali metal ions.[7a,b]

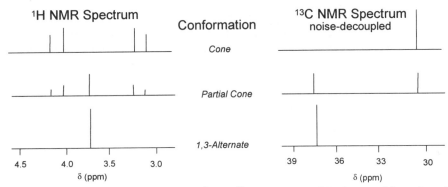

Figure 3: Patterns of the signals expected in the ^{1}H and ^{13}C NMR spectra of the three possible conformations of 1,3-dialkoxycalix[4]arene-crown ethers.

For calix[6]- and calix[8]arene derivatives the conformational mobility and the number of possible conformations increase substantially and therefore two-dimensional NMR techniques have been extensively used for clarifying dynamic processes and to elucidate the structure of the macrocycles and of their complexes. For example, these techniques were of fundamental importance in the understanding of the conformational behaviour of the 1,3,5-trimethoxy-2,4,6-trialkoxy-p-*tert*-butylcalix[6]arene (**3**).[8] Surprisingly these derivatives are conformationally more rigid than the hexaalkoxy-p-*tert*-butylcalix[6]arenes (**4**) which bear six bulky substituents at the lower rim.

Compounds **3** are mainly in a symmetric conformation which shows a very simple ^{1}H NMR spectrum: there is only an AX system ($\delta_A \approx$ 4.5-4.6 ppm, $\delta_X \approx$ 3.3-3.4 ppm, J = 12-14 Hz) for the methylene bridge protons and two singlets for the ArH protons ($\delta \approx$ 7.25 and 6.6 ppm), indicating that these compounds are mainly in a *flattened cone* conformation (C_{3v} symmetry). The presence in all compounds **3** of a singlet for the methoxy groups at unusually high field (δ = 2.3-2.5 ppm) indicates that these groups are shielded by the cavity of the macrocycle and that possibly CH/π interactions stabilise the *flattened cone* conformation. Depending on the Y substituent, however, an other minor conformer may exist in solution. With the aid of ROESY spectra at 273 K it was possible to deduce that the minor conformer (C_s symmetry) possesses a *1,2,3-alternate* structure. The differences in Gibb Free Energy ($\Delta G°$, kJ mol^{-1}, T = 303 K) for compounds **3** were calculated from the population of the two conformers in CDCl$_3$ using Boltzmann equation, and are reported in Table 1. The Activation Energies ($\Delta G^\ddagger$, kJ mol^{-1}, T = 328 K) were calculated from interconversion rate constants which were determined by integration of exchange peaks in EXSY experiments. Since from CPK model analysis one can conclude that the *p*-phenyl-benzyl substituent in compound **3f** is too bulky to pass through the annulus of the calixarene, the interconversion should take place *via* the *tert*-butyl through the annulus mechanism. On the other hand when $\Delta G^{\ddagger}$ is smaller than 82 kJ mol^{-1} the rate-limiting step in the interconversion is the passage of the phenoxy substituent through the annulus.

TABLE 1: Gibb Free Energy ($\Delta G°$, kJ mol^{-1}, T = 303 K) and Activation Energy ($\Delta G^{\ddagger}$, kJ mol^{-1}, T = 328 K) for the *flattened cone/1,2,3-alternate* interconversion in compounds **3** (CDCl$_3$).

Compound	Y	$\Delta G°$ (kJ mol^{-1}, T = 303 K)	$\Delta G^{\ddagger}$ (kJ mol^{-1}, T = 328 K)
3a	COONEt$_2$	7	87
3b	COOBu^t	6	78
3c	C$_6$H$_5$	3	70
3d	4-BrC$_6$H$_4$	2	86
3e	2-naphthyl	3	85
3f	4-C$_6$H$_5$C$_6$H$_4$	3	88

The conformational rigidity of these 1,3,5-trimethoxy-2,4,6-trialkoxy-p-*tert*-butylcalix[6]arenes (**3**) was exploited for the synthesis of receptors which show a high selectivity for guests having a C$_{3v}$ symmetry, such as the guanidinium cation[9] or the trianion of the benzen-1,3,5-tricarboxilic acid.[10]

References

1. For recent review articles on calixarenes, see: (a) Böhmer, V. (1995) *Angew. Chem., Int. Ed. Engl.* **34**, 713-745. (b) Pochini, A. and Ungaro, R. (1996) in *Comprehensive Supramolecular Chemistry*, Vögtle, F. Ed., Elsevier Science Ltd., Oxford, p. 103-142. (c) Ikeda, A. and Shinkai S. (1997) *Chem. Rev.* **97**, 1713-1734. (d) Casnati A. (1997) *Gazz. Chim. Ital.* **127**, 637-649.
2. Arduini, A. and Casnati, A. (1996) «Calixarenes» in *Macrocyclic Synthesis: a Practical Approach*, Parker D. Ed., Oxford University Press, Oxford, p. 145-173.
3. Groenen, L.C., van Loon, J.D., Verboom, W., Harkema, S., Casnati, A., Ungaro, R., Pochini, A., Ugozzoli, F. and Reinhoudt, D.N. (1991) *J. Am. Chem. Soc.* **113**, 2385-2392.
4. Casnati, A., Pochini, A., Ungaro, R., Ugozzoli, F., Arnaud, F., Fanni, S., Schwing, M.-J., Egberink, R.J.M., de Jong, F., and Reinhoudt, D.N. (1995) *J. Am. Chem. Soc.* **117**, 2767-2777.
5. Casnati, A., Pochini, A., Ungaro, R., Bocchi, C., Ugozzoli, F., Egberink, R.J.M., and Reinhoudt, D.N. (1996) *Chem. Eur. J.* **2**, 436-445.
6. (a) Gutsche, C. D. and Bauer, L. J. (1985) *J. Am. Chem. Soc.* **107**, 6052-6059. (b) Jaime, C., de Mendoza, J., Prados, P., Nieto, P. M. and Sànchez, C. (1991) *J. Org. Chem.* **56**, 3372-3376.
7. (a) Bocchi, C., Careri, M., Casnati, A. and Mori, G. (1995) *Anal. Chem.* **67**, 4234-4238. (b) Lugtenberg, R.J.W., Brzozka, Z., Casnati, A., Ungaro, R., Engbersen, J.F.J. and Reinhoudt, D. N. (1995) *Anal. Chim. Acta* **310**, 263-267.
8. van Duynhoven, J.P.M., Janssen, R.G., Verboom, W., Franken, S.M., Casnati, A., Pochini, A., Ungaro, R., de Mendoza, J., Nieto, P.M., Prados, P. and Reinhoudt, D.N. (1994) *J. Am. Chem. Soc.* **116**, 5814-5822.
9. Casnati, A., Minari, P., Pochini, A., Ungaro, R., Nijenhuis, W.F., de Jong, F. and Reinhoudt, D.N. (1992) *J. Isr. Chem. Soc.* **32**, 79-87.
10. Scheerder, J., Engbersen, J.F.J., Casnati A., Ungaro, R. and Reinhoudt, D.N. (1995) *J. Org. Chem.* **60**, 6448-6454.

STRUCTURE, MOLECULAR DYNAMICS, AND HOST–GUEST INTERACTIONS OF A WATER–SOLUBLE CALIX[4]ARENE

JÜRGEN H. ANTONY AND ANDREAS DÖLLE
Institut für Physikalische Chemie
Rheinisch–Westfälische Technische Hochschule Aachen
D–52056 Aachen, Germany

The influence of host–guest interactions between water–soluble *p*-sulfonatocalix[4]arene and various tetraalkylammonium cations on the molecular reorientational motions was studied in aqueous solutions by measurement of ^{13}C and ^{14}N NMR relaxation data.

1 Introduction

Calix[*n*]arenes belong to a particularly interesting class of compounds regarding their host–guest interactions [1]. Calix[*n*]arenes are cyclic compounds, easily obtained and modified according to the requirements for host molecules, which makes them superior to crown ethers and cyclodextrins. By sulfonation in *para* position to the phenolic hydroxyl group water–soluble calixarenes are obtained which are of great importance for potential applications in aqueous solutions.

Most calixarenes are flexible compounds which exist in different conformations being in solution transformed into each other. In aqueous solution the water–soluble *para*-sulfonatocalix[4]arenes exist predominantly in the cone conformation, although a permanent interconversion between the two cone conformers occurs. The interconversion process is accompanied by hydrogen bond breaking and forming between the calixarene and the surrounding water molecules.

The dynamic and conformational properties of the calixarenes are of great importance for their host characteristics when interacting with guests. In order to gain a deeper understanding of the host properties of calixarenes, the rotational dynamics of the *p*-sulfonatocalix[4]arene (**1**) were investigated in a previous study [2] for solutions of the acid form and the pentasodium salt of **1** by NMR relaxation methods. Host–guest interactions between *p*-sulfonatocalix[4]arene and alkyl- or alkylphenylammonium cations have already been studied by NMR chemical shifts [3]. In the present study, the rotational dynamics of **1** acting as host molecule for tetramethyl-, trimethylphenyl- (TMPA) and trimethyladamantanylammonium cations as guests were investigated. As an example only the results for the host–guest interaction between calix[4]arene-*p*-sulfonic acid (**1**) and TMPA are shown.

2 Results and Discussion

The ^{13}C spin–lattice relaxation times T_1 and the ^{1}H–^{13}C NOE factors η of the aliphatic (CH$_2$) and aromatic (CH) carbon atoms of calix[4]arene-p-sulfonic acid (**1**) without and with trimethylphenylammonium chloride (TMPA) were measured for a solution in D$_2$O and a temperature range from 270 to 345 K. The observed relaxation rates ($1/T_1$) and NOE factors are presented in Figure 1 as functions of the reciprocal temperature $1/T$.

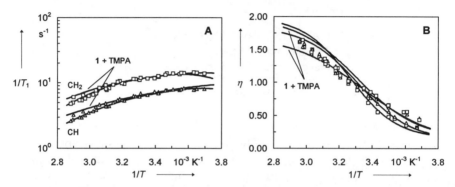

Figure 1. ^{13}C NMR relaxation rates $1/T_1$ (A) and {^{1}H}–^{13}C nuclear Overhauser factors η (B) of calix[4]arene-p-sulfonic acid (1) without and with TMPA in D$_2$O

The corresponding relaxation rates and NOE factors for the ^{13}C nuclei of the guest TMPA in a D$_2$O solution containing either phenol-p-sulfonate (PS) or calix[4]arene-p-sulfonic acid are shown in Figure 2. PS is added to the solution in the corresponding concentration in order to mimic the properties of the calixarene in the solution without exhibiting the host–guest interactions with the TMPA.

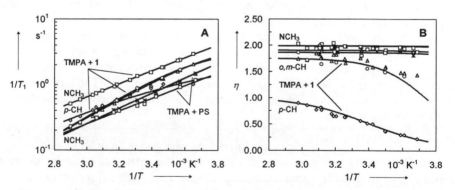

Figure 2. ^{13}C NMR relaxation rates $1/T_1$ (A) and {^{1}H}–^{13}C nuclear Overhauser factors η (B) of TMPA with phenol-p-sulfonate (PS) or calix[4]arene-p-sulfonic acid (1) in D$_2$O

Furthermore, the longitudinal ^{14}N NMR relaxation rates of TMPA were measured for a solution of TMPA with PS or the calixarene **1** in a temperature range from 271 to 323 K (Figure 3).

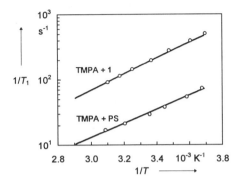

Figure 3. ^{14}N NMR relaxation rates $1/T_1$ of TMPA with phenol-p-sulfonate or calix[4]arene-p-sulfonic acid in D_2O

The relaxation data for the different solutions were fitted to the corresponding equations [2]. From the fit results the relaxation data were recalculated and plotted in the above Figures as lines. In Figure 4 the correlation times for reorientation of the ^{13}C—^{1}H vectors, which were calculated from the fits, are plotted as a function of reciprocal temperature.

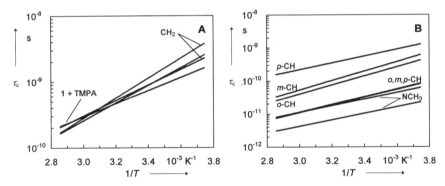

Figure 4. Reorientational correlation times of the ^{13}C—^{1}H vectors for calix[4]arene-p-sulfonic acid (1) without and with TMPA (A), and for TMPA with phenol–p–sulfonate (PS) or calix[4]arene-p-sulfonic acid (1) in D_2O (B, upper curves: TMPA with calixarene 1)

The results illustrate clearly that the effect of host–guest interactions is not very pronounced on the relaxation data and reorientational motion of the calixarene itself whereas it is significant for the guests. Particularly different are the ^{14}N spin–lattice relaxtion rates: They are by one order of magnitude larger for the guest TMPA interacting with the calixarene.

The larger correlation times for the guest molecules probably arise from a locking of the guest to the host calixarene which diminishes the velocity of reorientation. The small effects of the host–guest interactions on the rotational dynamics of the calixarene molecules can be explained by the fact that the guest molecules are stuck almost completely into the host molecules (the „calix") which results in a not much larger volume of the reorienting host–guest unit when compared to the calixarene molecule alone.

3 Conclusions

The relaxation data demonstrate that the investigated tetraalkylammonium cations exhibit host–guest interactions with the p-sulfonatocalix[4]arene in aqueous solution. Besides informations about the molecular dynamics also statements about the structure of the host–guest complexes in aqueous solution can be made.

4 Acknowledgements

Financial support by the Deutsche Forschungsgemeinschaft (Do 363/2) and Fonds der Chemischen Industrie is gratefully acknowledged, and the authors thank Prof. Dr. M. D. Zeidler for his support of this work.

5 References

1. (a) Gutsche, C. D. (1985) The Calixarenes, in Vögtle, F. and Weber, E. (eds.), *Host Guest Complex Chemistry. Macrocycles. Synthesis, Structures, Applications*, Springer-Verlag, Berlin-Heidelberg-New York-Toronto, pp 375-421. (b) Vicens, J. and Böhmer, V. (eds.) (1991) *Calixarenes: a versatile class of macrocyclic compounds*, in Davies, J. E. D. (ed.) *Topics in Inclusion Science*, vol. 3, Kluwer Academic Publishers, Dordrecht-Boston-New York. (c) Shinkai, S. (1993) Calixarenes as the Third Supramolecular Host, *Adv. Supramol. Chem.* **3**, 97-130. (d) Shinkai, S. (1993) Calixarenes – The Third Generation of Supramolecules, *Tetrahedron* **49**, 8933-8968. (e) Takeshita, M. and Shinkai, S. (1995) Recent Topics on Functionalization and Recognition Ability of Calixarenes: The 'Third Host Molecule', *Bull. Chem. Soc. Jpn.* **68**, 1088-1097.
2. Antony, J. H., Dölle, A., Fliege, T, and Geiger, A. (1997) Structure and Dynamics of p-Sulfonatocalix[4]arene and Its Hydration Shell. Nuclear Magnetic Relaxation Results, *J. Phys. Chem. A* **101**, 4517-4522.
3. Shinkai, S., Araki, K., Matsuda, T., Nishiyama, N., Ikeda, H., Takasu, I., Iwamoto, M. (1990) NMR and Crystallographic Studies of a p-Sulfonatocalix[4]arene–Guest Complex, *J. Am. Chem. Soc.* **112**, 9053-9058.

CATION COMPLEXATION AND STEREOCHEMISTRY OF MACROCYCLES STUDIED BY ^{13}C NMR RELAXATION-TIME MEASUREMENTS

ÇAKIL ERK
Technical University of Istanbul, Chemistry Department, Science Faculty, Maslak, 80626, Istanbul, Turkey.

Abstract

Cation binding by macrocycles was studied using ^{13}C NMR dipole-dipole relaxation measurements in solution. The association constants of [3n]Crown-n and its cyclohexyl, benzo and dioxo derivatives with Li^+, Na^+, K^+ and Ca^{2+} cations were studied in polar solvents. The relaxation times, T_1^{obs} and NOE factors (η) of ^{13}C atoms in the backbone of free and complex macrocycles were estimated in the extreme narrowing NMR conditions and gave T_0^{DD} and T_{10}^{DD} dipolar relaxation times. The internal motions coupled with the dipolar relaxation times are influenced by cation binding characterized by the association constants, K_a. The experimental results showed the effect of ionic radii and macrocycle size on ion selectivity as well as on the selectivity between non equivalent binding sites. The energy barriers of the preferential *a, ±g, a* conformational sequences present in the bound oxyethylene backbone were examined by molecular dynamics simulations using the MM+ force field in commercial software.

1. Introduction

The ion binding of cation receptors has been extensively studied by different analytical methods [1]. However, NMR techniques, including relaxation time measurements, are the most powerful methods now available to study ion-molecule interactions in solutions. Macrocycles ranging from podands and coronands to cyriptates and calixarenes have been studied in this way. Both partners in the interaction, the cation and the macrocycle can be examined by NMR. Cation quadrupolar relaxation rate studies have opened a new field that has provided information with which to test earlier theories and has led to new concepts on bio-chemical receptors [2]. Similarly, studies on **podands** have revealed the sensitivity of the method, even in the case of weak cation interactions [3]. Molecular motion within a complex is coupled to structural complementarity between complexed sides, which has been studied by intramolecular relaxation rate, $(1/T_1)_{intra}$ measurements. In particular, the $[^1H]$-^{13}C NMR method is reliable if it is conducted at various temperatures together with NOE factor measurements [4,5,7-9].

The stability of a cation-receptor system is determined by the exchange rate of both sides. However, the life time of a complex is usually more difficult to measure due to solvent and complex exchange rates [6]. The effect of cation binding is usually a decrease in the relaxation time that depends on the selectivity and the stability of the receptor [4,8,9].

The ability of the macrocycle to adopt a conformation with the oxygen dipoles oriented towards a cation of a given radius, necessary for cation encapsulation, may be examined via computer simulations using molecular dynamics [11,12].

2. Experimental

The relaxation times of free and complexed macrocycle backbone ^{13}C atoms, T_0, T_{10} and NOE factors were estimated at 50 MHz in appropriate polar solutions. Measurements were made using the inversion recovery method at various temperatures. Under these conditions extreme narrowing ($\omega_0 \tau_c \ll 1$) applies. The energy barriers on the internal motions of the free and cation complexed spins were calculated, together with the binding energy. The complexed macrocyclic ligand molar fractions, P_{10}, and the association constants, K_a, of the complexes were determined from the combination of the relationships of $1/T_1^{DD} = S\ P_0 / T_0^{DD} + P_{10} / T_{10}^{DD}$ and $1/\{K_a [L_o]\}^{n+m-1} = (1-nP_{10})^n (1-mP_{10})^m / P_{10}$ using the total cation, $[A_o^+]$ and macrocycle concentrations, $[L_o]$ in a complex solution. The computational molecular dynamics and theoretical energy barriers of $a, \pm g, a$ conformational sequences of free oxyethylene backbone, (Table 2), were calculated with MM+ methods using Hyper®chem, ver 4.0.

Representative results are displayed in Tables 1 and 2, Additional data can be found in [9].

3. Results

Comparison of dipolar relaxation rate measurements and the energy of optimised geometry as well as molecular dynamics calculations are giving us a better understanding of the basis of internal motions and pseudorotations of macrocycles [3,7,11].

Previous systematic work on this topic not only estimated the ion binding data of such compounds but also revealed the aspects of binding of diamagnetic cations of biochemical interest. However, we are now faced with the need to improve the analytical methods for the measurement of association constants (Table 1) [8,9].

Similar macrocycles were also studied using an equivalent analytical formalism with potentiometry and fluorescence spectroscopy [10].

The association constants of [3n]crown-n, benzo[3n]crown-n, cyclohexano[18]-crown-6, 2,6-dioxo[18]crown-6, and 2,6-dioxo-[15] crown-5 derivatives that complex with Li^+, Na^+, K^+ and Ca^{2+} cations were studied in polar solvents. The relaxation times of free and complexed ^{13}C spins, T_0 and T_{10} and NOE contributions of macrocycle backbones showed the effect of the cation radii and the ring size. The results were consistent with those reported elsewhere [3-5,7-10].

317

The T_1-NMR results, (Table 1) showed the reliability of the method. However, the roles of solvents and the concentration range of the macrocycle and the cation are not negligible.

Table 1. The 1:1 Association Constants of macrocycles found with T_1^{DD} measurements, Refs 8,9.

Cation	Macro cyc.ether	Temp.	Solvent	$LogK_a$ (1:1)
Li^+	[12]crown.4	25 °C	DOH	-.55
Na^+	[18]crown.6	25 °C	DOD	1.18
K^+	[18]crown.6	25 °C	DOD	2.47
Na^+	[18]crown.6	25 °C	Methanol-d	1.28
K^+	[18]crown.6	25 °C	Methanol-d	2.08
Na^+	DC[18]crown.6	25 °C	Methanol-d	1.44 (anti)
Na^+	DC[18]crown.6	25 °C	Methanol-d	2.19 (syn)
K^+	DC[18]crown.6	25 °C	Methanol-d	1.88 (anti)
K^+	DC[18]crown.6	25 °C	Methanol-d	3.19 (syn)
Ca^{2+}	DO[18]crown.6	25 °C	Methanol-d	1.8 - 2.8
Na^+	Bnz[18]crown.6	variable	Acetonitrile	varied
Na^+	Bnz[15]crown.5	variable	Acetonitrile	varied
Na^+	Bnz[12]crown.4	variable	Acetonitrile	varied

Table 2. NMR data and Molecular Dynamics of CC_n-OC of complexing of Benzo-18-crown-6

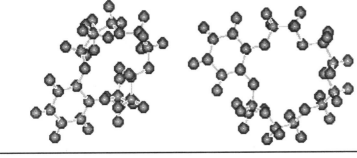

Relax Time	Barrier energy of (n)	C 1	C 2	C 3	C 4	C 5
T_{10}^{DD}	Experimental (a)	1.78	1.72	1.59	1.25	0.84
	anti,anti,anti (c)	13.14	12.58	10.84	10.58	6.10
T_0^{DD}	Experimental (b)	2.00	2.00	1.54	1.75	1.54
	anti,anti,anti (c)	10.35	9.94	9.08	9.81	7.45

(a) complexed molecule, $-E_b/R$ kJ mol^{-1}, (b) free molecule, 0.12 mol l^{-1},
(c) MM+ molecular dynamics results kcal mol^{-1} obtained from the Hyperchem.

4. References

1. Lisegang, G.W. Eyring, E.M., Synthetic Multidentate Macrocyclic Compounds, Izatt, R.M. and Christensen, J.J., Eds, (1978) Academic, New York , pp-254-260.
2. Forsen, S., Drakenberg, T. and Wennerström, H. (1987) NMR Studies of Ion Binding in Biological Systems, *Quarterly Rewiews of Biophysics,* **19(1/2)**, pp 83-114.
3. a. Detellier, C. and Laszlo, P. Topics in Molecular Interactions, Thomas, W.J.O-, Ratajczak, H. and Rao, C.N.R. Eds. (1985) *Studies in Physical and Theoretical Chemistry,* Vol 37, Elsevier, Amsterdam. pp 291-336. b. Hertz, H.G., Kratochwil, A. and Weingärtner, H. Topics in Molecular Interactions, Thomas, W.J.O-., Ratajczak, H. and Rao, C.N.R. Eds (1985*) Studies in Physical and Theoretical Chemistry,* Vol **37**, Elsevier, Amsterdam. pp 337-402.
4. Grootenhuis, P.D.J., van Eerden, J., Sudhölter, E.J.R., Reinhoudt, D.N., Roos, A., Harkema, S. and Feil, D. (1987*)* Complexation of Macrocyclic Polyethers Studied by Carbon-13 NMR Relaxation Time Measurements and Molecular Mechanics, *J.Amer.Chem.Soc.,***109**,4792-4797.
5. Echegoyen, L, Kaifer, A., Durst, H.D. Gokel, G.W. (1984) Carbon-13 Chemical Shifts and Relaxation Times as a Probe of Structural and Dynamic Properties in Alkali and Alkaline Earth Cryptate Complexes, *J.Org.Chem.***49**,688-690.
6. Srehlow, H. and Knohe, W. (1977) Fundamentals of Chemical Relaxation, VCH, Weinheim, pp 1-98.
7. a. Erk, Ç. (1986) The ^{13}C Dipole-Dipole Relaxation and Pseudorotation of some Macrocyclic Ethers and a Polyoxa-lactone, *Appl.Spectrosc.,***41**.100-103, see refs. b. Erk, G. (1989) ^{13}C Dipole-Dipole Relaxation Time and Internal Motions of Oligocyclic Ethers, *J.Mol.Liqs.,***40**,1-15.
8. a. Erk, Ç. (1985) Determination of The Association Constants of Macrocyclic Ethers by the Aid of ^{13}C Dipole-Dipole Relaxation Time Measurements. Part I, *Spectrosc.Lett.,***8(9)**,723-750. b. Erk, G. (1986) The Dipole-dipole Relaxation Time and Internal Motions of Cyclic Oligoethers. Part III, *Spectrosc.Lett.,***19 (10)**,1173 -1182. c. Erk, G. (1988) The Determination of Macrocyclic Ether Complexes by use of ^{13}C Dipole-Dipole Relaxation Time Measurements. Part. IV, *J. Mol. Liqs.,***37**,107-115.
9. a. Erk, Ç. (1988) Host-Guest Interaction Mechanisms with ^{13}C Dipole-Dipole Relaxation Time Measurements. Part V, Equilibrium Constants of Potassium and Sodium Acetates Complexing with 18.crown.6 in CH_3OD, *J.Mol.Liqs.,***40**,143-153. b. Erk, G. (1990) The ion-dipole Interaction Mechanism of Macrocyclic Ethers with ^{13}C Dipole-Dipole Relaxation Time Measurements, Part VI, *J.Phys.Chem.,***94**,8617-8621 c. Erk, G. (1996) Ion-Dipole Interactions of Macrocyclic Ethers using ^{13}C Dipole-Dipole Relaxation Time Measurements, Part XI, *J.Incl.Phen.,***26**,61-66 d. Erk, G. and Zeidler, M.D, (1997) ^{13}C NMR Relaxation Time of Benzo[18]crown-6 and Complexing in Acetonitrile with $NaClO_4,H_2O$, *J.Mol.Liqs.,* in press.
10. a. Gakir, Ü and Erk, G. (1991) The Complexation of 18.crown.6 with NaCl in Dioxan-Water and Determination of Equilibrium Constants with ISE, *Thermochim.Acta,***178**,67-73. b. GöÁmen, A. and Erk, Ç. (1993) Potentiometric Determination of Stability Constants of Macrocyclic Ethers using Linear Regression Techniques, *Fresenius J. Anal.Chem.,***347**,471-474. c. GöÁmen, A. and Erk, Ç. (1996) The Cation Binding of Benzo Crowns in Acetonitrile using Fluorescence Spectroscopy, *J. Incl.Phen.,***26**, 67-72.
11. Anet, F.A.L and Anet, R. (1975) Conformational Process in Rings, Jackman, L.M. & Cotton, F.A., Eds., *Dynamic NMR Sspecroscopy*, Academic, pp 543-616.
12. Troxler, L. and Piff, G. (1994) Conformation and Dynamics of 18.Crown.6, Cyritand 222 and their Cation Complexes in Acetonitrile Studied by Molecular Dynamic Simulations, *J.Am. Chem.Soc.,***96**,116,1468.

A COMPARATIVE STUDY ON LIQUID STATE CSA METHODS

GY. BATTA[1], K. E. KÖVÉR[2], J. KOWALEWSKI[3]
[1]*Research Group for Antibiotics of the Hungarian Academy of Sciences, H-4010 Debrecen, P. O. Box 70, Hungary*
[2]*Department of Chemistry, L. Kossuth University, H-4010 Debrecen, P.O.Box 20, Hungary*
[3]*Division of Physical Chemistry, Arrhenius Laboratory, University of Stockholm, S-106 91 , Stockholm, Sweden*

1. Introduction

There is a gaining interest in obtaining chemical shift anisotropy data of molecules in solution. Relaxation interferences between two different second rank tensorial relaxation mechanisms are manifested in differential multiplet relaxation [1] or line broadening [2], and this effect can be put to good use in structural and dynamics studies. We propose enhanced CSA/DD cross-correlated relaxation experiments for the measurement of chemical shift anisotropy in solution and demonstrate that their performance can be improved by spin-lock (SL) and/or field gradient pulses. Though the measured effects may be as small as 0.002 s^{-1}, the methods are often suitable at natural isotopic abundance. ^1H and heteronucleus CSA data were measured for triphenyl-silane (TPSi) (^{29}Si), cyclosporin A (^{15}N) and α-D-Trehalose, a symmetric disaccharide of glucose, (^{13}C). 1D data are evaluated both with the initial rate and the Redfield relaxation matrix approach providing cross-correlated relaxation rates and CSA values. 1D experiments and the 2D unbiased method (2D in the sense of magnetization modes too) were compared for our model compounds. The comparison gives experimental evidence on the equivalence of the 1 and 2D methods within the limits of two-spin approach. For the easy application of the 2D method double G-BIRD, X-filtered ^1H-^1H NOESY is suggested. It is shown that transversal and rotating frame (ortho-ROESY) experiments provide comparable ^{13}C-CSA effects, and both methods are suitable for macromolecules. Recent studies suggest that simultaneous use of longitudinal/transversal methods may yield additional dynamics/exchange information for proteins [3].

2. Improved longitudinal methods (X-detected, one-dimensional)

These techniques are most suitable for small to medium sized molecules in between the extreme narrowing regime and a few nanoseconds global correlation time. For example, the resolving power of ^{13}C NMR allows simultaneous ^{1}H and ^{13}C CSA measurements of all CH groups within a molecule at natural abundance. In the polarisation-transfer sequence [4] (Fig. 1) pure <zz> order creation and efficient decoupling is of vital importance, since one-spin order detection is achieved by decoupling the antiphase part from the signal. Spin-lock and/or gradient purging is essential, especially at short τ mixing times.

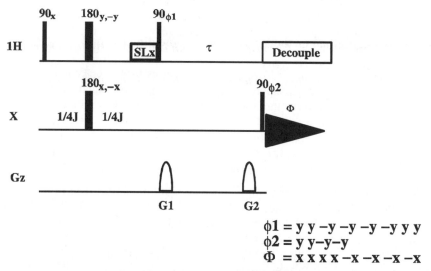

Figure 1. X-CSA Sequence, (Polarisation Transfer Enhanced <zz> --- <z>)

^{1}H CSA experiments are simple and efficient with the DQ-filtered transient heteronuclear NOE experiment [4] and refocussing of the antiphase signal allows ^{1}H decoupling during acquisition

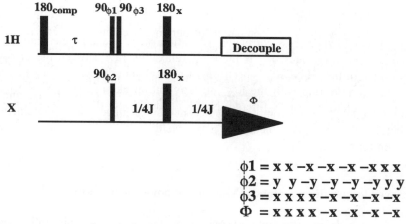

Figure 2. ^{1}H -CSA Sequence (DQF NOE, <z> --- <zz>)

^{1}H decoupling enhances signal to noise ratio, and initial rate measurements are feasible at ca. 1M concentrations even for aliphatic ^{1}H/^{13}C vectors (which have only moderate 5/30 ppm chemical shift anisotropy). ABX type strong coupling in the ^{1}H {^{13}C} satellite spectrum of carbohydrates may interfere with the signal arising from cross-correlated relaxation and potentially leads to misinterpretation off cross-correlated rates [5]. Such strong coupling is not present in oligopeptides or labeled proteins since the α protons are well separated from the NH region.

3. Improved longitudinal methods (^{1}H-detected, one-dimensional)

In the case of ^{15}N or ^{29}Si nuclei low sensitivity requires proton detection schemes. At natural abundance, very high suppression rate of background magnetization may be achieved (up to ca. 30.000). In the X-CSA experiment INEPT is utilized in preparing the initial -X_z magnetization and at this point background magnetization can be destroyed by strong orthogonal spin-lock fields or z-gradients. Then the sequence is analogous to DQF methods with optional refocussing and ^{1}H spin-lock purging before acquisition. The ^{1}H CSA terms can be measured from another sequence, where proton inversion is followed by DQ-filter, spin-lock purging (optional refocussing) and detection (optional X-decoupling). When the CSA/DD rates are very small (in case of TPSi the maximum interference signal is only half of the ^{13}C satellite of the central ^{29}Si line, i.e. ca. 0.25%) then X-decoupling is not recommended in the ^{1}H detected variant. To the contrary, longer correlation time and significant ^{15}N and ^{1}H CSA effects in unlabelled Cyclosporin A allow X-decoupling, and as a unique example ^{15}N CSA could be measured at natural abundance.

4. Two-dimensional (unbiased) method

It has been described [6] that three different pairs of semiselective experiments (^{1}H-NOESY, X-NOESY and HOESY), after suitable algebraic transformations give the full relaxation matrix R. For a two-spin system, the R_{31} and R_{32} elements are the proton and X nucleus CSA/DD cross-correlated relaxation rates. Diagonal elements of R are the intrinsic relaxation rates, R_{21} is the cross-relaxation rate between the proton and the X nucleus. In the ^{1}H-NOESY spectrum, at natural abundance, the long tails of the parent lines may overlap with the low intensity auto-peaks of the H-X doublet, and distort peak volumes. An X-filtered NOESY (double G-BIRD element) in fact eliminates this problem and gives more reliable proton CSA terms.

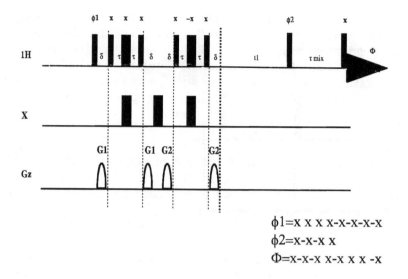

Figure 3. Double G-BIRD, X-filtered NOESY

A comparison of 2D unbiased and 1D initial rates for the case of a 1.7 M disaccharide (α-D- trehalose, τ_C = 0.5 ns at 300K) gave identical results for all elements in the R matrix within error limits. It is important, that initial rates are free from multispin effects. Simultaneous fit of the 1D ^{1}H and ^{13}C-CSA experiments with the Trudeau method slightly underestimates the ^{1}H CSA term, perhaps because proton-proton interactions are non-negligible at longer evolution times.

5. Transversal methods

Transversal CSA/DD cc. rates are sensitive to low frequency molecular tumbling and are roughly monotonic function of global reorientational correlation time. Sensitive, 2D methods were recently developed [8,9], mainly for ^{15}N labelled protein applications. However, these methods are less suitable for small or medium sized molecules, since the experiments require very short evolution time. The rotating frame "ortho-ROESY" [10] experiment is less sensitive to longer evolution times. Using the same disaccharide model a gradient variant was successfully tested, and the measured ^{13}C CSA value was identical with the longitudinal values.

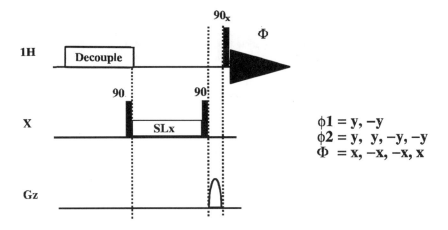

Figure 4. ^{13}C ortho-ROESY, ^{1}H detection, gradient variant

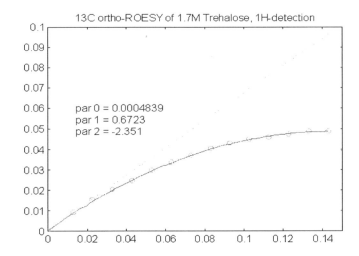

Figure 5. Build-up of antiphase magnetization in ortho-ROESY experiment

6. Conclusions

All the methods compared in this study gave nearly identical results for the requested CSA terms. Not surprisingly, at natural abundance, when X-detection scheme is used, the detection of ^{1}H- CSA terms is easier, while ^{1}H detection is less demanding for ^{13}C-CSA measurement. Theoretically, for the simplest cases, spin-locked and not locked transversal methods have the same sensitivity.

7. References

1. Werbelow, L. G. (1996) Relaxation Processes: Cross-correlation and interference terms, in *"Encyclopedia of Nuclear Magnetic Resonance"* (D. M. Grant and R.K. Harris, eds. Vol 6) pp .4072-4078, Wiley, London
2. a) Guéron, M., Leroy, J. L., and Griffey R. H. (1983) Proton nuclear magnetic relaxation of ^{15}N-labeled nucleic acids via dipolar coupling and chemical shift anisotropy, *J. Am. Chem. Soc.* **105**, 7262-7266. b) Goldman, M. (1984) Interference effects in the relaxation of a pair of unlike spin-1/2 nuclei *J. Magn. Reson.* **60**, 437-452.
3. Palmer, A.G. (1998) Spin relaxation methods for characterizing picosecond-nanosecond and microsecond-millisecond motions in proteins. NATO Advanced Research Workshop, Sitges, 5-9 May,
4. Batta, Gy. and Gervay, J. (1995) Solution-phase ^{13}C and ^{1}H chemical shift anisotropy of sialic acid and its homopolymer (colominic acid) from cross-correlated relaxation *J. Am. Chem. Soc.* **117**, 368-374.
5. Batta, Gy., Kövér, K. E., Gervay, J., Hornyák, M. and Roberts, G. M. (1997) Temperature dependence of molecular conformation, dynamics, and chemical shift anisotropy of α,α-Trehalose in D$_2$O by NMR relaxation, *J. Am. Chem. Soc.* **119**, 1336
6. Maler, L. and Kowalewski, J. (1992) Cross-correlation effects in the longitudinal relaxation of heteronuclear spin systems, *Chem. Phys. Lett.* **192**, 595-600.
7. Trudeau, J. D., Bohmann, J., and Farrar, T. C. (1993) Parameter estimation from longitudinal relaxation studies in coupled two-spin 1/2 systems using Monte Carlo simulations, *J. Magn. Reson A.* **105**, 151-166.
8. Tjandra, N., Szabo, A., and Bax A. (1996) Protein backbone dynamics and ^{15}N chemical shift anisotropy from quantitative measurement of relaxation interference effects, *J. Am. Chem. Soc.* **118**, 6986-6991.
9. Tessari, M., Mulder, F. A. A., Boelens, R., and Vuister, G. W. (1997) Determination of amide proton CSA in ^{15}N-labeled proteins using ^{1}H CSA/^{15}N-1H dipolar and ^{15}N CSA/^{15}N-^{1}H dipolar cross-correlation rates, *J. Magn. Reson.* **127**, 128-133.
10. Brüschweiler, R. and Ernst, R. R. (1992) Molecular dynamics monitored by cross-correlated cross relaxation of spins along orthogonal axes, *J. Chem. Phys.* **96**, 1758-1766.

HYDROGEN BONDING AND COOPERATIVITY EFFECTS ON THE ASSEMBLY OF CARBOHYDRATES

M. LÓPEZ DE LA PAZ, J. JIMÉNEZ-BARBERO AND C. VICENT*
Departamento de Química Orgánica Biológica, Instituto de Química Orgánica (CSIC)
Juan de la Cierva 3, E-28006 Madrid (Spain)

Hydrophobic interactions, hydrogen bonding and cation binding are at the origin of the recognition processes in which carbohydrates are involved. Their relative participations are related to the topology of the carbohydrate, to the particular orientation of the residues exposed to the interaction, and to the medium which surrounds the sugar. The study of carbohydrate OH···OH hydrogen bond energetics is of fundamental interest for understanding these recognition processes.

1. Role of intramolecular OH···OH hydrogen bonds on the self-assembly behaviour of carbohydrates

A relevant feature of multiple hydrogen-bonded complexes is the non additivity of molecular interactions, which has given rise to the concept of cooperativity, considered as the stregnthening of a first hydrogen bond (HB) between a donor and an acceptor when a second HB is formed between one of these two species and a third partner.[1] Theoretical[2] and experimental[3] methods have allowed the quantification of this effect. Evidences of intra- and intermolecular σ-cooperativity in carbohydrates has arisen from neutron diffraction data of crystalline structures of mono- and disaccharides and protein—carbohydrate complexes, confirming that the HBs length is shorter for those OHs involved in multiple intramolecular HBs than for isolated OH···OH bonds.[4] In solution, cooperativity in carbohydrates has been studied by NMR in polar solvents[5] and by FT-IR;[6] but the geometric requests and energetic implications of cooperativity in carbohydrate recognition have never been evaluated in detail.

Within a project to design self-assembled structures based on carbohydrate intermolecular OH···OH HBs, the energetic advantage of establishing cooperative intermolecular HBs for assembling simple carbohydrates, and the relative configuration of hydroxy groups that favours this process have been evaluated. Therefore the possibility of self-assembly of 1,2- and 1,3-diols and amidoalcohols, of different relative configuration and position with respect to the anomeric centre and different conformational flexibility, have been explored. The 1,3-diaxial diol **1** (figure 1), a 3-amido-1,6-anhydro-3-deoxy-β-D-glucopyranoside derivative, showed significant aggregation behaviour in $CDCl_3$. Diol **1** involves position 2 and 4 of the pyranose sugar ring. The proximity of OH-2 to the anomeric center makes it more acidic, therefore it is expected an intramolecular HB polarized in one direction (O2-H2→O4-H4).

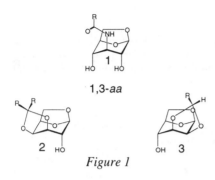

Figure 1

A detailed study of the intramolecular HB network for the monomer of **1** was carried out in diluted chloroform solution by ^{1}H-NMR. Additionally, monoalcohols **2** and **3** were used as models to study the influence of a second hydroxy group with a 1,3-diaxial orientation, as present in diol **1**, on intermolecular cooperativity. $^3J_{CHOH}$ analysis shows that all compounds have a chair conformation in CDCl$_3$ solution.

Monoalcohols **2** and **3** in CDCl$_3$ at low concentration show high $^3J_{CHOH}$ values (9Hz), consistent with a fixed conformation of the CHOH angle (larger than 150°), which can be attributed to an hydrogen bonded OH. For the 1,3-diaxial diol **1**, the OH−2 resonance follows the same trend (as expected for hydrogen bonded to OH−4 or O−5); but, in contrast, the OH−4 resonance has a medium size $^3J_{CHOH}$ value, indicating that it is not hydrogen bonded. Neither of OH−2 or OH−4 resonances achieve exchange decoupling at any accessible concentration, a characteristic feature of fixed OHs.[7]

Partial deuteration of **1** in CDCl$_3$ at low concentration show that the OH−2 resonance has a negative isotopic effect (−0.0165 ppm) consistent with OH−2 being donor.[8] Therefore, these results show that both monoalcohols have its hydroxy groups intramolecularly fixed by a HB to O−5. In contrast, for **1** the 1,3-diaxial orientation of both hydroxyls favours OH−2 to be hydrogen bonded, as a donor, to OH−4. These HBs must affect the self-assembly characteristics of the different compounds.

The characterization of the aggregates in solution was done by different methods.

^{1}H-NMR dilution experiments in CDCl$_3$ at 299K of **1-3**, allowed us to calculate the stability constant of the dimerization process. Dilution data were fitted to a dimerization process using an infinite non-cooperative model. Neither of the monoalcohols dimerize. In contrast, diol **1** presents a dimerization constant of 70 M^{-1} at 299 K in CDCl$_3$ ($\Delta G° = -2.5 \pm 0.1$ kcal mol^{-1}) and 150 M^{-1} in CDCl$_3$-CCl$_4$ (1-1.3) mixture ($\Delta G° = -3.0 \pm 0.1$ Kcal mol^{-1}). In order to check the self-assembly ability of a non intramolecular hydrogen-bonded hydroxy group, dilution experiments of ethanol in CDCl$_3$-CCl$_4$ were performed, showing a dimerization constant of 0.1 M^{-1} for the dimer of ethanol.

Vapour pressure osmometry measurements (VPO) in CDCl$_3$ suggested that, for a concentration range between 0.05-0.01 M, the monoalcohols **2** and **3** are monomers. On the other hand, **1** presents a molecular weight which corresponds to 1.6 times that of the monomer. This value is in agreement with the percentage of dimer which is present in solution according to the NMR-derived stability constant (80 %).

Chemical shifts, $^3J_{CHOH}$ values and temperature coefficients also indicate that the assembly is mediated by OH···OH hydrogen bonds. Table 1 shows that the temperature coefficients of OH−2 for **2** and OH−4 for **3** are not concentration dependent, in contrast with the observations made for **1**. The $^3J_{CHOH}$ values of **1-3** at high concentration indicate their involvement in HBs, with the exception of OH−4 of **1** which shows a $^3J_{CHOH}$ value of 5.4 Hz at all concentrations. The NH of **1** does not show any concentration dependence of NMR parameters. This experimental evidence is consistent with the amide not being involved in the self-association process. Additionally, a concentration dependent FT-IR experiment shows no evidence for intermolecular hydrogen bonded species involving the NH or the CO of the amide (v_{NH} = 3335 cm^{-1} and v_{CO} = 1772 cm^{-1} at high and low concentration).

The clear difference in the solution self-assembly behaviour of **1** respect to monoalcohols **2** and **3** indicates that the addition of one extra OH in a pyranoid ring, having a 1,3-*syn* diaxial orientation, accounts for an extra-stabilization of the dimer of 2.5 kcal mol^{-1} in CDCl$_3$, compared to the monoalcohols.

Table 1. ^{1}H-NMR Chemical Shifts, Coupling Constants, and Temperature Coefficients of OH Resonances of **1, 2, 3** at two Different Concentrations

Compound	C (mM)	OH-2			OH-4		
		δ (ppm)	Δδ/ΔT^a (ppb/K)	$^3J^b$ (Hz)	δ (ppm)	Δδ/ΔT^a (ppb/K)	$^3J^b$ (Hz)
1	110	4.00	−10.0	9.0	4.68	−16.0	5.4
	0.05	2.82	−2.3	10.2	2.73	−4.0	5.4
2	110	2.21	−5.5	8.4	—	—	—
	0.05	1.95	−2.2	8.7	—	—	—
3	110	—	—	—	2.53	−5.2	8.7
	0.05	—	—	—	2.30	−2.4	9.3

^{a}Measured between 297 K and 313 K ^{b}Data at 298 K

Thermodynamic parameters for the dimerization of **1** in a CDCl$_3$-CCl$_4$ (1-1.3) mixture were obtained from a Van't Hoff plot (from 296-318 K). $\Delta H°$ = −6.5 kcal mol^{-1} and $\Delta S°$ = −11.8 cal mol^{-1} K^{-1} were estimated. The stability constant of the dimer at 299 K was 150 M^{-1} ($\Delta G°$ = −3.0 ± 0.1 Kcal mol^{-1}).

To quantify the effect of cooperativity on the dimerization process, it is important to know the structure of the dimer. In principle, the structure of the dimer present in solution could be of two types (Figure 2): A (open dimer), with one cooperative intermolecular HB stabilizing the dimer, and B (closed dimer), with two intermolecular HBs established in a cyclic and cooperative way. In both cases, and according to the isotopic effects, the directionality of the intramolecular HB should be OH–2→OH–4. However, the implication of O–5 in the network cannot be excluded. NOESY experiments of a concentrated solution of **1** (0.1 M) were performed in order to obtain structural information about the dimer. Since two free OH groups are present in the molecule, regular experiments as well as MINSY-type spectra were recorded in order to exclude chemical exchange-mediated cross peaks.[9]

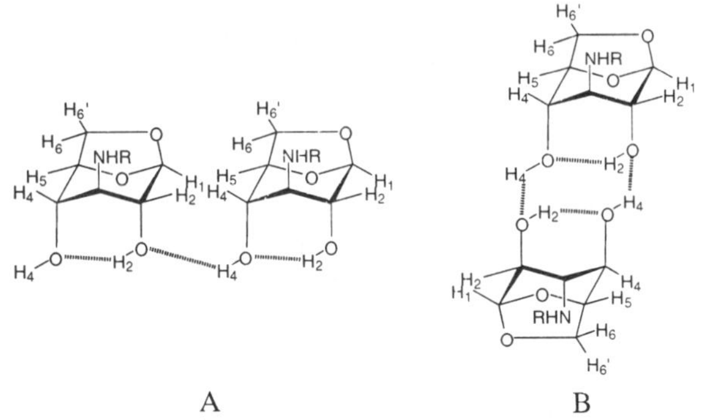

Figure 2

Thus, besides the regular NOESY spectrum, additional experiments saturating OH–2 and OH–4 hydroxyls were recorded. The obtained results unambiguosly indicate the presence of intermolecular NOEs. In particular, H–1/H–6exo, H–1/H–6endo, H–1/H–5, H–2/H–4, H–5/H–2 and H–1/H–4 cross peaks were detected with weak and medium-weak intensities. These NOEs are not seen when the experiments are carried out at low concentration, when no dimer is present in solution. These NOEs are only compatible with the structure of an open dimer. This open structure is also supported by the measured $^3J_{CHOH}$ values. MM2* calculations (carried out using MM2* with the GB/SA solvent model for chloroform) of this structure show that it is stable and account for the observed NOEs. This structure shows that the carbonyl group cannot be hydrogen bonded to any donor moiety.

Thus, in the assembly of carbohydrate **1** only one intermolecular cooperative OH⋯OH bond accounts for 2.5 kcal mol^{-1} in CDCl$_3$ and for 3 kcal mol^{-1} in CDCl$_3$-CCl$_4$.

2. Conformational restriction using intramolecular hydrogen bonds. Application to carbohydrate self-assembly

H-bonding involving neutral amidic-type NH and carbonyl groups has played a predominant role in the design of unique self-assembled supramolecular structures. The possibility of generating self-assembled supramolecular structures based on intermolecular H-bonding of hydroxyl groups has not been demostrated until very recently.[10]

We have used the 1,3-diaxial diol showed in figure 3, as an hydroxyl-based interaction unit and explored its ability to provide conformational control of self-recognition processes by intramolecular H-bonding.

The first step to achieve a well defined self-assembled structure is controlling the three dimensional structure of the monomeric unit. Receptor **4** (figure 3) has been designed for the purpose of OH⋯OH hydrogen bonding-mediated self-assembly. Molecular modeling of **4** showed that the linkage of the spacer to position 3 of the sugar, via an amide moiety, could allow the formation of intramolecular NH⋯O–1 hydrogen bond, thus fixing the conformation around the HN—C3 bond. This particular orientation places both hydroxyl groups (OH–2 and OH–4) pointing outwards, hence allowing the self-assembling of the carbohydrate receptor. Additionally, the formation of intramolecular NH⋯N$_{Py}$ HBs, involving the pyridine spacer,[11] allows complete control of the conformation of the molecule by locking the dicarboxylic moiety in a *cis-out* conformation.[12] In fact, MM2* calculations predicted this rotamer as the most energetically favourable with a NH⋯O–1 distance (2.3 Å) in agreement with the array of intramolecular HBs shown in figure 3.

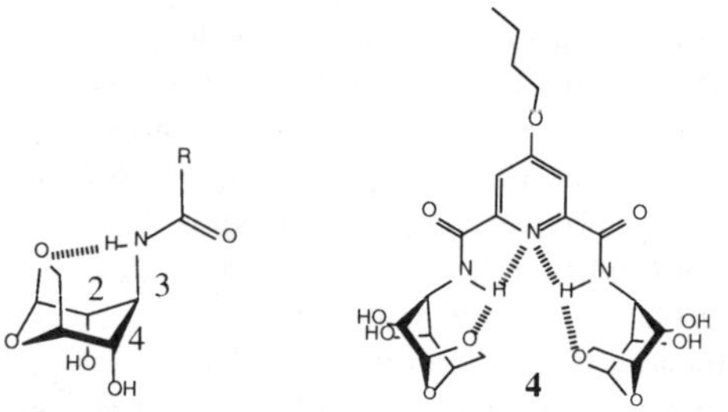

Conformational analysis of **4** by NMR in CDCl$_3$ showed that the pyranoid ring in solution is a $^1C_4$ chair. Experimental distances obtained from NOEs, and H-N-C3-H3

Figure 3

dihedral angles estimated from vicinal $^3J_{NHH3}$ values are in agreement with the MM2° calculated geometry of **4**.

Additional experimental evidences for the presence of the NH···O−1 HB in solution in the *interaction unit* came from FT-IR (CH_2Cl_2) and ^{1}H-NMR ($CDCl_3$) (variable temperature experiments and relative chemical shifts). Model compounds **7** and **8**, as well as protected amides **5** and **6** were used to proof its existence (Figure 4). FT-IR spectra showed the stretching NH stretching of **7** at 3449 cm^{-1} and at 3438 cm^{-1} for **5** (Δv = 11 cm^{-1}). Similar tendency is observed when the NH stretching vibration of **8** and **6** are compared (Δv = 26 cm^{-1}). This shifting might be attributed to the additional intramolecular NH···O−1 HB present in **5** and **6**. The ^{1}H-NMR chemical shifts (299 K, $CDCl_3$) of the NH resonances of **7** (6.18 ppm) and **5** (6.61 ppm), as well as those of the pyridin-based **8** (7.54 ppm) and **6** (8.50 ppm) are in agreement with the increasing involvement of the corresponding NH in intramolecular H-bonding. Thus, the combined FT-IR and NMR evidences suggest that the NH···O−1 HB is indeed present in solution, thus restricting the conformation around the HN-C3 linkage.

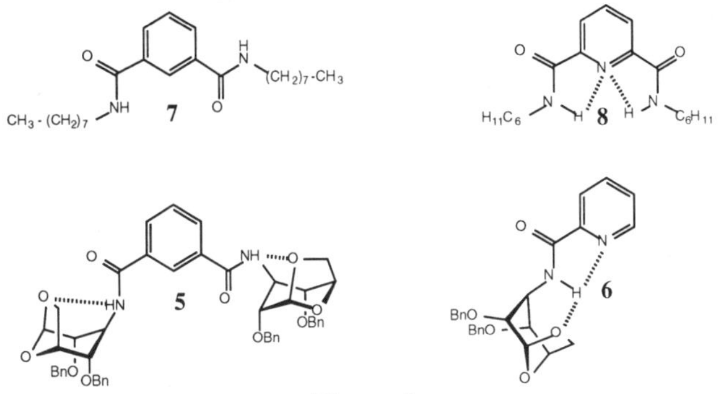

The self-assembly of **4** in $CDCl_3$ was studied by ^{1}H-NMR, monitoring the concentration and temperature dependence of the OHs and NH chemical shifts and coupling constants. OHs chemical shifts and temperature coefficients values of **4** clearly changed upon dilution (5 mM to 0.05 mM). On the other hand, the NH resonances were not modified (Table 2). ^{1}H-NMR variable temperature experiments (299-319 K) of **4-8** in $CDCl_3$ showed small temperature coefficients for the NH resonances. However, those measured for the OH−2 and OH−4 signals of **4** were larger (−0.0075 and −0.014 ppm/K, respectively) at the highest concentration. The size of these values are consistent with their involvement in intermolecular HBs.

Figure 4

Table 2. ^{1}H-NMR Parameters Dependence of NH and OHs Resonances of **4**

Signal	Concentration (mM)	δ (ppm)	Δδ/ΔT[a] (ppb/K)	^{3}J[b] (Hz)
OH2	5.00	3.34	−7.5	9.2
OH2	0.05	2.94	−2.7	10.3
OH4	5.00	3.66	−1.4	2.5
OH4	0.05	2.99	−5.2	5.1
NH	5.00	8.60	−2.9	7.8
NH	0.05	8.53	−2.4	8.4

[a]Measured between 297 K and 313 K   [b]Data at 298 K

Regarding the $^3J_{H,OH}$ couplings, OH−2 and OH−4 behave differently. $^3J_{H2,OH2}$ is independent of both concentration and temperature. The measured value (about 10 Hz) indicates a particular orientation of the hydroxyl proton, probably involved in intramolecular HB (with O−5 or/and O−4). On the other hand, at the highest concentration (5 mM), $^3J_{H2,OH2}$ varies from 2.5 Hz to 5.4 Hz between 297 K and 313 K. The measured value at low concentration (0.05 mM) is 5.0 Hz, regardless of temperature. This behaviour is consistent with a particular orientation of the OH−4 group at high concentration, probably involved in intermolecular HB.

These results indicate that the self-assembly process of the acyclic receptor **4** is mediated by intermolecular OH···OH H-bonding and not by the amide.

Acknowledgments: Financial support by DGICYT (PB96-0833) is gratefully acknowledged. M.L.P. thanks Comunidad de Madrid for a FPI Fellowship.

3. References and notes

[1] Frank, H. S. and Wen, W. Y. (1957) *Discuss. Faraday Soc.* **24**, 133.
[2] Sear, R. P. and Jackson, G. (1996) Thermodynamic Perturbation Theory for Association with Bond Cooperativity *J. Chem. Phys* **105**, 1113-1120.
[3] Maes, G. and Smets, J. (1993) Hydrogen Bond Cooperativity. A Quatitative Study Using Matrix-Isolation FT-IR Spectroscopy *J. Phys. Chem.* **97**, 1818-1825.
[4] Jeffrey, G. A. and Saenger, W. (1991) *Hydrogen Bonding in Biological Structures*, Springer-Verlag, Berlin.
[5] Christofides, J. C. and Davies, D. B. (1987) Co-operative Intramolecular Hydrogen Bonding in Glucose and Maltose *J. Chem. Soc. Perkin Trans. II*, 97-102.
[6] Zhbankov, R. G. (1992) Hydrogen Bonds and Structure of Carbohydrates *J. Mol. Struct.* **270**, 523-539.
[7] Pearce, C. M. and Sanders, J. K. M. (1994) Improving the Use of Hydroxyl Proton Resonances in Structure Determination and NMR Spectral Assignment: Inhibition of Exchange by Dilution *J. Chem. Soc. Perkin Trans. 1*, 1119-1123.
[8] Craig, B. N., Janssen, M. U., Wickersham, B. M., Rabb, D. M., Chang, P. S. and O'Leary, D. J. (1996) Isotopic Perturbation of Intramolecular Hydrogen Bonds in Rigid 1,3-Diols: NMR Studies Reveal Unusually Large Equilibrium Isotope Effects *J. Org. Chem.* **61**, 9610-9613.
[9] Massefski, W. Jr. and Redfield, A. G. (1988) *J. Mag. Res.* **78**, 150.
[10] Hanessian, S., Simard, M. and Roelens, S. (1995) Molecular Recognition and Self-Assembly by Non-amidic Hydrogen Bonding. An Exceptional "Assembler" of Neutral and Charged Supramolecular Structures *J. Am. Chem. Soc.* **117**, 7630-7645.
[11] Hamuro, Y., Geib, S. J. and Hamilton, A. D. (1994) Novel Molecular Scaffolds: Formation of Helical Secondary Structure in a Family of Oligoanthranilamides *Angew. Chem. Int. Ed. Engl.* **33**, 446-448.
[12] Names *cis-in*, *cis-out*, and *trans* were used to describe the rotation around the $C_{aromatic}$-CO bond depending on the situation of the carbonilyc oxigens relative to the N_{py} both cis to the N_{py} (cis-in), both trans to the N_{py} (cis-out), and one cis-one trans (trans).

INFLUENCE OF SUBSTITUENTS ON THE EDGE-TO-FACE AROMATIC INTERACTION: HALOGENS

G. CHESSARI, C. A. HUNTER, J. L. JIMENEZ BLANCO, C. R. LOW,[a] J. G. VINTER[a]
Krebs Institute for Biomolecular Science, Department of Chemistry, University of Sheffield, Sheffield S3 7HF United Kingdom
[a] *James Black Foundation, 68 Half Moon Road, Dulwich, London SE24 9JE United Kingdom*

1. Introduction

Weak intermolecular interactions provide the key to understanding the relationship between structure and function in chemical and biological systems (1). Although hydrogen bonding is the most commonly used interaction for the control of supramolecular architecture (2), the non-covalent interaction between aromatic centres, the π-π interaction, is also of some interest. Aromatic interactions play an important role in determining the fine structure of the DNA double helix and the tertiary and quaternary structure of proteins, as well as the properties of many synthetic supramolecular systems (3). In an effort to quantify, understand and develop reliable computer models for these interactions, we have developed a new approach for studying intermolecular interactions based on molecular zippers and the double mutant cycle concept (4) (Fig. 1).

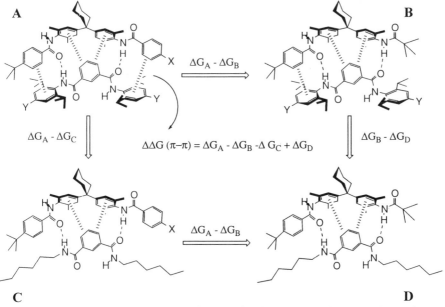

Figure 1 Chemical double-mutant cycle for measuring the terminal π-π interaction in complex **A**.

The zippers are 1:1 complexes of amide oligomers held together by hydrogen bonds and edge-to-face π-π interactions. The molecules are essentially rigid so that experiments are not complicated by losses of conformational mobility on complexation. This rigidity also ensures that the π-π interactions in complex **A** (Fig. 1) are fixed in an edge-to-face orientation by the covalent structures of the molecules and the hydrogen bonding. The magnitude of the terminal π-π interaction in the complex **A** could be estimated by removing one of the interacting aryl groups, as in complexes **B** and **C**. However, these single chemical mutations not only remove the interaction of interest, but also remove other secondary interactions with the core of the complex. The change in the secondary interactions can be quantified by using the double mutant, complex **D**. Therefore, the magnitude of the terminal π-π interaction in the absence of secondary effects is the difference of the ΔΔGs for the two horizontal mutations of Fig 1. The value of ΔG for complexation can be experimentally determined by NMR titration in CDCl$_3$, and the limiting complexation-induced changes in ^{1}H NMR chemical shift can also be used to determine the structures of the complexes.

We have previously used the chemical double mutant cycle approach to quantify the effects of nitro and dimethylamino substituents in the edge-to-face aromatic interaction in complex **A** (5). Changing the substituents X and Y in Fig. 1 causes remarkable changes in the magnitude of the aromatic interaction (Fig. 2), and the results correlate well with the Hammett substituent constants, σ (Equ. 1).

$$\Delta\Delta G(\pi\text{-}\pi) = 5.2\ \sigma_X\ \sigma_Y - 1.9\ \sigma_X + 1.4\sigma_Y - 1.5 \qquad (1)$$

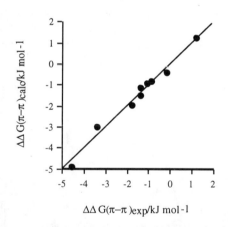

Figure 2 Correlation between aromatic interaction energies from experiment and calculated using Equ. 1.

2. Results and Discussion

Although the correlation in Figure 2 suggests that it will be possible to make general quantitative predictions about the magnitudes of edge-to-face aromatic interactions, it is based on a fit of four variables to nine experimental data points. In order to investigate the generality of these results, we have therefore measured the interaction energies for a quite different set of substituents, the halogens. Fluorine, chlorine, bromine and iodine have rather similar Hammett substituent constants but are quite different in size and polarisability α, and these experiments will therefore allow us to determine whether van der Waals factors play a significant role in our system. A set of halogenated amides was therefore synthesised, and ^{1}H NMR titrations were performed on the appropriate complexes. The limiting complexation-induced changes in chemical shift show that the structures of the complexes are not affected by halogenation.

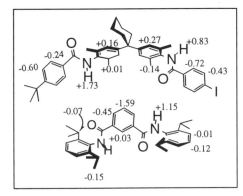

Figure 3. Limiting complexation-induced changes in chemical shift for the t-butyl and iodine complexes.

The X-ray crystal structures of **1** and **2** also clearly show that substitution with iodine has no impact on the geometry of the edge-to-face interaction in this system (the dimers of these simple amides found in the solid state are effectively half of complex **A** with the same set of hydrogen bonds and edge-to-face aromatic interactions).

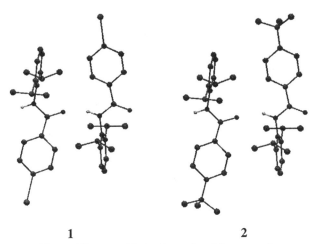

Figure 4 X-ray crystal structures of model compounds.

Table 1 and Figure 5 show the aromatic interaction energies measured for the set of four halogen substituents and a comparison of these results with the predictions of Equ. 1.

TABLE 1. Aromatic interaction energies and properties of the substituents

Halogens	σ_p	α^a	$\Delta\Delta G(\pi-\pi)_{exp}$	$\Delta\Delta G(\pi-\pi)_{calc}$
F	0.15	0.56	-2.3 ± 0.3	-2.0
Cl	0.24	2.18	-1.9 ± 0.3	-2.2
Br	0.26	3.05	-2.1 ± 0.2	-2.2
I	0.28	5.35	-2.5 ± 0.3	-2.2

[a] unites of 10^{-24} cm^3

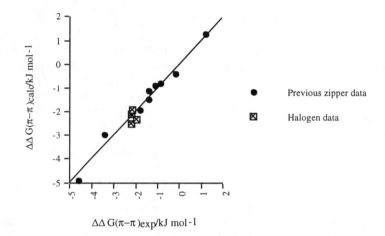

Figure 5 Correlation between experimental and calculated interaction energies.

The agreement is excellent and indicates that it is electrostatics, which dominate the behaviour of this system. The halogens are mildly electron withdrawing and so increase the net positive charge on the hydrogens of the 'edge' ring which leads to an increased electrostatic interaction with the π-electron density of the 'face' ring. Neither changes in polarisability nor direct van der Waals interactions with the halogen substituents play a thermodynamically significant role.

3. Acknowledgements

We thank the Lister Institute (CAH), the BBSRC (GC), the James Black Foundation (GC) and the Spanish Government (JLJB) for funding.

4. References

1. Hunter, C. A. (1993)*J. Mol. Biol.* **230**, 1025-1030. Petsko, G. A. and Burley, S. K. (1988) *Adv. Protein Chem.* **39**, 125-189. Lehn, J.-M. (1988) *Angew. Chem., Int. Ed. Engl.* **27**, 89-112. Cozzi, F., Cinquini, M., Annunziata, R., Dwyer, T. and Siegel, J. S. J. (1992) **114**, 5729-5733.
2. Desiraju, G. R. (1995) *Angew. Chem., Int. Ed. Engl.* **34**, 2311-2327.
3. Hunter, C. A. (1991) *J. Chem. Soc. Chem. Commun.* **11**, 749-751.
4. Adams, H., Carver, F. J., Hunter C. A., Morales, J. C. and Seward E. M. (1996) *Angew. Chem., Int. Ed. Engl.* **35**, 1542-1544.
5. Carver, F. J., Hunter C. A. and Seward, E. M. (1998) *Chem. Commun.* **7**, 775-776.

SUBJECT INDEX

^{2}H-NMR	304
^{7}Li-NMR	94, 143
^{14}N-NMR	311
^{23}Na-NMR	94, 143
^{31}P-NMR	28, 158
^{39}K-NMR	143
^{131}Xe-NMR	83
^{133}Cs-NMR	143
Actinides	20
Adenylate kinase	164
Aggregation	19, 248, 303, 325
Anion recognition	267
Anisotropic environments	81, 92ff, 255
Annexin	274
Antibodies	117
Arginine kinase	164
Aromatic interactions	331
Association constants	3
ATP-utilizing enzymes	155
β-Galactosidase	99
β-sheet	62
Binding, non specific	161
Bipyridinium	3
BIRD sequence	84
Bound Water	239, 304
C-lactose	103
Calbindin D$_{9k}$	175
Calixarene	45, 71, 307, 311
Cap binding protein	247
Capsules	47
Carbohydrate	101, 191, 328
Carr-Purcell-Meiboom-Gil	171, 216
Catenane	2ff
Cation extraction	22
Cavitand	20
CHAPS	247
Chelates	135
Chemical shift anisotropy	319
CMP(O) cavitands	19
Coalescence temperature	3
Conformational dynamics	75, 171, 212, 256
Contrast agents for MRI	136, 146
Cooperativity	325
Coordination	139
Correlation time	71, 148, 173ff, 314
Coupling constant	202, 206
CP-MAS	43
Creatine kinase	164
Cyclodextrin	142
Cyclophane	3
Daisy chains	14
Dendrimers	42
Density operator	95
Desymmetrization	69, 71
Deuterated molecules	118ff, 249
Deuterium exchange	69
Diels-Alder reaction	41
Diffusion	301ff
Diffusion anisotropy	174
Dihydrofolate reductase	282
Dumbbell	5
eIF4E	247
Entropic contribution to binding	171
Enzyme bound substrates	155
Europium(III) extractants	19
Exchange relaxation	158, 184, 213
Exchange rates	171

EXSY	30, 56, 73, 307	NAD	85
		Nanotubes	61ff
		NMRD	149
Fv fragment	118	NOE, heteronuclear	171, 213, 233, 312, 315
Glycerate kinase	164		
gp120	118	NOESY	27, 41, 192
Guanidino-carboxilate int.	268, 287		
		NOESY Difference	118ff
HAWAII experiment	240	NOESY-HSQC	230
Helical superhelices	269	NOESY-TOCSY	197
Helicity score	270	Nuclear fuel cycle	20
Hevein	111		
HMBC	57, 80, 192	Order Parameter	171
HSQC-TOCSY	196	p-sulfonatocalix[4]arene	213
Hydrogen bonding	325	P53	273
Hydrogen exchange	69, 74, 235	Palindrome	68
		Paramagnetic cations	133ff, 156
Hydrosteroid dehydrogenase	275	Peptide	61, 68, 117, 255
Ion channels	64		
Isotope editing	118	Periplasmic glucans	191ff
Isotope filtering	239	Phosphotransferase system	260
Isotopic labeling	166, 203, 229	Physisorbtion	92
		Porphyrin dimers	39
Isotopic shifts	71	Porphyrin trimers	40
Itinerant phosphoryl	168	Preorganization	20
		Procarboxypeptidase B	277
Lactate dehydrogenase	85	Protein assignment	229ff, 249
Lanthanide	19, 133ff		
Lanthanide Induced Relax.	133ff	Protein structure	227ff, 260
Lanthanide Induced Shifts	133		
Lectins	99	Protein-protein interaction	247, 262, 281
Lysyl t-RNA synthetase	274		
		Pseudorotaxanes	3ff
Metalloporphyrins	37	PUREX	20
Metalloproteinases	291	Pyruvate decarboxylase	273
Micelles	248	Pyruvate kinase	165
Model free formalism	171ff		
Molecular dynamics	259	Quadrupolar coupling	93
Molecular weight	303	Quadrupolar nuclei	83
MQF-NMR	92, 304	QUIET-NOESY	84
MRI contrast agents	146		
Multiple conformations	290	Rate constants	5, 59
Myosin regulatory domain	275	Relaxation	171ff, 213, 311
N-acetyl glucosamine	112	Relaxation dispersion	149, 186

Relaxation interference	319	Serine protease PB92	227
Relaxation matrix	162, 321	Shuttle, molecular	11
Relaxation, paramagnetic	133ff, 157	Slippage	5
		Solvent inclusion	52
Relaxation, quadrupolar	94	Spin diffusion	84
Relaxation, rotating frame	171	Sping-pong	69
Relaxation, T1	29, 72, 171, 213, 312, 315	Surface recognition	267
		Surfaces	92
		Symmetry	45, 67ff, 309
Relaxation, T2	171, 213		
Relaxivity	133ff		
Resorcinarene	20	Tenascin	171
Ribonuclease H	171	TOCSY-NOESY	69, 197
Ribonucleotide reductase	276	Torsion angle dynamics	236
Ricin	104	Transferred NOE	83ff, 99ff, 118, 157
Ring current shifts	39		
ROESY	27, 41		
ROESY, off-resonance	205	Transferred ROE	101
Rotaxane	2ff	Trefoil proteins	282
		Triarylmethane	79
Saturation transfer	11, 233	TRUEX	20
Scytalone dehydratase	277		
Second rank tensor	95	Water soluble calixarene	311
Self-assembly	1, 45, 61		
Self-complexation	23	X-ray	25, 49